Reducing Production Costs

Other books by Donald J. Wheeler:

Understanding Variation–the Key to Managing Chaos

Twenty Things You Need to Know

Making Sense of Data

Understanding Statistical Process Control

Advanced Topics in Statistical Process Control

The Six-Sigma Practitioner's Guide to Data Analysis

EMP III: Evaluating the Measurement Process and Using Imperfect Data

Understanding Industrial Experimentation

Tables of Screening Designs

Short Run SPC

Beyond Capability Confusion

Normality and the Process Behavior Chart

Range Based Analysis of Means

SPC at the Esquire Club

Reducing Production Costs

Donald J. Wheeler
Fellow American Statistical Association
Fellow American Society for Quality

SPC Press
Knoxville, Tennessee

SPC Press
5908 Toole Drive, Suite C
Knoxville, Tennessee 37919
(865) 584–5005
Fax (865) 588–9440
www.spcpress.com

ISBN 978-0–945320–70-8
290 + *xii* pages
134 figures
97 examples
18 data sets
12 reference tables
appendix with 8 additional tables
glossary
index

1 2 3 4 5 6 7 8 9 0

Contents

About the Author *ix*
Acknowledgments *x*
Preface *xi*

PART ONE: INTRODUCTION

Chapter 1 The Nature of Data 1
1.1 Cause-and-Effect Relationships 1
1.2 Identifying Control Factors 5
1.3 Assignable Causes and Common Causes 7
1.4 Two Ways to Improve a Process 10
1.5 The Fork in the Road to Process Improvement 12
1.6 Summary 14

Chapter 2 The Analysis of Observational Data 15
2.1 Observational Studies versus Experimental Studies 15
2.2 Process Behavior Charts 17
2.3 Average and Range Charts 20
2.4 The Chart for Individual Values and Moving Ranges 24
2.5 Caution regarding Software 29
2.6 How Process Behavior Charts Work 29
2.7 Using Process Behavior Charts for Continual Improvement 31

Chapter 3 The Opportunities 41
3.1 Performance and Potential 41
3.2 Capability Indexes 45
3.3 Performance Indexes 47
3.4 The Gaps 49
3.5 The Two Definitions of Trouble 50
3.6 The Four Possibilities 52
3.7 Summary 56

Chapter 4 The Costs of Using Conforming Product 57
4.1 The Problem of Production 57
4.2 Tighten the Specifications: The Center Cut Approach 60
4.3 Random Assembly with the Original Specifications 61
4.4 Tweaking the Target Value for the Bearing Diameter 63
4.5 Pre-Sorting the Parts 65
4.6 The Traditional Approaches to Production 66
4.7 The Effects of Reducing Variation at the Source 68
4.8 Summary 73

PART TWO: USING EXCESS COSTS

Chapter 5	**From Capabilities to Excess Costs**	**75**
	5.1 The Effective Cost of Production and Use	75
	5.2 The Gaps between Performance and Potential	80
	5.3 The Batch Weight Data	82
	5.4 The Methanol Percentages	87
	5.5 Tokai Rika Revisited	89
	5.6 The Cover Examples	91
	5.7 The Tables for Two-Sided Specifications	92
Chapter 6	**Economic Operation**	**93**
	6.1 The Excess Costs of Production	93
	6.2 The Excess Costs of Use	94
	6.3 The Effective Cost of Production and Use	95
	6.4 The Zone of Economic Operation	98
	6.5 The Spurious 1.5 Sigma Shift	99
	6.6 Summary	100
Chapter 7	**Excess Costs for One-Sided Specifications**	**101**
	7.1 One-Sided Safety Specifications	101
	7.2 One-Sided Process Specifications	104
	7.3 One-Sided Product Specifications	108
	7.4 Excess Costs of Use for One-Sided Product Specifications	111
	7.5 Using One-Sided Product Specifications	114
	7.6 Reassessing the Run-Out Data	114
	7.7 The Contamination Data	116
	7.8 Summary	119
Chapter 8	**Excess Costs for Targets at Boundaries**	**121**
	8.1 The Excess Cost Function	121
	8.2 Using Tables Nine and Ten	123
	8.3 The Run-Out Data	124
	8.4 The Contamination Data	126
	8.5 Summary	128
Chapter 9	**When Nominal Costs Change with Aim**	**129**
	9.1 One-Sided Product Specifications without a Target	129
	9.2 One-Sided Product Specifications with a Target Value	136
	9.3 Target at Boundary	141
	9.4 Summary	144
Chapter 10	**Excess Costs for Count Data**	**145**
	10.1 Specifications and Count Data	145
	10.2 Effective Costs of Production for Counts of Items	147
	10.3 Effective Costs of Production for Counts of Events	148

PART THREE: TABLES OF EXCESS COSTS

Chapter 11	**The Formulas for the Excess Costs**	**151**
	11.1 The Structure of the Effective Cost of Production and Use	151
	11.2 The Average Costs of Scrap, Rework, and Use	152
	11.3 Finding the Effective Cost of Production and Use	157
	11.4 One-Sided Process Specifications	159
	11.5 The Excess Cost of Production with One-Sided Product Specs	160
	11.6 The Excess Cost of Use with One-Sided Product Specs	161
	11.7 One-Sided Product Specifications with Targets at Boundaries	162

Tables for Two-Sided Specifications		**169-216**
Table One	When All Nonconforming Units Are Scrapped	170-176
Table Two	When All Nonconforming Units Are Reworked	178-184
Table Three	When the Average Is on the Scrap Side of the Target Value and the Cost of Rework Is Equal to the Cost of Scrap	186-192
Table Four	When the Average Is on the Scrap Side of the Target Value and the Cost of Rework Is 0.1% the Cost of Scrap	194-200
Table Five	When the Average Is on the Rework Side of the Target Value and the Cost of Rework Is Equal to the Cost of Scrap	202-208
Table Six	When the Average Is on the Rework Side of the Target Value and the Cost of Rework Is 0.1% the Cost of Scrap	210-216

Tables for One-Sided Specifications		**218-225**
Table Seven	Excess Costs of Scrap and Excess Costs of Use when Nonconforming Product Is Scrapped	218-221
Table Eight	Excess Costs of Rework and Excess Costs of Use when Nonconforming Product Is Reworked	222-225

Tables for One-Sided Specifications with Target at Boundary		**226-240**
Table Nine	When Nonconforming Product Is Scrapped	226-232
Table Ten	When Nonconforming Product Is Reworked	234-240

APPENDIX

I. Formulas for Measures of Dispersion 243

II. The Coverage of Three-Sigma Limits 244

III. The Transformation of Data 248

Table A.1 Bias Correction Factors for Measures of Dispersion 252

Table A.2 *XmR* Charts: Charts for Individual Values 254

Table A.3 Average and Range Charts Based on the Average Range 256

Table A.4 Average and Range Charts Based on the Median Range 258

Table A.5 Average and Standard Deviation Charts Based on the Average Standard Deviation Statistic 260

Table A.6 Average and Standard Deviation Charts Based on the Median Standard Deviation Statistic 262

Table A.7 Average and Root Mean Square Deviation Charts Based on the Average Root Mean Square Deviation Statistic 264

Table A.8 Average and Root Mean Square Deviation Charts Based on the Median Root Mean Square Deviation Statistic 266

Table A.9 Normal Theory Values of the Integrals **IA**, **IB**, **IC**, and **ID** 268

Table A.10 Burr and Normal Values for the Integrals **IC** and **ID** 277

Glossary of Symbols 283

Index 287

About the Author

Donald J. Wheeler is a consulting statistician who had the good fortune to work with Dr. W. Edwards Deming and David S. Chambers. Dr. Wheeler graduated from the University of Texas, Austin, with a Bachelor's degree in Physics and Mathematics, and holds M.S. and Ph.D. Degrees in Statistics from Southern Methodist University. From 1970 to 1982 he taught in the Statistics Department at the University of Tennessee, Knoxville, where he was Associate Professor. Between 1981 and 1993 he periodically assisted Dr. Deming with his four-day seminars. He is a Fellow of the American Statistical Association and a Fellow of the American Society for Quality. He has consulted with over 250 organizations and has conducted over one thousand seminars in sixteen countries on five continents. He is author or co-author of 25 books and over 200 articles. Through these seminars and books Dr. Wheeler has had a profound impact on companies and organizations around the world.

Acknowledgments

I am indebted to all of those who took the time to read and comment on this book. First among these is James Beagle who not only reviewed the text and content, but also undertook to verify Tables One to Ten.

Next must come Dr. Andrew C. Palm who made some very helpful suggestions regarding the organization of the book. All readers of this book owe him a debt of gratitude. Finally we have Dr. Jose Ramirez, Raymond Phillips, and Dr. Henry Neave whose comments and suggestions proved useful in transforming the manuscript into the finished book. Writing a book can be a very solitary occupation, and the considered opinions of competent reviewers are a pure gift when one is trying to communicate complex ideas to a wider audience. I have been very fortunate to have the advice and assistance of such a capable review board.

Finally, I must acknowledge the help and support of my wife, Fran Wheeler, who has been my editor and publisher for 26 years. Without her gentle insistence that "things make sense" I fear that my books would have been much more difficult to read.

Preface

The purpose of this book is to describe a proven methodology for reducing production costs. This methodology does not require troubleshooters with special skill sets, but it can make use of them. This methodology does not require special projects, but it can be used as part of them. This methodology does not require a realignment of the organization, but it may result in changes in how the organization works. Rather than being structural or organizational, this methodology is built upon a consistent understanding and interpretation of the data generated by your processes. With this approach you will no longer have to reengineer a process in order to make improvements. With this approach you can discover how to get the most out of your existing process. With this simple understanding of the data it turns out that virtually everyone in the organization can contribute to the reduction of production costs. Moreover, the techniques given in this book allow you to actually quantify these cost reductions in a straightforward manner.

While much of the material in this book has been available for years, there has not been any serious effort to integrate these techniques within a conceptual framework that will both allow and promote the reduction of production costs. All too often the techniques have been taught without any viable conceptual framework for how to use the techniques. As a result many have come to the mistaken conclusion that the techniques do not work.

On the other hand, for the many who were taught the conceptual framework, the techniques have become the basis for doing business. They have simply become the way that everyone does their job. And the spectacular results posted by the organizations using this approach are the proof that the methodology given here does, indeed, work in practice.

This book is intended to breathe a breath of life into the "dry bones" of the computations and techniques of SPC, and to take you to the point where you understand how simple and easy it is to continually and relentlessly reduce production costs.

Deming's First Theorem:

"No one gives a hoot about profits—
if they did they would be interested
in learning better ways to make them."

Chapter One

The Nature of Data

In the classroom, two plus two is always equal to exactly four. Yet when a manufacturer tries to lay down a two micron film on a two micron substrate, the result is seldom equal to exactly four microns. About the best that we can hope for is that the result will be four microns thick on the average. And this is the basic difference between *numbers* and *data*.

Numbers inhabit the mathematical plane, where 1 is always 1, 2 is always 2, and there is no uncertainty. Here it can truly be said that two numbers that are not the same are different.

On the other hand, data belong to the real world. They are generated by some underlying process. As the process varies, the data will also vary. This variation introduces things that we never dreamed of in arithmetic. For example, when we use numbers to express our data, we suddenly find that two numbers that are different may, in fact, represent the same thing! So when it comes to using data, we have to understand how to deal with variation or we are liable to end up being misled or confused.

Since it is the underlying process that produces the variation, we will need a model for the structure of an underlying process in order to gain some insight into the origins of variation within our data. This chapter will provide such a model.

1.1 Cause-and-Effect Relationships

No matter what your process, no matter what your data, all data display variation. Any measure you can think of that will be of interest to your business will vary over time. The reasons for this variation are many. There are all sorts of causes that have an impact on your process and the process outcomes. It is not unrealistic to think that your processes and systems will be subject to dozens, or even hundreds, of cause-and-effect relationships. And this multiplicity of causes has two consequences: it makes it easy for you to pick out an explanation for why the current value is so high or so low; and it makes it very hard for you to know if your explanation is even close to being right. In order to make any headway in understanding data we need to start with a discussion of these cause-and-effect relationships.

To this end we need to perform a thought experiment. Think about any product characteristic you may choose. We will call this Characteristic No. 1. Next list all of the cause-and-effect relationships that you can think of that will have an impact upon Characteristic No. 1.

In most cases you will have dozens of cause-and-effect relationships. For purposes of this illustration we will assume that after naming the factors and their possible interactions you end up with a total of 21 cause-and-effect relationships. Because of these cause-and-effect relationships, we would expect the values for Characteristic No. 1 to vary about some average value in the manner shown in Figure 1.1.

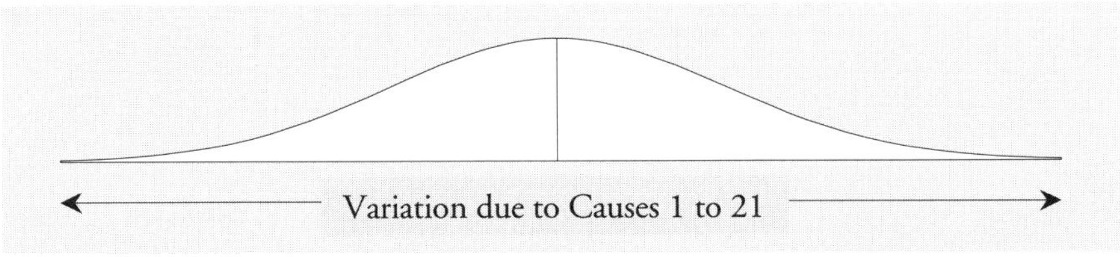

Figure 1.1: The Original Distribution of Characteristic No. 1

Now some of these causes will have large effects upon our characteristic, while others will have smaller effects. In Figure 1.2 we use bars to show the effects of each of the 21 causes upon Characteristic No. 1. As each cause varies within some range of interest the bar shows the amount by which Characteristic No. 1 varies. Thus, the height of each bar shows how much of the variation in Characteristic No. 1 can be attributed to each individual cause. With perfect knowledge about the effects of each cause upon our product characteristic, we could arrange these causes in order of descending impact to create a Pareto diagram.

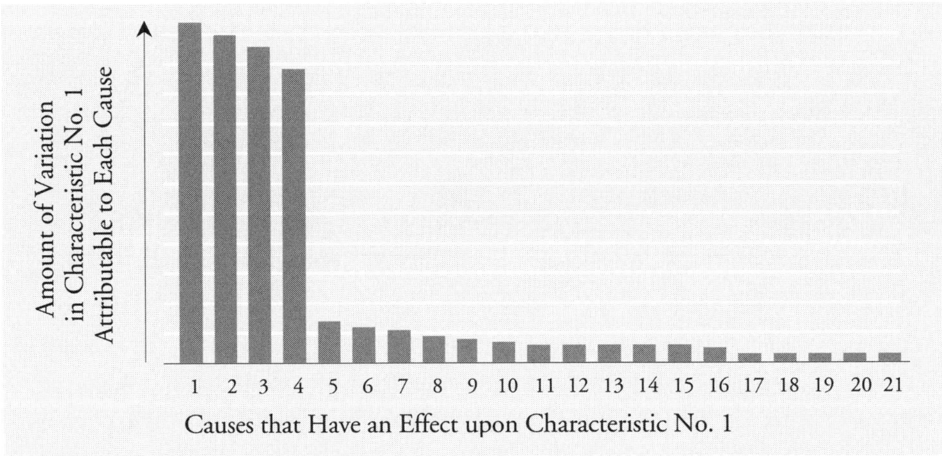

Figure 1.2: A Pareto Diagram of Causes and Their Effects for Characteristic No. 1

In Figure 1.2 it is the causes with the large bars (the dominant causes) that will provide the greatest leverage for changing the average value. Causes with small bars will provide little opportunity to change the average of Characteristic No. 1. Therefore, the producer will typically select the levels of the dominant causes to obtain the desired average level for the product characteristic.

Moreover, when the levels for the dominant causes are held constant these causes will no longer contribute to the variation in the product characteristic. Here, Causes 1, 2, 3, and 4 account for 79% of the total variation, while Causes 5 through 21 account for the remaining 21%. By holding Causes 1, 2, 3, and 4 at fixed levels the producer not only determines the average value for Characteristic No. 1, but also removes 79% of the total variation in this product characteristic. This can be seen in Figures 1.3 and 1.4.

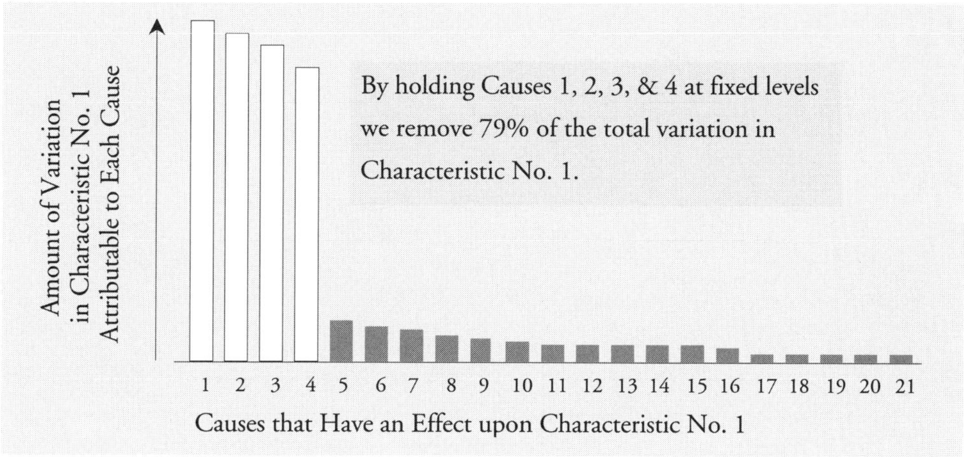

Figure 1.3: An Economic Production Process for Characteristic No. 1

Because of the way variation works, the 21% of the variance that remains in Figure 1.3 shows up in Figure 1.4 as a distribution with a spread that is $\sqrt{0.21} = 0.458$ or 46% of the spread of the original distribution.

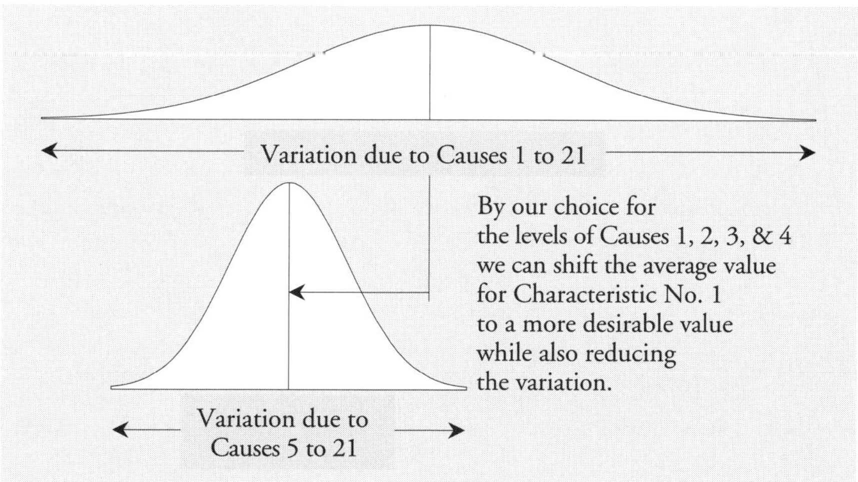

Figure 1.4: The Effects of Controlling Causes 1, 2, 3, & 4 for Characteristic No. 1

Finally, once the first four causes have been controlled, it is unlikely to be economical to attempt to control any further causes. For example, Causes 5, 6, 7, and 8 only contribute 9% of the total variation in Characteristic No. 1. While it will probably cost about the same to control these four causes as it does to control the first set of four causes, the payback for controlling these additional four causes will only be about one-ninth as big as it was for the first four causes. This diminishing return may be seen in Figure 1.5.

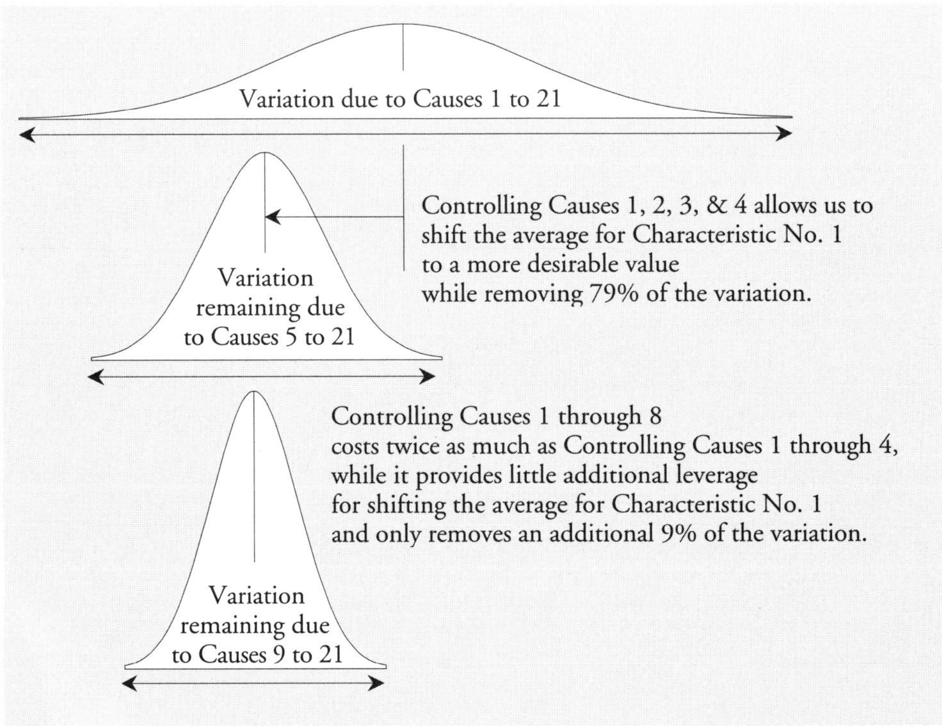

Figure 1.5: The Diminishing Return of Controlling Causes 5, 6, 7, & 8 for Characteristic No. 1

Economic production requires that we distinguish between those causes with dominant effects and the many causes with small effects. Controlling dominant cause-and-effect relationships will be economical. Controlling the lesser cause-and-effect relationships will not be economical. The diminishing returns shown in Figure 1.5 make it unreasonable to devote resources to controlling any cause-and-effect relationships for Characteristic No. 1 beyond the first four. Thus, Causes 5 to 21 will make up the set of Uncontrolled Factors. It is this group that creates virtually all of the variation in the product characteristic.

Thus, the first benefit of having a conceptual model for the nature of data is that it provides a reasonable explanation for why it is not economical to try to control all of the inputs to a production process. The dominant causes are the only ones that we want to include in the set of Control Factors.

1.2 Identifying Control Factors

Of course, before we can draw the Pareto diagram of Figure 1.2 we will need perfect knowledge about *all* of the cause-and-effect relationships for Characteristic No. 1. In the absence of perfect knowledge, what can we do? We generally start off as shown in Figure 1.6 for Characteristic No. 2 and ask R&D to sort out the gaps in our knowledge. To this end R&D will need to find the amount of variation attributable to each cause. Next they will need to identify the Control Factors, and then for each of these they will need to establish the appropriate levels for these Control Factors that should be used in production.

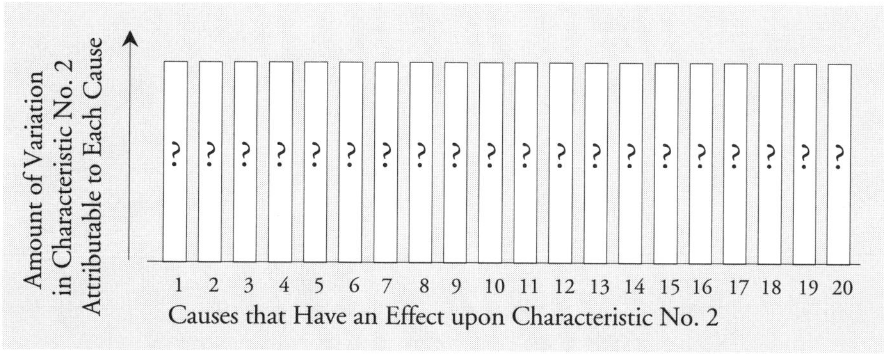

Figure 1.6: What We Usually Know about Causes and Effects

Unfortunately, due to pressures of time and money, it is a rare thing for R&D to ever study all of the causes on the list. Generally they will use both theory and experience to redraw Figure 1.6 to look like Figure 1.7, and then they will tackle the cause-and-effect relationships that are thought to be the more critical (Causes 1 through 10 in this case). The remaining

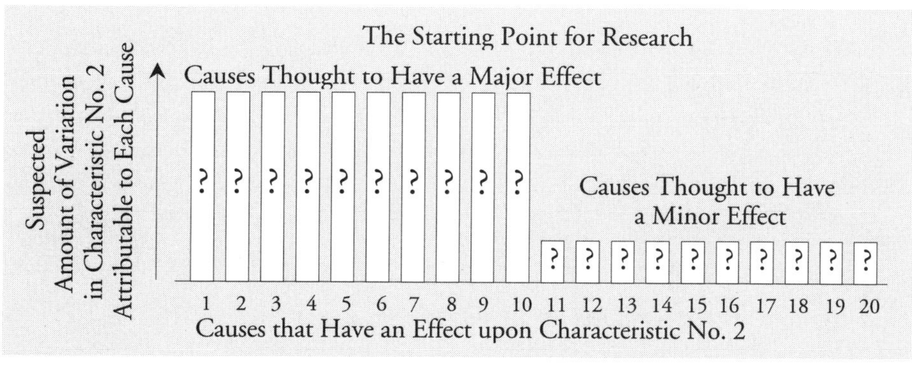

Figure 1.7: The Starting Point for Research

causes will not be studied—they will either be ignored, or randomized, or held constant during the course of the experiments performed by R&D.

Figure 1.8 shows the results reported by R&D for Characteristic Number 2. Among the ten causes studied R&D found Causes 5, 1, and 7 to be the dominant factors, collectively accounting for 77% of the observed variation. Thus, they told the Production Department that they needed to control Causes 5, 1, and 7, and they also defined the appropriate levels to use with each of these Control Factors.

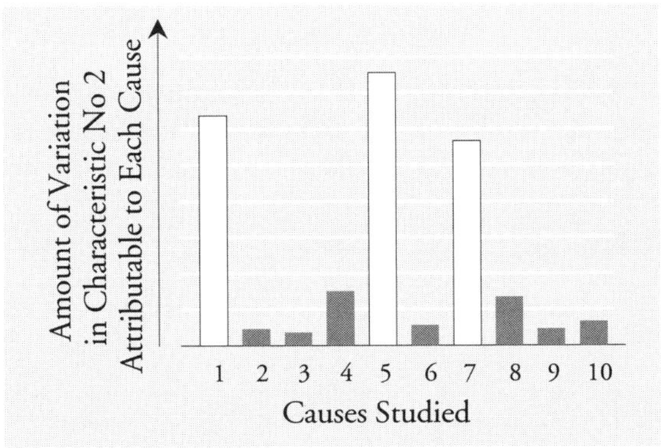

Figure 1.8: Causes Used to Control Characteristic No. 2

So the production process was set up using Causes 5, 1, and 7 as Control Factors for Characteristic No. 2. However, at the start of production they immediately had problems with too much variation in Characteristic No. 2. This resulted in a high scrap rate. So they decided to also control Cause 4. While this cost extra money, it was expected to remove an additional 6% of the variation found by R&D. Unfortunately, as seen in Figure 1.9, controlling Cause 4 had virtually no impact upon the process outcomes.

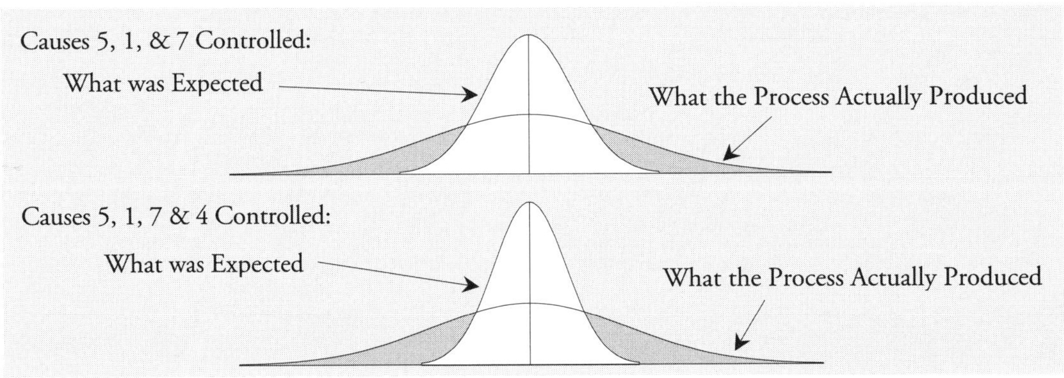

Figure 1.9: Expected and Observed Outcomes for Characteristic No. 2

As they fell further and further behind the production schedule, and as the mountain of scrap increased, they began to talk about the "skill" that it took to make this product. Words like "art" and "magic" were used. Inspection and rework facilities were expanded, and soon the production department had settled down to a routine that W. Edwards Deming called the Western approach to production—"burn the toast and scrape it."

Of course, the problem was not with the set of Control Factors—by virtue of their being controlled these causes ceased to be sources of variation for the product stream. Likewise the problem was not with the other six causes studied by R&D—these causes all had small effects upon Characteristic No. 2. Nothing was wrong with the information provided by R&D, except that it was incomplete—the problem was with the factors *not* studied by R&D.

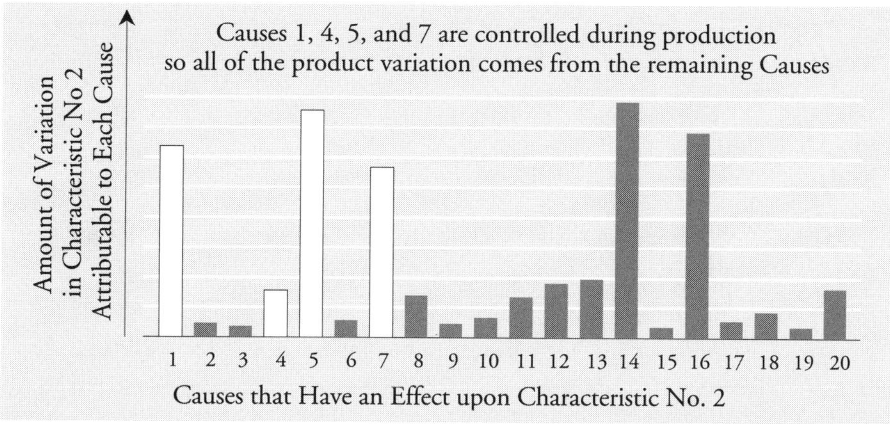

Figure 1.10: The Rest of the Story for Characteristic No. 2

As seen in Figure 1.10, Causes 14 and 16 are dominant cause-and-effect relationships for Characteristic No. 2. They are as big, or bigger than Causes 5, 1, and 7. But they were not studied. Although the manufacturer was not aware of the impact of Causes 14 and 16, the process continued to be under their influence. Since Causes 14 and 16 had not been studied, and were *thought* to be part of the lesser causes back in Figure 1.7, the manufacturer was not exerting any control over the levels of these two factors. As a result, when the levels of either one of these causes changed, it would result in a corresponding change in the product characteristic. While the manufacturer remained unaware of impact of Causes 14 and 16, he still suffered the consequences of their effects.

So while Causes 5, 1, and 7 need to be in the set of Control Factors, the producer needs to have Causes 14 and 16 in the set of Control Factors as well. Having Cause 4 in the set of Control Factors is a mistake caused by the incomplete information provided by R&D.

1.3 Assignable Causes and Common Causes

Whenever the set of Uncontrolled Factors contains a dominant cause-and-effect relationship, this dominant, uncontrolled cause will haunt the process, creating exceptional variation, excess costs, and general grief. Walter Shewhart called such dominant Uncontrolled Factors

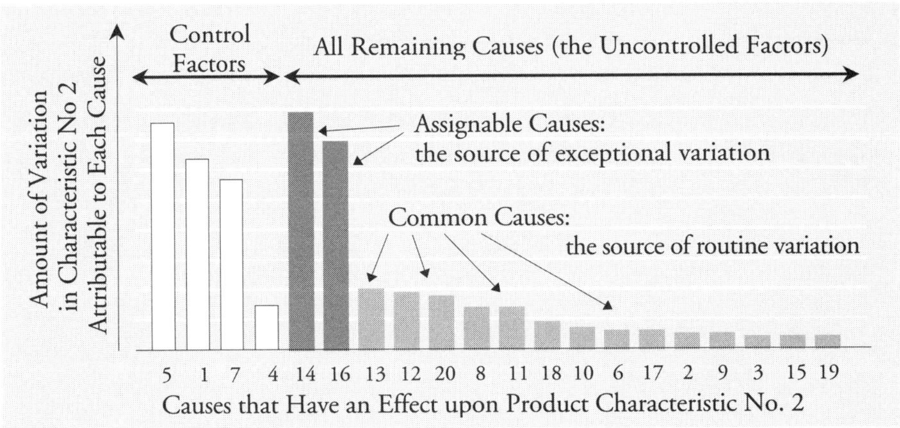

Figure 1.11: Three Types of Cause-and-Effect Relationships

Assignable Causes. W. Edwards Deming called the remaining group of lesser Uncontrolled Factors Common Causes.

Thus, we have Control Factors, Assignable Causes, and Common Causes. While it is the levels of the Control Factors that determine the average level for the product characteristic, the fact that the levels of the Control Factors are held constant during production means that they contribute little or no variation to the product characteristic. Virtually all of the variation in the product characteristic will come from the set of Uncontrolled Factors.

This is why *effort spent trying to fine tune the Control Factors will be of marginal benefit as long as there are Assignable Causes present.* It does not matter what levels of Factors 5, 1, 7 and 4 we choose to use as long as we are doing nothing about Factors 14 and 16! The variation introduced by the Assignable Causes will more than obliterate any fine tuning of the average level we might achieve by adjusting the Control Factors.

When the set of Uncontrolled Factors contains Assignable Causes, it will be worthwhile to identify those Assignable Causes and take steps to make them part of the set of Control Factors. By doing so we will be removing a major chunk of the variation in the product characteristic, while, at the same time, we will also be gaining added leverage for controlling the average level of the product characteristic. For this reason, effort spent in identifying Assignable Causes will have a high payback and should always be given priority.

You cannot optimize any system
when some of the dominant cause-and-effect relationships remain unidentified.

When all of the dominant cause-and-effect relationships are included in the set of Control Factors and only Common Causes remain in the set of Uncontrolled Factors we will have a picture like Figure 1.12. Here the variation in the product characteristic is the result of *many different cause-and-effect relationships where no one cause-and-effect relationship is dominant.* The resulting process variation will be the routine, run-of-the-mill variation that is inherent in every production process.

Since there is almost always a cost associated with controlling a cause-and-effect relationship, there will always be some point at which the variation removed by controlling additional causes is no longer sufficient to justify the expense involved. This point of diminishing returns for Characteristic No. 2 is shown in Figure 1.12.

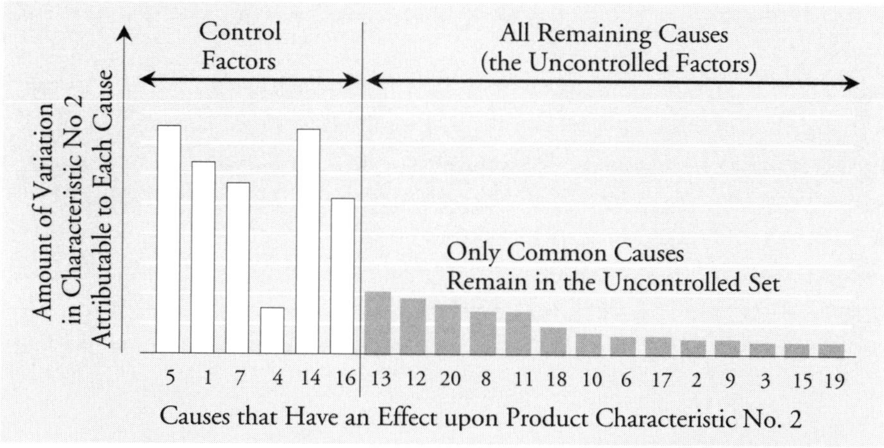

Figure 1.12: An Economic Equilibrium for Characteristic No. 2

Seeking to identify and control Common Causes will generally prove to be a low-payback strategy. When all of the dominant causes are contained in the set of Control Factors you will have reached an economic limit on finding and removing causes of variation. At this point of economic equilibrium the product stream will have the greatest degree of uniformity that is consistent with economic production. Controlling additional factors will be uneconomical.

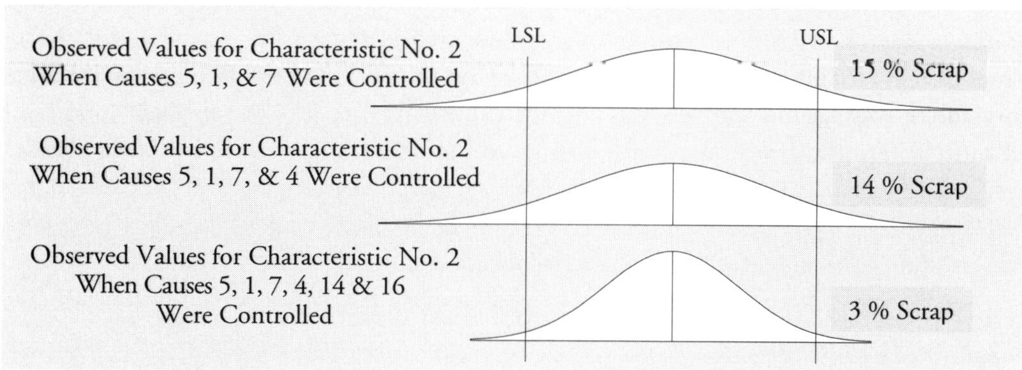

Figure 1.13: Outcomes for Characteristic No. 2 corresponding to Figures 1.8, 1.10, and 1.12

As shown in Figure 1.13, this economic equilibrium may not be perfect, but it is the best we can expect to achieve without changing the process in some major way. In order to achieve greater consistency for Characteristic No. 2, and to thereby eliminate the 3% scrap, we will have to upgrade or reengineer this production process. This will typically require the introduction of new technology, new equipment, new procedures, or new materials into the process. *The introduction of these new elements effectively adds new cause-and-effect relationships to our list.* Of course, with these new factors to consider we end up returning to the stage represented by Figure 1.7. While we may have partial knowledge about our upgraded process, we will still have to learn how the new cause-and-effect relationships fit in with the original set of cause-and-effect relationships in order to once again select the dominant causes for our set of Control Factors.

Thus, when you upgrade or reengineer a production process you are changing the set of cause-and-effect relationships and returning to the starting point defined in Figure 1.7. On the other hand, when you identify Assignable Causes and make them part of the set of Control Factors you are simply moving a production process toward its point of economic equilibrium. This distinction between these two completely different approaches to process improvement is the second major benefit of having a conceptual model for the nature of data. Since these two ways to improve a process have different objectives, different costs, and different outcomes it is important to understand how these two approaches fit together.

1.4 Two Ways to Improve a Process

To identify these two different approaches to process improvement we shall use the objective of each approach as a label. One approach seeks to get a process to operate up to its full potential, while the other approach is concerned with operating to meet requirements.

OPERATING AT FULL POTENTIAL

When your set of Uncontrolled Factors contains one or more Assignable Causes it will be economic to identify the Assignable Causes and make them part of the set of Control Factors. When this is done you will be gaining leverage for manipulating the process average while you are removing major chunks of variation from the product stream. By moving Assignable Causes into the set of Control Factors you can take any process to its point of economic equilibrium where the product stream will have the minimum variation that is consistent with economic production. At this point the process can be said to be operating up to its full potential.

As we have seen, any failure to include all of the dominant cause-and-effect relationships in the set of Control Factors will leave the process operating short of its full potential. Such a process will have excess variation. This increased variation will result in increased costs. Some of these costs may be the direct costs of scrap and rework, and some of these costs may be the increased cost of using the conforming material in subsequent operations, but either way, increased variation always results in increased costs. This is why economic operation requires that you operate your processes at their full potential.

Identifying the Assignable Causes and making them part of the Control Factors is simply the only way to operate any process up to its full potential, and this will only happen when observational studies are used to finish up the job begun by R&D. Experience with this approach to process improvement has typically and repeatedly shown reductions in process variation. These reductions range from 50% to over 90% compared to that achieved by using the following technique alone.

OPERATING TO MEET REQUIREMENTS

When you have a process that is not meeting requirements you will usually have scrap or rework costs, and you will naturally want to minimize these excess costs. Here we have to talk about two different ways that a process can fail to meet requirements.

One way a process can fail to meet requirements is by being off-target. When this happens it is natural to seek to move the process average closer to the target value. To this end, your only steering mechanism will consist of the levels of the Control Factors. By your choices of these levels you can change the average value for the process outcomes. However, if you are not operating your process predictably, your set of Control Factors will be incomplete—some Assignable Causes will still remain in the set of Uncontrolled Factors. Any experiments you may perform to determine the appropriate levels of the Control Factors will not include these Assignable Causes, and in consequence, your experimental results will be incomplete. So, when you need to adjust the process average, but are not operating your process predictably, you will be handicapped in your efforts. Moreover, changes in the Assignable Causes may change the process average and thereby undermine your efforts to move the process closer to the target. Hence, operating the process at full potential will be the first step in operating your process on target.

The other way that a process can fail to meet requirements is to have too much variation. When this happens it is natural to seek to reduce the process variation. As we saw in Figure 1.13, operating a process up to its full potential will reduce the variation to the economic minimum. So, if your process is not yet at this point of economic equilibrium, you will need to identify the Assignable Causes that are keeping you from achieving this minimum variation state. Experiments with the Control Factors will not help here. Observational studies are required in order to identify the Assignable Causes (see Chapter Two).

On the other hand, if the process is already being operated up to its full potential, and like the process in Figure 1.13, it is still not meeting requirements, then the only thing left is to upgrade or reengineer the process. As you introduce new materials, new equipment, new technologies, or new procedures into your process you will be introducing additional cause-and-effect relationships. If any of these new causes have dominant effects they will belong in the set of Control Factors, and until they are included in the set of Control Factors, the upgraded process will not be operating up to its full potential, even though it may be doing better than the old process. Here you may well start out with some experimental studies, but you will still need observational studies to complete the process of bringing the upgraded process up to full potential.

Thus, while it is important to operate to meet requirements, operating at full potential is a key element in meeting requirements in an economic manner. Both approaches to process improvement are necessary. Neither is sufficient by itself.

1.5 The Fork in the Road to Process Improvement

Everyone understands the need to operate a process so that it will meet requirements: If the specifications given by the customer are not met, then the customer may go elsewhere. This economical imperative is clear, yet the way to achieve this objective is hardly ever clear. Conventional wisdom tells us to upgrade our bad processes and leave our good processes alone. And the usual definition of a bad process is the percentage of nonconforming product. The conventional wisdom of this approach to improvement is shown schematically in Figure 1.14.

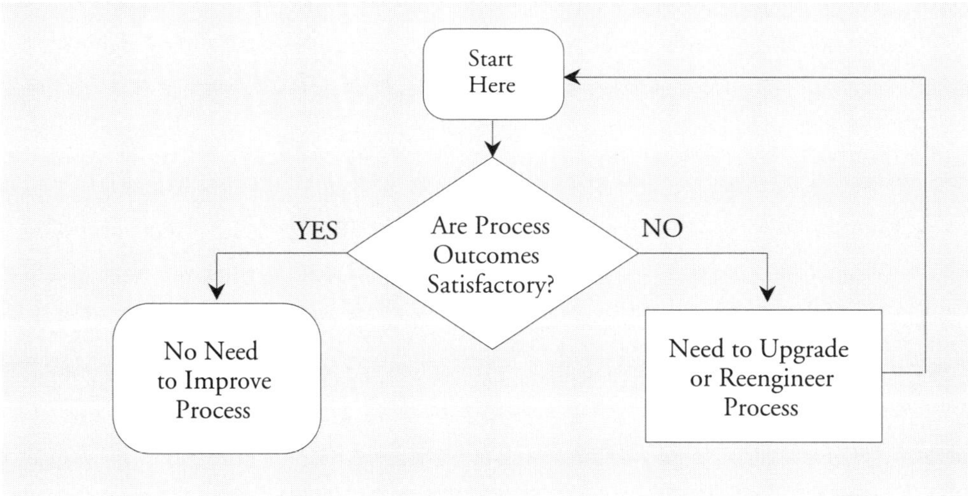

Figure 1.14: Conventional Wisdom about Process Improvement

As noted earlier, this habit of always upgrading your bad processes continually takes you back to Figure 1.7, where you only have partial knowledge about the system of cause-and-effect relationships that affect your process outcomes. Consequently, your process upgrades seldom work as well as they were expected to work. That this happens should not be a surprise. After all, if you cannot operate your current process up to its full potential, then how will you ever operate a new, and upgraded, process up to its full potential? While you might hope that the new process will operate up to its full potential, conventional wisdom offers no methodology for achieving this elusive goal. The only methodology for operating a process up to its full potential is the methodology provided by the process behavior chart. It defines the second path to process improvement.

Thus, there is a fork in the road to process improvement, and the fork you need to take will depend upon what type of behavior your process displays. As shown in Figure 1.15, if you pick the wrong improvement strategy, your efforts are likely to be wasted. Upgrading an unpredictable process is a waste of time and effort, while looking for root causes of variation

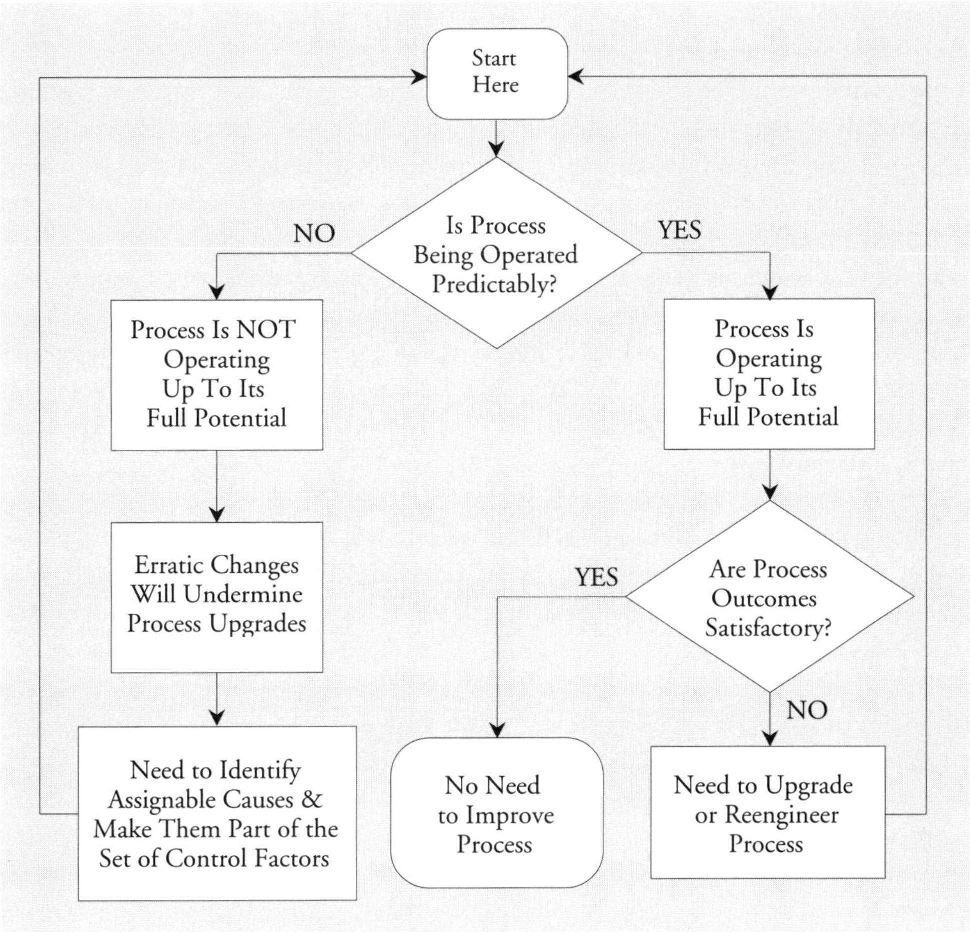

Figure 1.15: The Fork in the Road to Process Improvement

in a predictable process will lead to frustration. And the only way to answer the question in the first decision diamond of Figure 1.15 will be to use a process behavior chart.

Since it has proven to be easier and cheaper to operate existing processes up to their full potential than to upgrade and reengineer processes, any attempt to discuss process improvement that does not take advantage of process behavior charts to operate processes predictably is inherently flawed and incomplete. Shewhart's fundamental observation that some processes are operated predictably, while others are operated unpredictably, is demonstrably true. It has been proven over and over again. Ignore this fact of life and your process improvement efforts are likely to be ineffectual.

Since the process behavior chart is the way to answer the question in the first decision diamond of Figure 1.15, the next chapter will discuss the fundamentals of using process behavior charts. Since the question in the second decision diamond requires an assessment of capability, Chapter Three will provide a summary of this important topic.

1.6 Summary

The conceptual model for the nature of data given here explains why it is not economical to try to control all of the cause-and-effect relationships that affect any given product characteristic. By defining the economic equilibrium for a process this model also shows how to move a process toward that state of equilibrium. Since operating a process at full potential requires a different approach than reengineering a process, this model effectively defines a second path to process improvement and shows how the two paths are related. Some of the implications of this model for how we use data are the following:

The first implication is that trusting in the results of a series of "experimental studies" can never be completely satisfactory. Experiments are very successful at answering specific questions and confirming suspected relationships. They are of limited utility when we do not know what questions to ask.

The very nature of an experiment demands that you study some relationships while ignoring other relationships. For this reason, what you learn from experiments is likely to be true, but it is also likely to be incomplete. Thus, the second implication of the model above is that you need to complement and complete the work of R&D by using observational studies where you listen to your process data when *all* of the cause-and-effect relationships are present. This is the only reliable way to determine if your process contains any Assignable Causes like Factors 14 and 16. Observational studies are a necessary step in learning from the process data and developing a complete understanding of our processes.

The third implication of the model above is that you should start any investigation of an existing process with an observational study. If you try to adjust, improve, or optimize an existing process without first checking for the presence of Assignable Causes you may end up playing with less than a full deck. As a result you may find that you are erroneously adjusting the level of Factor 4 instead of working to remove the effects of Factors 14 and 16.

Thus, in addition to the tools and techniques that allow you to analyze experimental data, you will also need a way to analyze data from observational studies. And the primary tool for use with observational studies are process behavior charts.

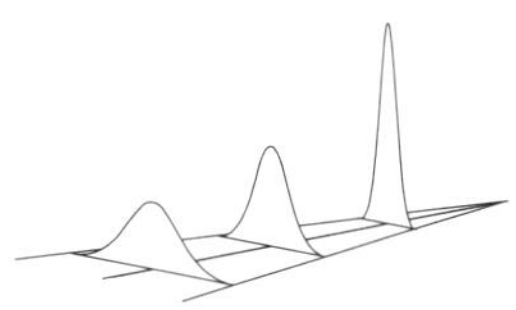

Control Factors Set the Average
Uncontrolled Factors Create the Variation

Chapter Two

The Analysis of Observational Data

When you perform an experiment, you are essentially interrogating our process by asking a series of questions of the form: "What will happen if I do this?" As long as you ask the right questions you can obtain useful information. But if you do not know enough to ask *all* of the important questions, then you may well miss out on some useful information.

With an observational study you are simply listening to your process. Instead of trying to figure out what questions to ask, you let the process tell you what is important.

As our model in the previous chapter showed, there is a time for both kinds of studies. This chapter will focus on how to analyze and interpret data from an observational study. Since this will differ from the way we analyze and interpret data from an experimental study, we need to begin by clarifying the differences between these two ways of obtaining data.

2.1 Observational Studies versus Experimental Studies

Observational studies are studies where the data are obtained as a by-product of some continuing operation. These data may be deliberately and intentionally collected, but they are obtained while the process is operated in an ordinary manner. In other words, in an observational study, the data track the operation of the process. Thus, observational data will consist of a stream of values which should contain no signals. Here you will need to analyze and interpret each value (or each set of values) *as it becomes available.* This will require a sequential analysis technique.

On the other hand, experimental data are deliberately collected under different conditions, with the intention of determining if the different conditions result in different outcomes. Here you will end up with a fixed amount of data that should contain some signals. Such data are usually analyzed all at once in a single act of analysis.

Hence, there are structural differences between observational data and experimental data, and these differences require different types of analytical techniques.

When analyzing data there are two mistakes that can be made. One of these is detecting a signal when no signal is present. Such mistakes are called false alarms. The other is missing a signal when it is present. In performing any analysis it is possible to reduce the occurrence of one type of mistake by increasing the chance of making the other mistake. Analyses with a small risk of a false alarm are said to be conservative. Analyses with larger risks of a false alarm are said to be traditional or exploratory.

Because experimental data are collected under special conditions they tend to be more expensive than observational data. Moreover, because of the way they are obtained, we expect experimental data to represent the differences between the special conditions. Thus, when we analyze experimental data we are looking for signals that we have paid good money to create and which we believe are contained within the fixed amount of data. In order to avoid missing these signals within our experimental data we will tend to use a traditional analysis, or even an exploratory analysis with a larger risk of a false alarm. (Any false alarms will be sorted out fairly quickly in practice, but if we fail to detect a difference we may have to perform more experiments at additional expense.)

When analyzing data from an observational study, the fact that all of the data were supposedly collected under the same conditions will mean that we do not expect to find any differences within the data. Furthermore, since any differences that we may find will be interpreted as indications of unplanned changes in the process, we will want to be sure about any differences we find. With additional data available, we can afford to play a waiting game with observational studies. This means that our analysis will need to be conservative, with a small risk of a false alarm, so that we will have strong evidence of a process change before taking action.

These differences between observational studies and experimental studies are summarized in Table 2.1.

Table 2.1: Observational Studies versus Experimental Studies

Observational Studies	*Experimental Studies*
One Condition Present	Two or More Conditions Present
Data Should Contain No Signals	Data Should Contain Some Signals
Additional Data Available	Fixed Amount of Data Available
Sequential Analysis Procedure	All Data Analyzed at One Time
Conservative Analysis Used	Traditional or Exploratory Analysis Used

Thus, with an observational study, we have a different attitude to the analysis, the analysis itself is done differently, and the nature of what we are looking for changes. This is why many who are used to the analysis of experimental data have trouble understanding the analysis of observational studies.

Process behavior charts were expressly created for the analysis of data from an observational study.

2.2 Process Behavior Charts

In the 1920s a physicist named Walter Shewhart noted that production data behaved differently than data coming from physical processes such as radioactive decay and Brownian motion. As he thought about this discrepancy he came to the conclusion that variation comes in two flavors: While every process will display *routine variation*, some processes will also display *exceptional variation*. The model given in Chapter One is a paraphrased version of Shewhart's explanation for this observed fact.

(Shewhart's label for routine variation was "controlled variation" and his label for exceptional variation was "uncontrolled variation." However, in both cases he was talking about the variation from the set of Uncontrolled Factors. Therefore, in the interest of clarity, we shall continue to use routine variation and exceptional variation.)

Exceptional variation may be thought of as the product of the Assignable Causes, while the routine variation may be thought of as the result of the Common Causes. The purpose of a process behavior chart is to detect any exceptional variation that may be present.

ROUTINE VARIATION

When the set of Uncontrolled Factors for a production process consists of nothing but Common Causes, that production process will display routine variation. The fact that the Common Causes are Uncontrolled Factors means that their levels will be constantly changing. However, since there are generally a large number of Common Causes, and since their

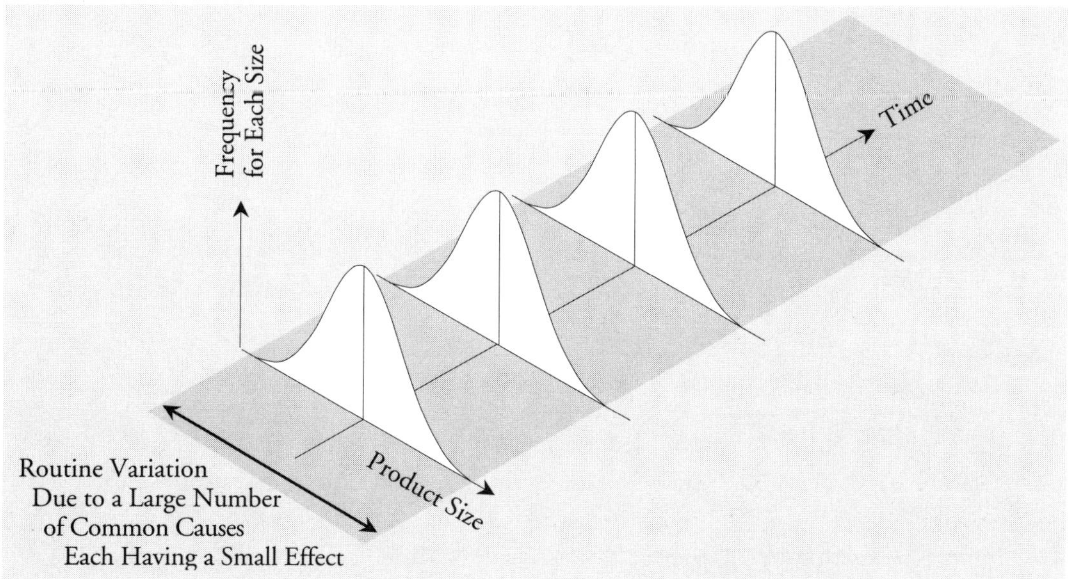

Figure 2.1: Routine Variation Results in a Predictable Process

effects are all small and all about the same size, the result of all of these changes will tend to look like a steady-state level of variation that can best be described as background noise. Here the production process will be operating at its economic equilibrium, and the variation in the product characteristic will be at its minimum. Such a process will be consistent over time, as shown in Figure 2.1.

EXCEPTIONAL VARIATION

On the other hand, when one or more Assignable Causes is present the production process will display exceptional variation. Over time, as the levels of the Uncontrolled Factors change, the many Common Causes will once again produce a steady-state level of background noise. However, because of the dominant nature of their effects, whenever an Assignable Cause changes level we will often see a definite change in the process outcomes. As a result the stream of process outcomes will appear to be erratic and changing. These changes may appear as shifts in level or as changes in dispersion, or both, as shown in Figure 2.2.

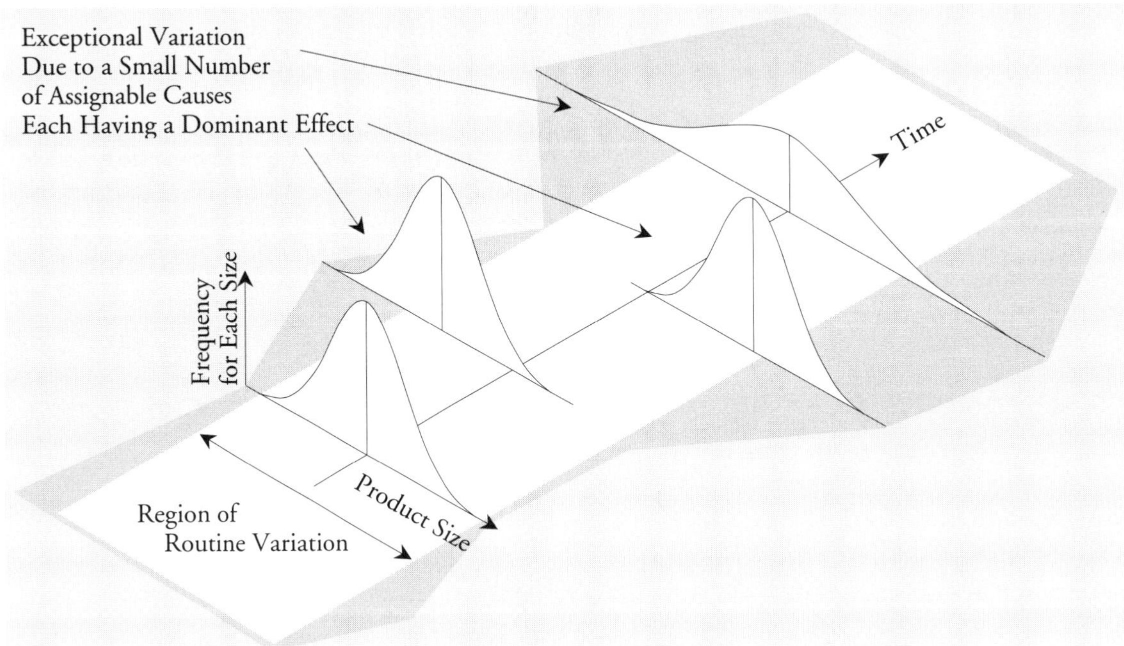

Figure 2.2: Exceptional Variation Results in an Unpredictable Process

Careful thought about our model from Chapter One will show that the only critical difference between an Assignable Cause and a Common Cause is the magnitude of their effects upon the process outcomes. As shown in Figures 2.1 and 2.2, this provides us with a clue on how we might make a practical distinction between routine variation and exceptional variation. While predictable processes will possess routine variation, unpredictable processes will

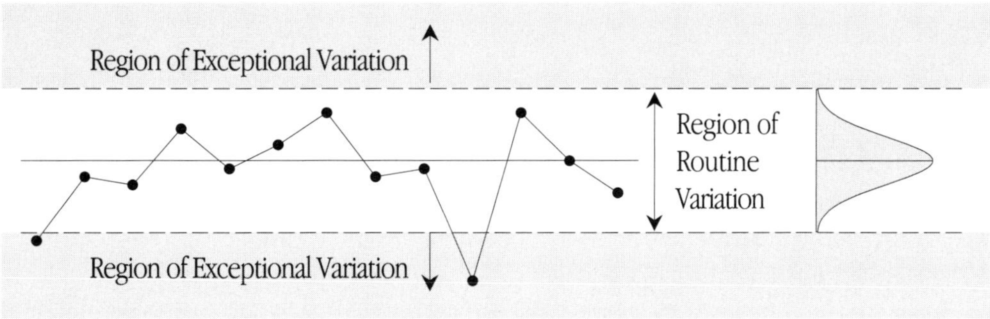

Figure 2.3: Separating Routine Variation from Exceptional Variation

possess both routine variation and exceptional variation. Thus, if we can filter out the routine variation, anything left over is likely to be exceptional variation. As seen in Figure 2.3, the question is where to draw the lines separating routine variation from exceptional variation.

If we place the limits too close together we will get *false alarms* when routine variation causes a point to fall outside the limits. Given that we are looking at a sequential analysis procedure, and that we want a conservative analysis, we will want to have very few false alarms.

On the other hand, if we place the limits too far apart we will *miss some signals* of exceptional variation. In an observational study we are not trying to detect every signal regardless of its size, but are concerned with detecting those signals that are large enough that it will economical to take action based upon the signal.

THREE - SIGMA LIMITS

Thus, the trick is to strike a balance between false alarms and missing signals, and this is exactly what Walter Shewhart did when he created the *control chart.** In his book, *Economic Control of Quality of Manufactured Product,* Shewhart found that "three-sigma" limits would effectively balance the economic consequences of getting false alarms with those of missing some signals. Three-sigma limits separate the exceptional variation from the routine variation by bracketing approximately 99% to 100% of the routine variation. (This is true regardless of the shape of the curve in Figure 2.3. For more on this see Appendix II.) As a consequence of having filtered out virtually all of the routine variation, we can be reasonably sure that a point that falls outside the limits is the result of exceptional variation. It is this conservative nature of the three-sigma limits that provides a useful guide on when to take action in practice. Points outside the limits represent potential signals of Assignable Causes and need to be investigated as such. Such points represent opportunities to move the process closer to its economic equilibrium.

* Because of the baggage associated with the word "control" this name can be quite misleading. Therefore we shall use the more descriptive name of process behavior chart to describe this versatile technique.

CAPTURING ROUTINE VARIATION

In computing the three-sigma limits it is essential to use a measure of dispersion that will capture the routine variation, but which will, at the same time, avoid (or minimize) contamination by any signals of exceptional variation that may be present. To this end Shewhart chose to use a computation that is the foundation of all modern techniques of statistical analysis—the within-subgroup variation. When three-sigma limits are computed correctly they will always depend upon a summary within-subgroup measure of dispersion.

As a consequence of this need to capture and use the routine variation in computing the limits there are correct and incorrect ways of computing a measure of dispersion for use with a process behavior chart. This means, among other things, that you cannot use a global measure of dispersion, such as the standard deviation statistic computed using all of the data, to find limits for a process behavior chart. This is why the limits are referred to as three-sigma limits rather than three-standard-deviation limits.

The twin computational foundations of Shewhart's approach to the analysis of observational data are (1) the use of the within-subgroup variation and (2) the computation of three-sigma limits. Shewhart's approach not only has strong theoretical underpinnings, but it also has a proven track record. Practitioners should avoid tampering with this successful combination. (See Tables A.1 to A.8 for the correct formulas and the notation used.)

There are two basic types of process behavior charts. The original form used subgroups of data "collected under essentially the same conditions." This chart is known as an Average and Range Chart. A few years later a second form, known as a Chart for Individual Values and Moving Ranges, was developed for use with single values. These two types of charts will be sufficient for the analysis of almost all observational data.

2.3 Average and Range Charts

The Average and Range Chart organizes a stream of process measurements into rational subgroups of size n. These subgroups are said to be *rational* because they depend upon a *judgment* about the conditions under which the values were obtained—*when two values may be judged to have been collected under essentially the same conditions, they may be placed in the same subgroup.* (Conversely, when two values *may have been collected under different conditions,* they will need to be placed in different subgroups.) Thus, the question of how to organize the data into rational subgroups will always require process knowledge. Any subgrouping created without regard for the context of the data will be arbitrary rather than rational.

Having grouped together values that were collected under essentially the same conditions, we can use the variation within the subgroups to obtain an estimate of the routine variation. This need for homogeneity within the subgroups will tend to favor smaller subgroup sizes, with subgroup sizes of $n = 4$ to $n = 10$ being the most common.

To illustrate the Average and Range Chart we shall use a data set given by Walter

Table 2.2: 204 Measurements of Electrical Resistance in Megohms

Subgroup	-1-	-2-	-3-	-4-	-5-	-6-	-7-	-8-	-9-	-10-	-11-
Value 1	5045	4290	3980	3300	5100	4635	4410	4725	4790	4110	4790
Value 2	4350	4430	3925	3685	4635	4720	4065	4640	4845	4410	4340
Value 3	4350	4485	3645	3463	5100	4810	4565	4640	4700	4180	4895
Value 4	3975	4285	3760	5200	5450	4565	5190	4895	4600	4790	5750
Average	***4430.0***	***4372.5***	***3827.5***	***3912.0***	***5071.25***	***4682.5***	***4557.5***	***4725.0***	***4733.75***	***4372.5***	***4943.75***
Range	***1070***	***200***	***335***	***1900***	***815***	***245***	***1125***	***255***	***245***	***680***	***1410***
Subgroup	-12-	-13-	-14-	-15-	-16-	-17-	-18-	-19-	-20-	-21-	-22-
Value 1	4740	4170	4170	4175	2920	4090	4640	4215	4615	4700	4095
Value 2	5000	3850	4255	4550	4375	5000	4335	4275	4735	4700	3940
Value 3	4895	4445	4170	4450	4375	4335	5000	4275	4215	4700	3700
Value 4	4255	4650	4375	2855	4355	5000	4615	5000	4700	4095	3650
Average	***4722.5***	***4278.75***	***4242.5***	***4007.5***	***4006.25***	***4606.25***	***4647.5***	***4441.25***	***4566.25***	***4548.75***	***3846.25***
Range	***745***	***800***	***205***	***1695***	***1455***	***910***	***665***	***785***	***520***	***605***	***445***
Subgroup	-23-	-24-	-25-	-26-	-27-	-28-	-29-	-30-	-31-	-32-	-33-
Value 1	4445	4560	5000	4850	4855	4100	4050	4430	3075	4425	4840
Value 2	4000	4700	4575	4850	4160	4340	4050	4300	2965	4300	4310
Value 3	4845	4310	4700	4570	4325	4575	4685	4690	4080	4430	4185
Value 4	5000	4310	4430	4570	4125	3875	4685	4560	4080	4840	4570
Average	***4572.5***	***4470.0***	***4676.25***	***4710.0***	***4366.25***	***4222.5***	***4367.5***	***4495.0***	***3550.0***	***4498.75***	***4476.25***
Range	***1000***	***390***	***570***	***280***	***730***	***700***	***635***	***390***	***1115***	***540***	***655***
Subgroup	-34-	-35-	-36-	-37-	-38-	-39-	-40-	-41-	-42-	-43-	-44-
Value 1	4700	4450	3635	4340	5000	4770	4850	5000	5075	4925	5075
Value 2	4440	4450	3635	4340	4850	4500	4700	4700	5000	4775	4925
Value 3	4850	4850	3635	3665	4775	4770	5000	4500	4770	5075	5250
Value 4	4125	4450	3900	3775	4500	5150	5000	4840	4570	4925	4915
Average	***4528.75***	***4550.0***	***3701.25***	***4030.0***	***4781.25***	***4797.5***	***4887.5***	***4760.0***	***4853.75***	***4925.0***	***5041.25***
Range	***725***	***400***	***265***	***675***	***500***	***650***	***300***	***500***	***505***	***300***	***335***
Subgroup	-45-	-46-	-47-	-48-	-49-	-50-	-51-				
Value 1	5600	4325	4500	4850	4625	4080	5150				
Value 2	5075	4665	4765	4930	4425	3690	5250				
Value 3	4450	4615	4500	4700	4135	5050	5000				
Value 4	4215	4615	4500	4890	4190	4625	5000				
Average	***4835.0***	***4555.0***	***4566.25***	***4842.5***	***4343.75***	***4361.25***	***5100.0***	*Grand Average = 4498.18*			
Range	***1385***	***340***	***265***	***230***	***490***	***1360***	***250***	*Average Range = 658.6*			

Shewhart on page 20 of *Economic Control of Quality of Manufactured Product*. These data consist of 204 measurements of the resistance of an insulator used in telephone wire.

Shewhart arranged these data into 51 subgroups of size 4, based on the time order for the 204 measurements. The idea being that successive measurements are more likely to be obtained under similar conditions than values that are obtained at wider intervals of time. These subgroups, along with their averages and ranges, are shown in Table 2.2.

An Average and Range Chart will consist of two graphs: the subgroup averages will be on one chart and the subgroup ranges will be on the other chart. The central line for the Average Chart is typically the grand average, while the central line for the Range Chart is typically the average range.

To compute three-sigma limits for the Average Chart we will need to multiply the average range by the scaling factor A_2. From Table A.3, for $n = 4$, $A_2 = 0.729$:

$$\textit{Limits for Averages} = \textit{Average}(\bar{X}) \pm 3\ \textit{Sigma}(\bar{X}) = \textit{Grand Average} \pm A_2 \times \textit{Average Range}$$
$$= \bar{\bar{X}} \pm A_2\, \bar{R} = 4498.2 \pm 0.729\,(658.6) = 4018 \text{ to } 4978$$

To compute three-sigma limits for the Range Chart we again use the within-subgroup variation, as summarized by the average range, to characterize both location and dispersion. Using the bias correction factors of d_2 and d_3 along the way, we end up with scaling factors known as D_3 and D_4. For $n = 4$ the values for D_3 and D_4 are −0.282 and 2.282. However, since the ranges are non-negative, the boundary value of zero replaces the lower limit.

$$\textit{Limits for Ranges} = \textit{Average(R)} \pm 3\ \textit{Sigma(R)}$$
$$= \textit{Average(R)} \pm 3\frac{d_3}{d_2}\ \textit{Average(R)} = \left[1 \pm 3\frac{d_3}{d_2}\right] \textit{Average(R)}$$
$$= D_3\ \textit{Average(R)}\ \textit{to}\ D_4\, \textit{Average(R)} = 0\ \textit{to}\ 2.282\ (658.6) = 0\ \textit{to}\ 1503$$

These formulas and scaling factors convert the average range directly into those three-sigma limits that will filter out virtually all of the routine variation. In Figure 2.4 we find two subgroup ranges and 10 subgroup averages outside their respective three-sigma limits—clear evidence that Assignable Causes are affecting this process.

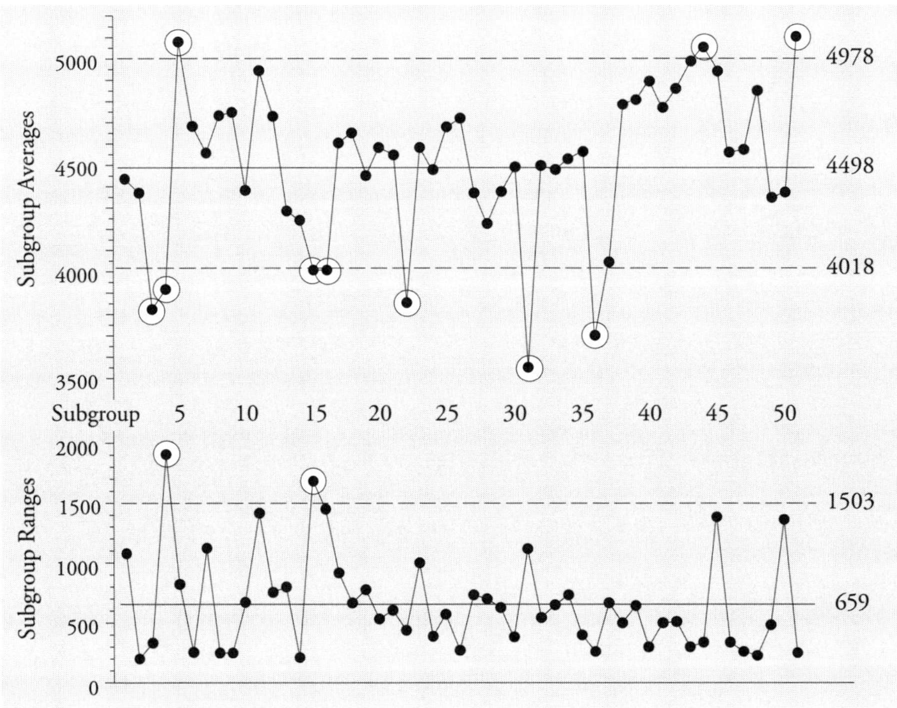

Figure 2.4: Average and Range Chart for Table 2.2

Here we do not need to be concerned with estimates of the process average or estimates of the process dispersion simply because, when a process is subject to Assignable Causes, there is no well-defined process average and there is no well-defined process dispersion. The question is *not* one of estimation, but rather one of taking action. As Shewhart wrote regarding Figure 2.4:

"Further research was instituted at this point to find these [Assignable] causes of variability. Several were found, and after these had been eliminated another series of observed values gave the results in fig. 7-b." [shown on the right side of Figure 2.5.]

The data for the right side of Figure 2.5 are given in Table 2.3. The grand average for these data is 4417.8, and the average range is 367.5. So what was the effect of finding the Assignable Causes and making them part of the set of Control Factors? This may be seen by comparing the width of the two sets of limits. In interpreting the right side of Figure 2.5 Shewhart wrote:

"Here we see that all of the points lie within the limits. We assumed therefore, upon the basis of this test, that it was not feasible for research to go much further in eliminating causes of variability. Because of the importance of this particular experiment, however, considerably more work was done, but it failed to reveal causes of variability. Here then is a typical case where the criterion [three-sigma limits] indicates when variability should be left to chance [Shewhart's term for the Common Causes]."

As we saw in Chapter One, once you reach the point of economic equilibrium for a process, further efforts at identifying causes of variation will always have a low payback.

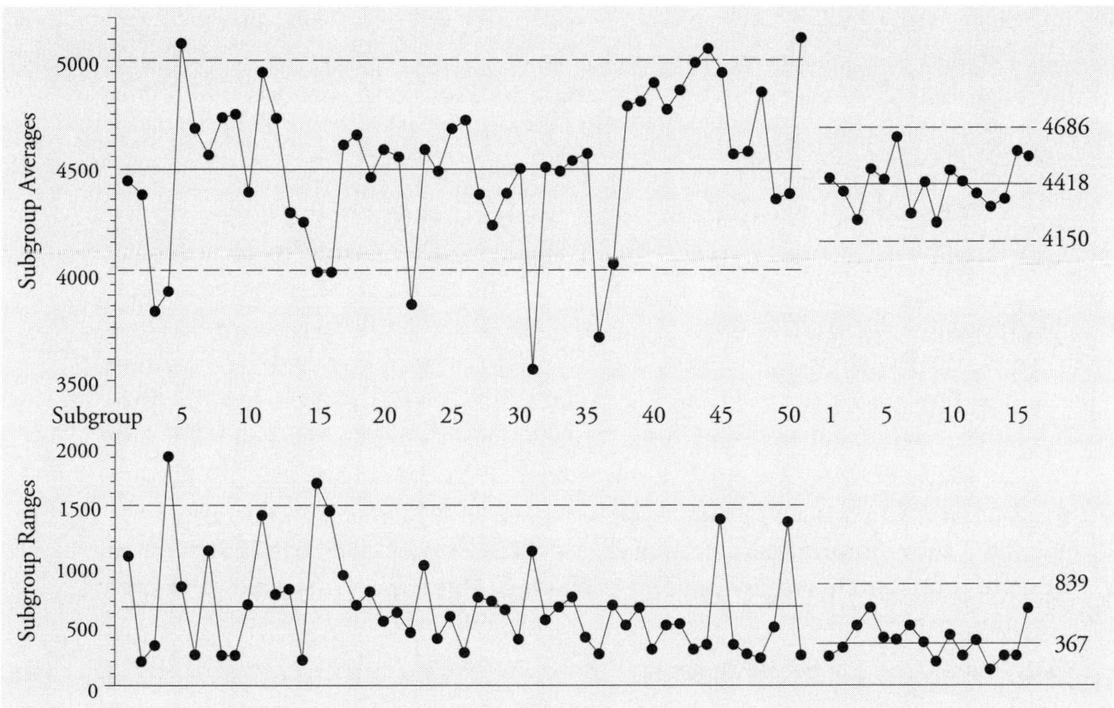

Figure 2.5: Average and Range Charts for Tables 2.2 and 2.3

Table 2.3: 64 Additional Measurements of Electrical Resistance in Megohms

Subgroup	-1-	-2-	-3-	-4-	-5-	-6-	-7-	-8-
Value 1	4400	4385	4340	4445	4465	4675	4260	4460
Value 2	4565	4545	4500	4775	4375	4800	4000	4360
Value 3	4325	4235	4000	4615	4660	4410	4360	4265
Value 4	4495	4370	4165	4125	4260	4675	4500	4635
Average	***4446.25***	***4383.75***	***4251.25***	***4490***	***4440***	***4640***	***4280***	***4430***
Range	***240***	***310***	***500***	***650***	***400***	***390***	***500***	***320***

Subgroup	-9-	-10-	-11-	-12-	-13-	-14-	-15-	-16-
Value 1	4240	4520	4320	4340	4310	4490	4475	4210
Value 2	4350	4545	4570	4350	4250	4240	4725	4540
Value 3	4220	4650	4380	4600	4315	4350	4540	4600
Value 4	4150	4230	4450	4220	4380	4320	4565	4850
Average	***4240***	***4486.25***	***4430***	***4377.5***	***4313.75***	***4350***	***4576.25***	***4550***
Range	***200***	***420***	***250***	***380***	***130***	***250***	***250***	***640***

2.4 The Chart for Individual Values and Moving Ranges

The second major type of process behavior chart is the Chart for Individual Values and Moving Ranges. Instead of arranging the original data into subgroups and plotting the averages and ranges, this chart uses the original data for one of the graphs. This means that every time a new value is obtained it is plotted, analyzed, and interpreted. The simplicity of this approach is particularly advantageous whenever data are constrained to appear one value at a time.

But how are we to obtain a within-subgroup measure of dispersion when there are no subgroups? As Shewhart wrote on page 314 of *Economic Control of Quality of Manufactured Product*:

> "Obviously, if the cause system is changing, the sample [subgroup] size should be as small as possible so that the averages of samples do not mask the changes. In fact single observations would be the most sensitive to such changes. Why then do we not use a sample size of unity? The answer is that if we do, we are faced with the difficulty of choosing the standard deviation to be used in the control charts. ... In fact, the sensitivity of the test [the process behavior chart] will increase, in general, with the decrease in subgroup size until the size of the subgroup is such that the data in any given subgroup come from a constant system of chance [Common] causes. In the absence of any *a priori* information making it possible to divide the data into rational subgroups, there would be some advantage therefore in reducing the subgroup size to unity."

Thus, from the very beginning, the philosophy behind the process behavior chart has been pushing us in the direction of the Chart for Individual Values. Initially, the question of how to measure the within-subgroup variation prevented us from reaching this goal. While Shewhart found a way around this problem for certain types of count-based data (*np*-charts, *p*-charts, *c*-charts and *u*-charts are all Charts for Individual Values), his solution does not generalize to other types of data.

In 1942, W. J. Jennett provided an answer to the problem posed by Shewhart. In his work with the MO Valve Company Ltd., Jennett proposed using a technique that had been used in studying the ballistics of artillery shells in the 1890s. This technique was known as the method of successive differences. By using the differences between successive values as a measure of dispersion, Jennett effectively created the Chart for Individual Values and a Moving Range, or *XmR* Chart.

Table 2.4 contains the same 204 resistivity values given in Table 2.2, along with their 203 moving range values. The average is still 4498.18, while the average moving range is 318.81.

Table 2.4: 204 Measurements of Electrical Resistance in Megohms

X	*mR*	*X*	*mR*	*X*	*mR*	*X*	*mR*	*X*	*mR*
5045		4740	1010	4445	795	4700	130	5600	685
4350	695	5000	260	4000	445	4440	260	5075	525
4350	0	4895	105	4845	845	4850	410	4450	625
3975	375	4255	640	5000	155	4125	725	4215	235
4290	315	4170	85	4560	440	4450	325	4325	110
4430	140	3850	320	4700	140	4450	0	4665	340
4485	55	4445	595	4310	390	4850	400	4615	50
4285	200	4650	205	4310	0	4450	400	4615	0
3980	305	4170	480	5000	690	3635	815	4500	115
3925	55	4255	85	4575	425	3635	0	4765	265
3645	280	4170	85	4700	125	3635	0	4500	265
3760	115	4375	205	4430	270	3900	265	4500	0
3300	460	4175	200	4850	420	4340	440	4850	350
3685	385	4550	375	4850	0	4340	0	4930	80
3463	222	4450	100	4570	280	3665	675	4700	230
5200	1737	2855	1595	4570	0	3775	110	4890	190
5100	100	2920	65	4855	285	5000	1225	4625	265
4635	465	4375	1455	4160	695	4850	150	4425	200
5100	465	4375	0	4325	165	4775	75	4135	290
5450	350	4355	20	4125	200	4500	275	4190	55
4635	815	4090	265	4100	25	4770	270	4080	110
4720	85	5000	910	4340	240	4500	270	3690	390
4810	90	4335	665	4575	235	4770	270	5050	1360
4565	245	5000	665	3875	700	5150	380	4625	425
4410	155	4640	360	4050	175	4850	300	5150	525
4065	345	4335	305	4050	0	4700	150	5250	100
4565	500	5000	665	4685	635	5000	300	5000	250
5190	625	4615	385	4685	0	5000	0	5000	0
4725	465	4215	400	4430	255	5000	0		
4640	85	4275	60	4300	130	4700	300		
4640	0	4275	0	4690	390	4500	200		
4895	255	5000	725	4560	130	4840	340		
4790	105	4615	385	3075	1485	5075	235		
4845	55	4735	120	2965	110	5000	75		
4700	145	4215	520	4080	1115	4770	230		
4600	100	4700	485	4080	0	4570	200		
4110	490	4700	0	4425	345	4925	355		
4410	300	4700	0	4300	125	4775	150		
4180	230	4700	0	4430	130	5075	300		
4790	610	4095	605	4840	410	4925	150		
4790	0	4095	0	4840	0	5075	150		
4340	450	3940	155	4310	530	4925	150		
4895	555	3700	240	4185	125	5250	325		
5750	855	3650	50	4570	385	4915	335		

Once we have these two summary values, Table A.2 defines the three-sigma limits to be:

Natural Process Limits $= Average(X) \pm 3\ Sigma(X)$

$= \bar{X} \pm E_2\ \overline{mR} = 4498.2 \pm 2.660\ (318.8) = 3650 \text{ to } 5346$

Upper Range Limit $= Average(R) + 3\ Sigma(R)$

$= D_4\ \overline{mR} = 3.267\ (318.8) = 1042$

The resulting *XmR* Chart is shown in Figure 2.6.

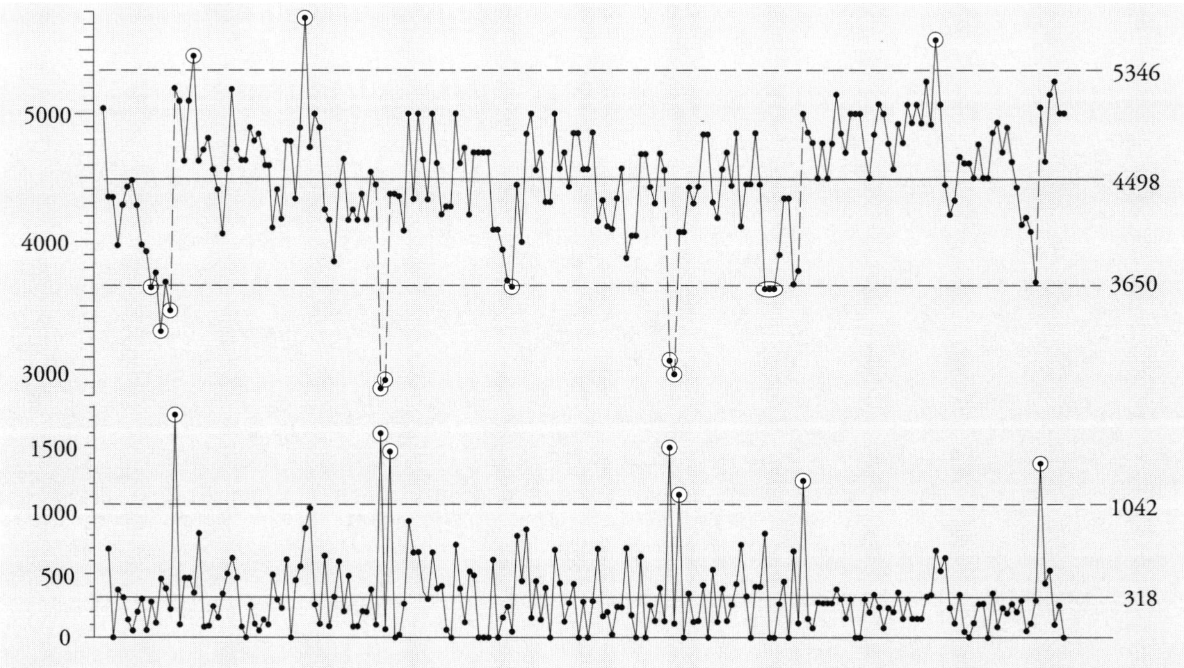

Figure 2.6: *XmR* Chart for Resistivity Values of Table 2.4

Fourteen individual values and seven moving ranges fall outside the computed limits. This is compelling evidence that this process is subject to the exceptional variation created by Assignable Causes. The seven out-of-limit moving ranges identify seven abrupt changes in level for this process. While these seven large values may have slightly inflated the average moving range and the limits, these limits are still sufficient to characterize the process behavior and identify the points where many of the changes occurred. The computations are robust, and as a result the technique is sensitive.

While the overall impression given by Figures 2.4 and 2.6 is the same, the Chart for Individual Values shows more detail. Moreover, since the values plotted are the actual observations, *X* Charts are generally easier to interpret. In the words of one shop foreman, "Nothing can hide on the *XmR* Chart."

The data from Table 2.3 are listed as individual values in Table 2.5 along with their mov-

Table 2.5: 64 Additional Measurements of Electrical Resistance in Megohms

X	*mR*	*X*	*mR*	*X*	*mR*	*X*	*mR*
4400		4465	150	4240	395	4310	90
4565	165	4375	90	4350	110	4250	60
4495	70	4660	285	4220	130	4315	65
4325	170	4260	400	4150	70	4380	65
4385	60	4675	415	4520	370	4490	110
4545	160	4800	125	4545	25	4240	250
4235	310	4410	390	4650	105	4350	110
4370	135	4675	265	4230	420	4320	30
4340	30	4260	415	4320	90	4475	155
4500	160	4000	260	4570	250	4725	250
4165	335	4360	360	4380	190	4540	185
4000	165	4500	140	4450	70	4565	25
4125	125	4460	40	4340	110	4210	355
4445	320	4360	100	4350	10	4540	330
4775	330	4265	95	4600	250	4600	60
4615	160	4635	370	4220	380	4850	250

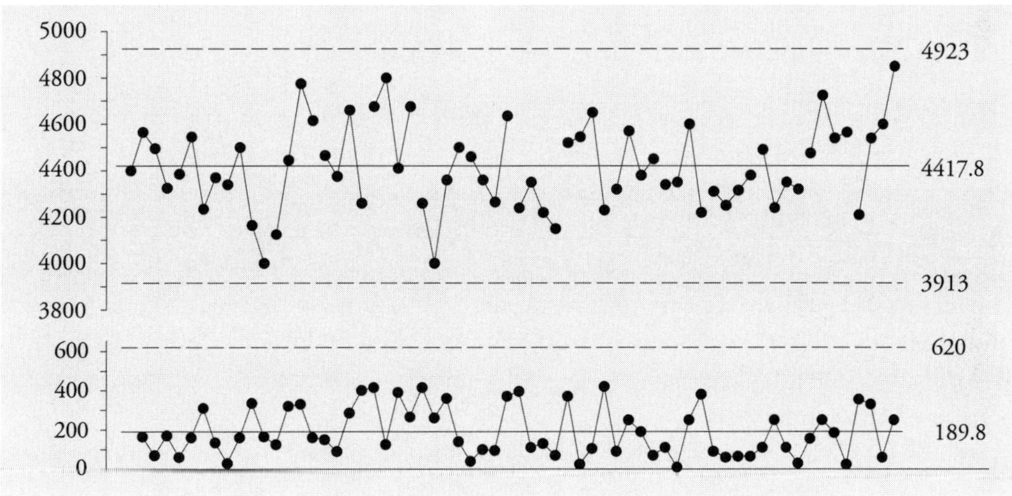

Figure 2.7: *XmR* Chart for Resistivity Values of Table 2.5

ing ranges. The average is 4417.8 and the average moving range is 189.8. The *XmR* Chart is shown in Figure 2.7.

As before, the Average and Range Chart on the right side of Figure 2.5 and the *XmR* Chart of Figure 2.7 tell the same story. After making the Assignable Causes part of the set of Control Factors they had a much more predictable process with substantially reduced variation. While Figure 2.7 is drawn to a different scale than Figure 2.6, with a little work you can see that the limits there are only 60% as wide as those in Figure 2.6.

While the previous examples serve to show how the two different types of charts deliver the same message, the real strength of the *XmR* Chart is that it allows us to work with data where the *only* rational subgroup size is $n = 1$. As an example of this type of data, consider the percentages of methanol in successive batches of alcohol shown in Table 2.6. Here we get one number per batch, and we want to analyze and interpret each value as they become available. Waiting until we have four or five batches before we plot a point is completely unsatisfactory. The average is 4.93 while the average moving range is 0.23.

Table 2.6: Methanol Percentages in 15 Successive Batches of Alcohol

X	4.6	4.7	4.3	4.7	4.7	4.6	4.8	4.8	5.2	5.0	5.2	5.0	5.6	5.5	5.2
mR		0.1	0.4	0.4	0.0	0.1	0.2	0.0	0.4	0.2	0.2	0.2	0.6	0.1	0.3

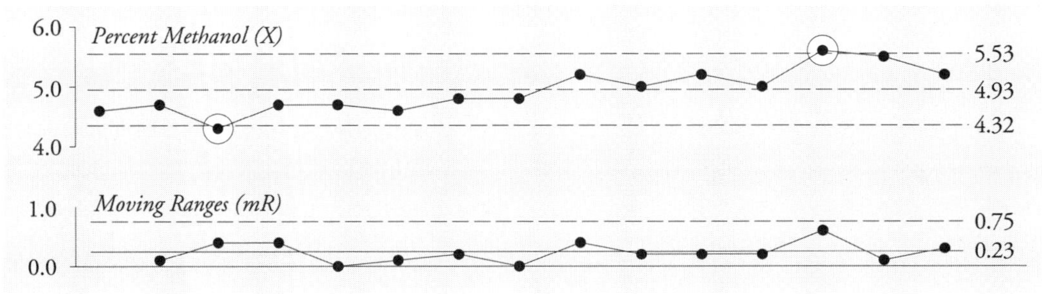

Figure 2.8: An *XmR* Chart for the Methanol Data

The two points outside the Natural Process Limits in Figure 2.8 confirm that the apparent upward trend in the percentages of methanol is indeed real. This production process was changing when it was thought to be operating in a steady state. The Assignable Cause of this trend will need to be identified and controlled before this process can operate up to its full potential.

So what would you predict for the next batch? Do you think it will have 4.9 percent methanol? The fact that this process is changing over time makes all predictions little more than wishful thinking. Having characterized this process as unpredictable, we have made all questions of estimation moot.

There are two basic requirements to be met when creating an *XmR* Chart. The first requirement is that *successive values need to be logically comparable.* This means that you cannot use a single *XmR* chart to compare apples and oranges. Your data must consist of a stream of values that represent the same process or operation. In Figure 2.8 we have satisfied the first requirement by comparing the same measurement on successive batches coming from the same production process.

The second requirement is that *the moving ranges need to isolate and capture the local, short-term, routine variation that is inherent in the process generating the data.* In Figure 2.8 we have satisfied the second requirement by using the differences between successive batches to make allowance for the routine batch-to-batch variation. Since the observed long-term variation exceeds the limits based on this short-term variation we can be confident that the production process in Figure 2.8 is changing.

While the use of the time-order sequence of the data will usually be sufficient to satisfy these two requirements of the *XmR* Chart, there are times when a careful consideration of the structure in the data will require something other than a time-ordered sequence.

(For example, if the data of Table 2.6 represented batches produced by two separate distillation units, say Unit *X* and Unit *Y*, and the values in time order formed the sequence *X*, *Y*, *X*, *Y*, *X*, *Y*, *X*, *Y*, *X*, *Y*,... then the moving ranges would represent the difference between the two units rather than the routine variation of either unit. Here the data would need to be arranged in time order *by unit* before creating two separate charts for the two units.)

2.5 Caution regarding Software

Software companies are not in the business of teaching people the right and wrong ways of computing limits for process behavior charts. They are in the business of selling software. So when a customer who doesn't know any better insists on computing the limits incorrectly (say by using a global measure of dispersion), they will give the customer what he wants. For this reason many programs that create process behavior charts include options that allow you to compute the limits incorrectly.

- Good software does not allow the limits to be computed incorrectly.
- Acceptable software will default to the correct computations, with a warning on the options that allow the incorrect computations.
- Bad software does not guide the user to the correct computations nor warn when the wrong computations are used.

Some software that correctly computes the limits for the Average and Range Chart will compute limits for the *XmR* Chart incorrectly. Some software will use or allow the use of something other than three-sigma limits. So how can you figure out which options to use? The simplest way to verify the computations in any program is to use a data set for which you know the correct limits. (This data set should contain points outside the three-sigma limits on both the chart for location and the chart for dispersion.) Use only those options that give you the correct limits. Tables A.2 to A.8 list virtually all of the correct ways to compute limits for *X* Charts and Average Charts and provide data sets to use in testing software.

There are no shades of gray here. This is not a matter of opinion, but a mathematical fact of life. Limits are either computed correctly, or they are computed incorrectly. And limits computed incorrectly are a sign of either ignorance or dishonesty. Let all users of software beware.

2.6 How Process Behavior Charts Work

To clarify the way that process behavior charts allow us to operate a process at or near its point of economic equilibrium, we need to consider what the various elements of a chart represent. In his second book, *Statistical Method from the Viewpoint of Quality Control,* Shewhart showed how a process behavior chart (1) defines an *Ideal,* (2) provides a *Methodology* for moving toward that Ideal, and (3) allows us to make a *Judgment* about how close to the Ideal we have come.

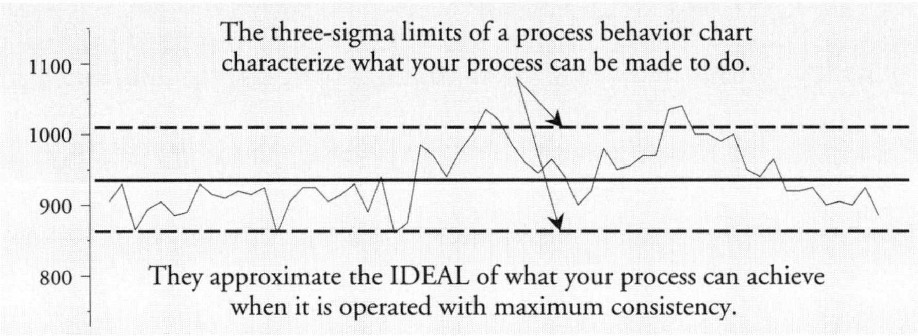

Figure 2.9: The Process Potential

Specifically, in Figure 2.9 the limits on a chart define the *process potential*. They define the range of values that you are likely to see from a predictable process, and they *approximate* the range of values that you could expect to see when you learn how to operate your unpredictable process closer to its full potential. To use Shewhart's word, the three-sigma limits define the *Ideal* of what the process can do when it is operated up to its full potential. They define what any process can accomplish.

The running record defines the actual *process performance*. As seen in Figure 2.10, whenever a point goes outside the three-sigma limits it identifies a departure from the routine, a change in the process, and the presence of an Assignable Cause of exceptional variation. By identifying these points the process behavior chart provides us with a methodology for when to look for Assignable Causes. When we can identify an Assignable Cause and move it from the set of Uncontrolled Factors to the set of Control Factors we will be removing a significant source of variation from the product stream. By removing sources of variation from the product stream we are not merely maintaining the status quo, but are rather tightening up on the process variation and improving both the predictability of the process and the consistency of the process outcomes. So the process behavior chart gives us a *Methodology* for actually moving our process toward the *Ideal*.

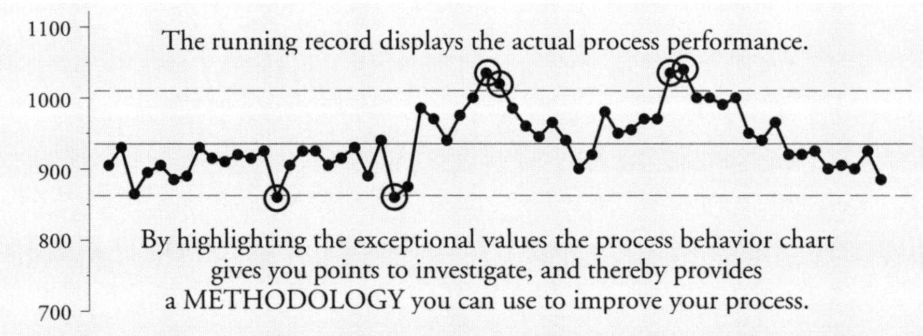

Figure 2.10: The Process Performance

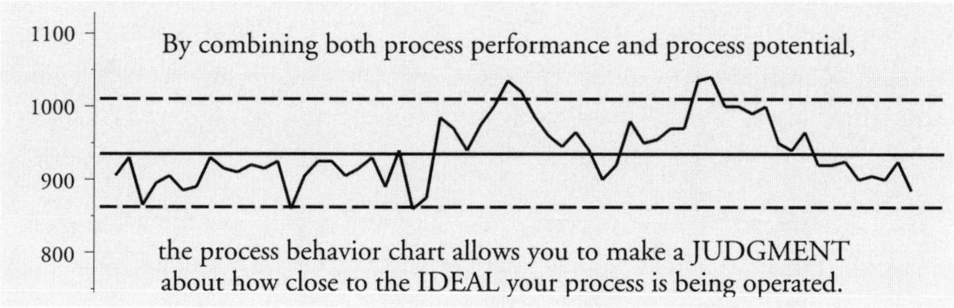

Figure 2.11: The Comparison of Performance with Potential

Finally, in Figure 2.11, by combining both the process potential and the process performance on the same graph, the process behavior chart allows us to make a *Judgment* about how close to full potential our process is operating. The absence of points outside the limits tells of a reasonable degree of predictability. For an unpredictable process, the number of points outside the limits, and the extent to which they fall outside the limits, will quantify the degree of unpredictability.

Hence, a process behavior chart is in itself an operational definition of how to get the most out of any process. It defines the Ideal, it provides a Method for moving toward that Ideal, and it allows a Judgment about how close the process is to its point of economic equilibrium. It is this complete package that makes the process behavior chart the locomotive of continual improvement.

2.7 Using Process Behavior Charts for Continual Improvement

As an example of how the charts may be used on a continuing basis to improve a production process, we use the production of the outer shell of an automotive cigar lighter by the Tokai Rika company. The dimension tracked is that of the distance from the flange to the detent as shown on the left side of Figure 2.12.

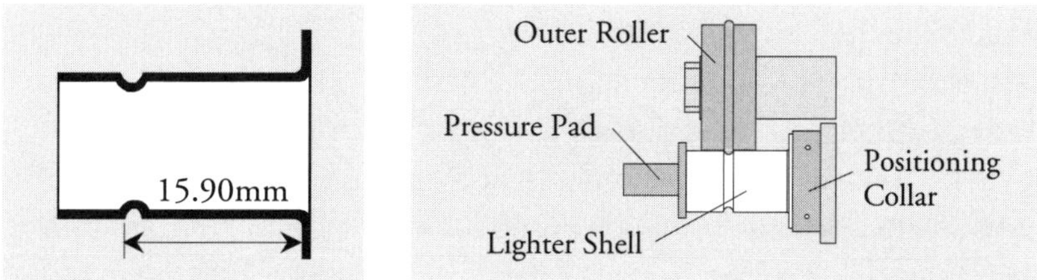

Figure 2.12: Cigar Lighter Detent Dimension and the Rolling Operation

Production involves a die stamping operation to form the cylinder with a flange on one end followed by a rolling operation which forms the detent. The data from the first 24 days of production for a new part number are given in Table 2.7. The values shown are the detent dimension expressed as the number of hundredths of a millimeter in excess of 15.00 mm. Thus, the target value of 15.90 mm for this dimension shown in Figure 2.12 corresponds to a recorded value of 90 and the specifications are 80 to 100. Four times each day (at 10 am, 11 am, 2 pm, and 4 pm) a single part is selected out of the product stream and measured. These four values are then combined into a single subgroup for that day and the average and range are computed.

Table 2.7: Tokai Rika Lighter Detent Data for Days 1 to 24

Day	-1-	-2-	-3-	-4-	-5-	-6-	-7-	-8-	-9-	-10-	-11-	-12-
10 am	90	91	88	88	87	90	88	90	91	90	92	91
11 am	88	87	91	89	91	91	92	90	89	91	89	88
2 pm	91	89	90	90	91	92	91	88	89	88	90	89
4 pm	89	89	89	90	90	89	90	89	87	90	90	88
Average	89.50	89.00	89.50	89.25	89.75	90.50	90.25	89.25	89.00	89.75	90.25	89.00
Range	3	4	3	2	4	3	4	2	4	3	3	3

Day	-13-	-14-	-15-	-16-	-17-	-18-	-19-	-20-	-21-	-22-	-23-	-24-
10 am	89	91	91	90	91	92	90	89	88	91	91	92
11 am	89	88	89	92	92	89	91	92	89	92	91	91
2 pm	91	88	89	92	89	89	90	92	90	90	90	88
4 pm	89	92	87	89	92	90	89	88	90	89	90	88
Average	89.50	89.75	89.00	90.75	91.00	90.00	90.00	90.25	89.25	90.50	90.50	89.75
Range	2	4	4	3	3	3	2	4	2	3	1	4

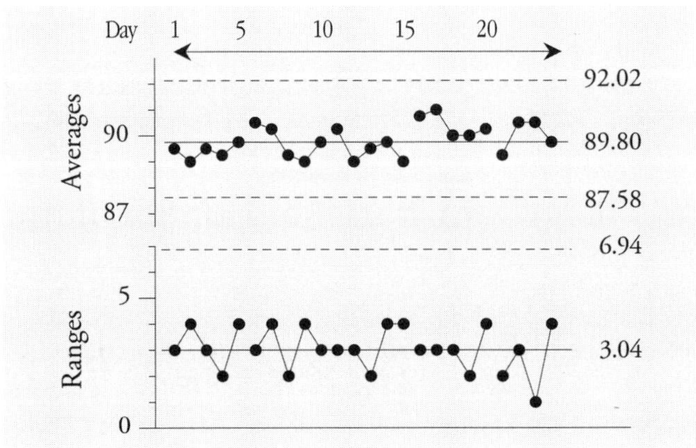

Figure 2.13: Tokai Rika Lighter Detent Dimension: Average and Range Chart for Days 1 to 24

At Tokai Rika they computed limits using the first 24 subgroups. For these 24 subgroups the grand average is 89.80 and the average range is 3.04. These values result in the limits shown in Figure 2.13. Interpreting the Average and Range Chart for this baseline period, we would conclude that this process shows a reasonable degree of predictability. The homogeneity of these data means that the parts we have measured may be taken as being representative of the parts we have not measured, and we may reasonably and realistically characterize the whole process stream using these 96 data. We do this by computing limits for individual values according to the formula from Table A.3:

$$\textit{Natural Process Limits} = \textit{Grand Average} \pm E_2 \ \textit{Average Range}$$
$$= 89.80 \pm 1.457\,(3.04) = 85.4 \text{ to } 94.2$$

With a predictable process such as this we would expect to find virtually all of the individual values within the Natural Process Limits. Here these limits range from 86 to 94 inclusive. Since we do not have access to the parts not measured during the baseline period, we can only compare these limits against the observed values obtained from this process after Day 24. We do this with the histogram in Figure 2.14 which shows the 244 data from the 61 day period of Day 25 to Day 85.

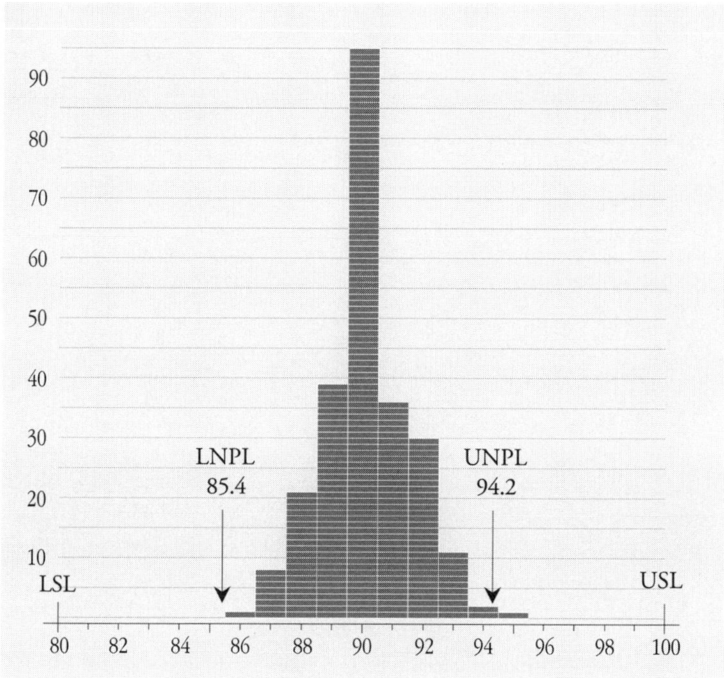

Figure 2.14: Histogram of Detent Values from Days 25 to 85

As expected, we find 243 out of 244 values in the interval of 86 to 94. It is important to note that the limits shown were not obtained from the data shown. We have *extrapolated* from the product produced in the baseline period (Days 1 to 24) to characterize the product produced and measured after that baseline period (Days 25 to 85). Moreover, if we can successfully extrapolate to characterize the product not yet produced, we can also extrapolate to characterize the product not measured as well. Thus, we can reasonably say that virtually all of the product produced on Days 1 to 85 fell in the range of 86 to 94 inclusive. At a production rate of 17,000 parts per day, this amounts to 1.45 million parts.

How can we justify using 96 data to characterize 1.45 million parts? The only justification for such an extrapolation is the predictability displayed by the underlying process in Figure 2.13. When a process has been operated predictably in the past, it is reasonable to assume that, barring any deliberate changes, it will continue to be operated predictably in the foreseeable future. On the other hand, when a process has been operated unpredictably in the past, it is unreasonable to assume that it will spontaneously do better in the future.

Table 2.8: Tokai Rika Lighter Detent Data for Days 25 to 125

-25-	-26-	-27-	-28-	-29-	-30-	-31-	-32-	-33-	-34-	-35-	-36-	-37-	-38-	-39-
91	90	89	90	89	90	90	90	91	92	92	93	90	90	90
92	92	90	90	88	91	91	92	92	92	93	94	90	90	90
90	89	92	89	92	92	93	90	92	91	92	94	90	90	90
93	89	92	89	92	89	90	90	93	92	93	95	90	90	91
91.50	90.00	90.75	89.50	90.25	90.50	91.00	90.50	92.00	91.75	92.50	94.00	90.00	90.00	90.25
3	3	3	1	4	3	3	2	2	1	1	2	0	0	1

-40-	-41-	-42-	-43-	-44-	-45-	-46-	-47-	-48-	-49-	-50-	-51-	-52-	-53-	-54-
90	89	90	90	91	90	90	90	90	90	90	90	90	90	90
91	90	91	90	90	91	90	90	90	90	90	90	90	88	91
90	90	90	91	91	90	90	91	90	90	91	90	89	91	88
90	91	91	89	90	90	90	90	90	90	90	90	90	90	90
90.25	90.00	90.50	90.00	90.50	90.25	90.00	90.25	90.00	90.00	90.25	90.00	89.75	89.75	89.75
1	2	1	2	1	1	0	1	0	0	1	0	1	3	3

-55-	--56-	-57-	-58-	-59-	-60-	-61-	-62-	-63-	-64-	-65-	-66-	-67-	-68-	-69-
90	90	87	88	87	86	87	88	89	90	89	89	87	89	89
89	91	88	90	88	87	87	89	90	91	87	89	90	89	88
90	90	88	89	88	89	89	89	88	88	88	90	89	89	89
89	90	89	89	89	87	90	88	89	89	89	90	89	88	88
89.50	90.25	88.00	89.00	88.00	87.25	88.25	88.50	89.00	89.50	88.25	89.50	88.75	88.75	88.50
1	1	2	2	2	3	3	1	2	3	2	1	3	1	1

-70-	-71-	-72-	-73-	-74-	-75-	-76-	-77-	-78-	-79-	-80-	-81-	-82-	-83-	-84-
88	91	93	91	92	91	91	90	89	88	88	90	90	93	93
89	92	90	89	91	92	90	92	90	90	92	91	90	92	90
88	90	91	91	92	90	90	89	91	91	91	91	93	91	91
88	93	92	92	90	92	92	90	92	89	90	92	90	90	91
88.25	91.50	91.50	90.75	91.25	91.25	90.75	90.25	90.50	89.50	90.25	91.00	90.75	91.50	91.25
1	3	3	3	2	2	2	3	3	3	4	2	3	3	3

-85-	-86-	-87-	-88-	-89-	-90-	-91-	-92-	-93-	-94-	-95-	-96-	-97-	-98-	-99-
91	90	90	91	93	90	91	92	92	92	93	93	93	93	92
90	91	92	92	92	92	93	93	90	91	93	92	93	91	93
92	90	92	90	92	90	90	91	90	91	92	93	92	92	92
92	90	91	93	92	90	90	91	91	90	92	91	93	93	93
91.25	90.25	91.25	91.50	92.25	90.50	91.00	91.75	90.75	91.00	92.50	92.25	92.75	92.25	92.50
2	1	2	3	1	2	3	2	2	2	1	2	1	2	1

-100-	-101-	-102-	-103-	-104-	-105-	-106-	-107-	-108-	-109-	-110-	-111-	-112-	-113-	-114-
92	93	94	93	94	94	94	93	93	93	94	93	93	94	95
93	92	93	94	93	94	93	93	94	93	93	93	93	93	94
92	93	93	94	94	94	93	94	94	94	93	93	95	95	94
92	93	94	94	95	95	94	93	94	93	94	93	94	95	93
92.25	92.75	93.50	93.75	94.00	94.25	93.50	93.25	93.75	93.25	93.50	93.00	93.75	94.25	94.00
1	1	1	1	2	1	1	1	1	1	1	0	2	2	2

-115-	-116-	-117-	-118-	-119-	-120-	-121-	-122-	-123-	-124-	-125-
94	95	93	92	95	93	93	94	94	94	95
94	94	94	95	95	95	95	95	95	94	94
94	94	94	94	94	94	94	94	96	95	93
94	94	93	95	94	95	95	95	95	92	95
94.00	94.25	93.50	94.00	94.50	94.25	94.25	94.50	95.00	93.75	94.25
0	1	1	3	1	2	2	1	2	3	2

Past behavior is always the best predictor of future behavior.

Thus, the characterization of a process as either predictable or unpredictable is a fundamental dichotomy for data analysis. However, even the best process will still be subject to the effects of entropy. While we may expect our process to continue as before, we need to be aware of changes that may occur without warning. As this example continues we will see how the operators at Tokai Rika used the Average and Range Chart not only to track this process, but also to improve it.

But first we need to address two questions that commonly occur at this point: "Are four pieces out of seventeen thousand enough?" and "Is one subgroup per day adequate?" Tokai Rika let the process behavior chart answer both of these questions. As we will see, four pieces per day is sufficient to detect changes that occur in this process. *As long as you are detecting changes in the process when they occur you have enough data.*

Moreover, since changes in this process tend to develop slowly over a period of several days, one subgroup per day is adequate. You only need enough subgroups to detect changes in your process. So the key question regarding subgroup frequency is, "How quickly can your process change?"

The limits from Figure 2.13 were extended forward and used to track and evaluate the production process. (This extension of limits from a baseline period is shown here by dashed lines. Baseline periods are shown by the arrows under the scale of days, and limits for baseline periods are shown by dotted lines.) As may be seen in Figure 2.15, they found evidence of a process change on Days 35 and 36.

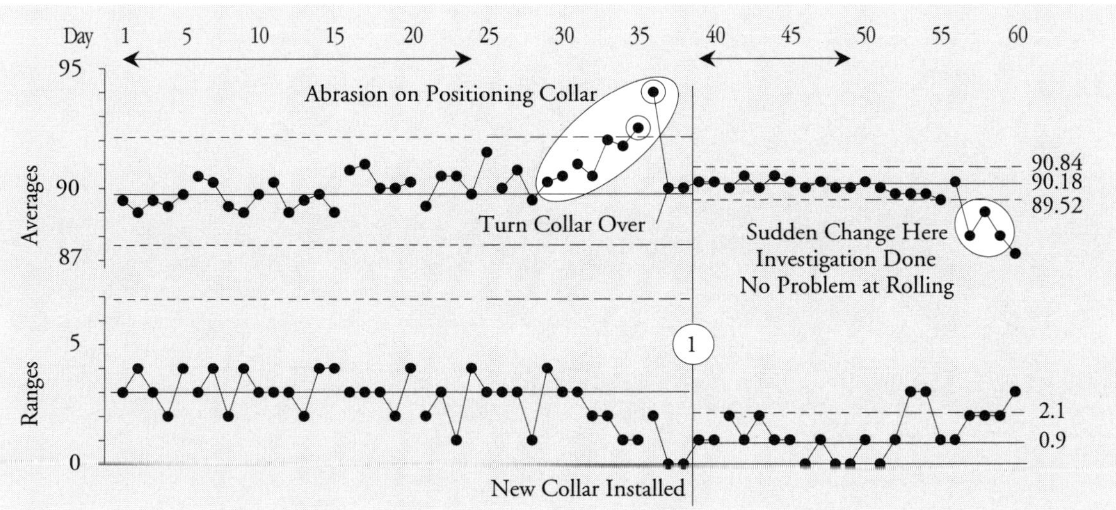

Figure 2.15: Tokai Rika Lighter Detent Dimension: Days 1 to 60

They knew they had a problem with Days 35 and 36. Looking back to the last time the process crossed the central line they decided that this problem could have begun as early as Day 29. Upon investigation, they found that the positioning collar had worn down and needed to be replaced (Figure 2.12 shows that a worn collar would probably cause the measurements to increase). Recognizing this as a problem of tool wear they did two things: they ordered a new positioning collar, and they turned the old collar over to get back on target while waiting for the new collar to arrive. This is indicative of a desire to operate right at the target value whenever possible. The new collar was installed on Day 39 and they wrote Intervention Report Number One detailing what was found and want was done.

Following this intervention they decided to compute new limits for the process. They ran without limits for Days 39 to 49 and used this period as their new baseline. With a grand average of 90.18 and an average range of 0.91 the new limits were considerably tighter than the previous limits. As they used these limits to track the process they soon found evidence of another process change.

Days 57, 58, and 59 are all below the lower limit and it is fairly clear that there has been a shift in the process. (They did not make any notes about the ranges falling above the upper limit a few days earlier.) Moreover, unlike the excursion on Days 29 to 36, there is no gradual change leading up to the first point outside the limit. Hence they noted that this was a sudden change and began to look for something broken. They began their investigation with the rolling operation. When no problems were found at rolling, they turned to the blanking operation.

As shown in Figure 2.16, after the better part of two weeks had passed, they finally discovered what was making the detent dimensions smaller—there was a very small wrinkle on the flange due to a defect in the die.

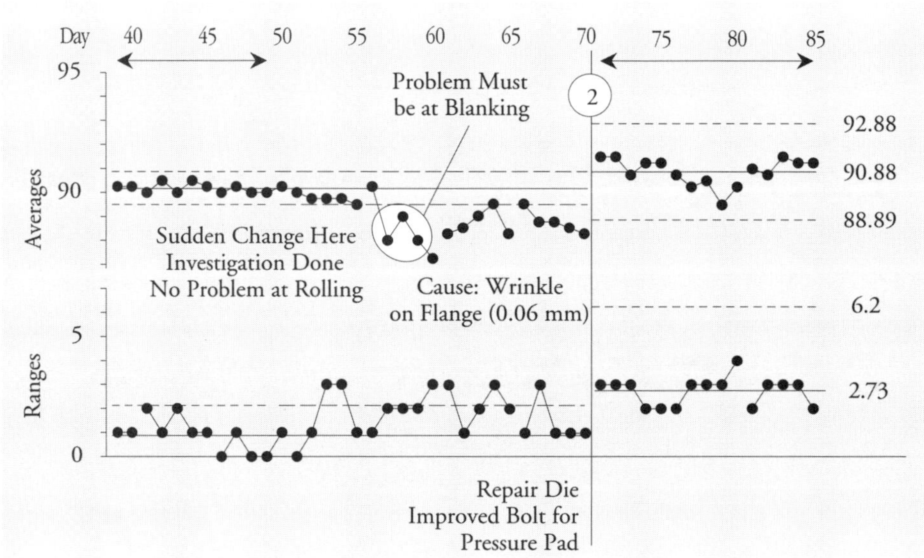

Figure 2.16: Tokai Rika Lighter Detent Dimension: Days 39 to 85

Thus, they scheduled a repair for the die on the weekend between Days 70 and 71. At the same time they modified the bolt holding the pressure pad since they had found that it was coming loose. Following these changes they wrote up Intervention Report Number Two and proceeded to collect data for a new set of limits. The process average went up to 90.88, which is probably due to the fact that this positioning collar already had 32 days of wear prior to this new baseline period.

To understand how quickly and efficiently they are tracking and responding to the upsets that occur in this process it might be helpful to look back at the histogram in Figure 2.14. That histogram contains the data from Days 25 to 85. During this period this process displayed two excursions and was subject to one tool change and one process upgrade and repair. In spite of the fact that the histogram contains the data from these excursions, we find all the values but one within the Natural Process Limits computed from Days 1 to 24! The signals of change were clear and unequivocal on the Average Chart, which allowed those who ran this process to respond in a timely manner. In the end, their interventions were sufficient to keep the product stream essentially homogeneous in spite of the process changes.

Note also that the purpose of the limits is to tell the story of what is happening in the process. It is not a matter of computing a single set of limits and using them forever. It is also not a matter of waiting until we have some minimum amount of data before we compute our limits. The baseline period of Days 39 to 49 was essentially chosen based on the fact that Day 49 was the last point on that sheet of paper. If a process is operated predictably, any baseline will tell the same story as another baseline. If a process is operated unpredictably, different baselines may tell different stories, but the point will not be one of "How good are the limits?" but rather "What can we do about the Assignable Causes?"

Table 2.9: Tokai Rika Lighter Detent Data for Days 126 to 185

-126-	-127-	-128-	-129-	-130-	-131-	-132-	-133-	-134-	-135-	-136-	-137-	-138-	-139-	-140--
91	91	93	89	91	90	90	92	93	92	89	90	92	92	89
90	90	90	90	90	91	89	90	90	89	90	93	91	90	90
91	90	91	90	91	90	89	91	91	90	90	92	92	89	88
92	91	90	90	89	93	91	92	92	89	90	92	92	89	90
91.00	90.50	91.00	89.75	90.25	91.00	89.75	91.25	91.50	90.00	89.75	91.75	91.75	90.00	89.25
2	1	3	1	2	3	2	2	3	3	1	3	1	3	2
-141-	-142-	-143-	-144-	-145-	-146-	-147-	-148-	-149-	-150-	-151-	-152-	-153-	-154-	-155--
90	88	91	89	89	92	90	92	91	92	89	90	91	91	90
91	92	90	90	91	92	89	91	90	90	91	91	92	92	92
90	90	91	90	92	91	90	92	92	90	90	91	91	91	90
90	91	91	89	92	91	90	92	92	89	90	92	92	90	92
90.25	90.25	90.75	89.50	91.00	91.50	89.75	91.75	91.25	90.25	90.00	91.00	91.50	91.00	91.00
1	4	1	1	3	1	1	1	2	3	2	2	1	2	2
--156-	-157-	-158-	-159-	-160-	-161-	-162-	-163-	-164-	-165-	-166-	-167-	-168-	-169-	-170-
92	90	90	89	92	91	91	92	91	89	90	91	92	89	91
89	91	89	93	91	90	90	92	90	91	90	92	90	91	92
88	92	89	91	91	90	92	90	90	92	89	91	90	90	90
89	90	89	91	91	89	92	91	90	90	90	93	88	91	90
89.50	90.75	89.25	91.00	91.25	90.00	91.25	91.25	90.25	90.50	89.75	91.75	90.00	90.25	90.75
4	2	1	4	1	2	2	2	1	3	1	2	4	2	2
-171-	-172-	-173-	-174-	-175-	-176-	-177-	-178-	-179-	-180-	-181-	-182-	-183-	-184-	-185--
90	92	93	90	91	91	92	90	91	92	90	91	90	89	89
90	91	90	92	92	89	91	91	90	93	89	90	89	90	90
91	91	94	90	90	93	90	94	93	90	91	90	89	90	90
89	92	94	91	91	90	93	95	94	89	90	89	90	90	91
90.00	91.50	92.75	90.75	91.00	90.75	91.50	92.50	92.00	91.00	90.00	90.00	89.50	89.75	90.00
2	1	4	2	2	4	3	5	4	4	2	2	1	1	2

Figure 2.17, on page 38, shows the limits from Days 71 to 85 extended forward.

The points for Days 102 to 105 are above the upper limit. This excursion appears to have begun on Day 95. It was interpreted as tool wear on the positioning collar. The first positioning collar lasted less than 40 days. This collar lasted about 55 days. Based on this experience, and possibly on experience with previous cigar lighter shells, they decided upgrade to a tungsten coated positioning collar.

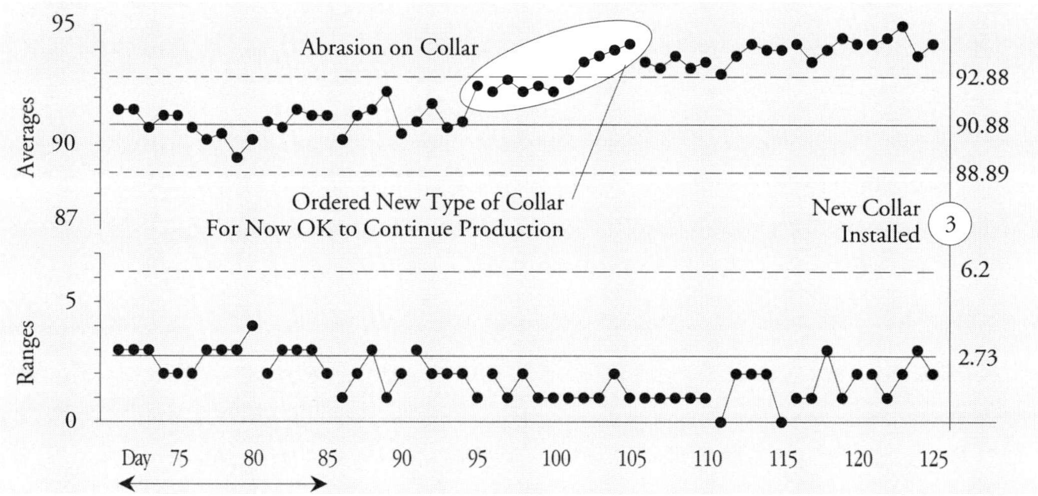

Figure 2.17: Tokai Rika Lighter Detent Dimension: Days 71 to 125

Since a tungsten coated collar was a special order part, they had to wait a month to get the replacement collar. Since this was too long a time for the trick of turning the collar over like they did on Days 37 and 38, and since the specifications of 80 to 100 provided them with a lot of elbow room, they decided that they could continue to operate with the worn positioning collar until the replacement arrived. While they frequently did whatever was needed to operate this process right at the target value, here they made a conscious and deliberate decision that they would allow this process to operate off-target for a limited period of time. Based on their experience with this process they were confident that as long as the process average did not go above 95 they would not be in danger of making any nonconforming parts. The highest daily average shown is for Day 123, which is 95.00.

Then, on Day 126, the tungsten coated positioning collar was installed and Intervention Report Number Three was written. Following this they used Days 126 to 140 to compute new limits. With a grand average of 90.57 and an average range of 2.13 they got the limits shown in Figure 2.18.

Between Day 1 and Day 140 they replaced and then upgraded the positioning collar, repaired the die, and improved the bolt for the pressure pad. While these are fairly routine types of maintenance operations, the cumulative effect of all of these things was a reduction in the process variation. By dealing with these issues in a timely manner they have actually reduced the variation in this process by 52 percent. (Because of the nature of variation, a 52 percent reduction in the variance will look like a 30 percent reduction in the width of the histogram since: $\sqrt{0.48} = 0.693$.)

Figure 2.19 shows the histogram of the data from Days 126 to 185 along with the Natural Process Limits based on the limits from Days 126 to 140. Since this histogram includes the data from Days 172 to 180, where the process suffered another excursion, we do find five values just outside the Natural Process Limits. Nevertheless, the limits in Figure 2.19 are only 70

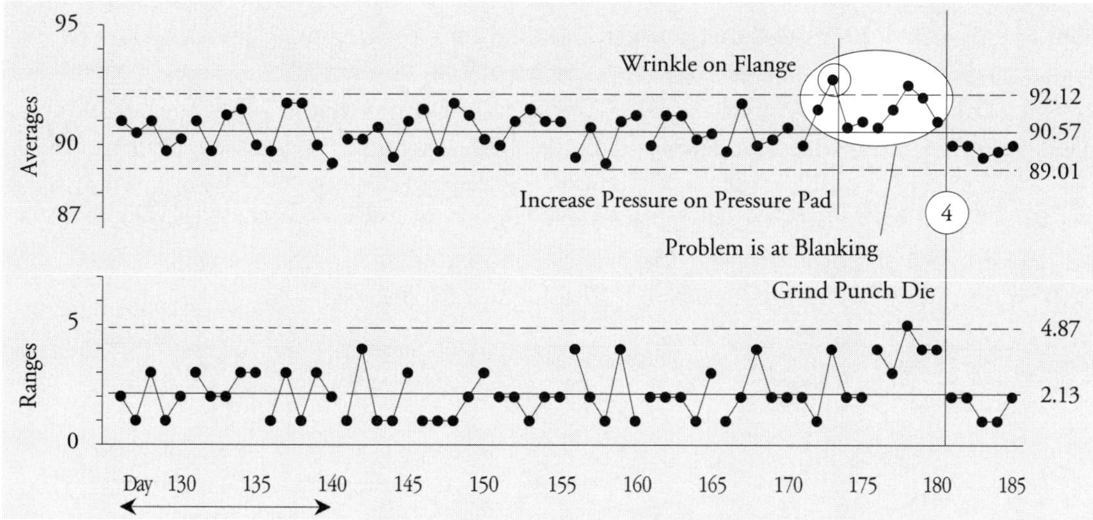

Figure 2.18: Tokai Rika Lighter Detent Dimension: Days 126 to 185

percent as wide as those in Figure 2.14. Over the course of the next year Tokai Rika maintained this reduced level of variation. Since 17,000 parts per day would meet the combined needs of all automotive manufacturers in Japan at the time of this chart, it would seem that Tokai Rika was the sole supplier for this part.

This process is about as predictable as any process will ever be. You simply cannot fix a process and then expect it to stay fixed. In nine months of operation there were four process excursions followed by four interventions. The process behavior chart called their attention

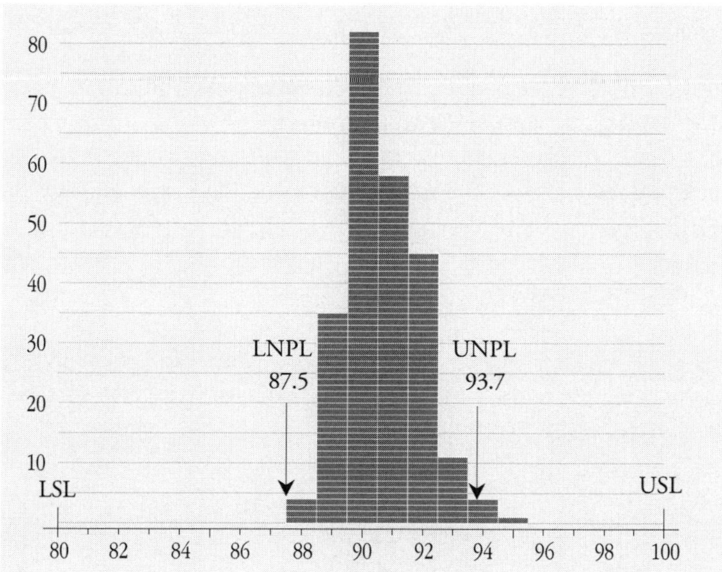

Figure 2.19: Histogram for Days 126 to 185

to the unplanned process changes and allowed them to evaluate their deliberate process interventions. They improved this process and then they maintained it so that it would continue to operate up to its full potential. This can be seen in Figure 2.20 which contains the histogram of the 780 measurements for Days 186 to 380. During this 195 day period they produced over 3.3 million Lighter Shells and they had two more process excursions. After adjusting for a change they made in the measurement process on Day 275, only 10 of these 780 values fall outside the Natural Process Limits computed from the baseline of Days 126 to 140! For the rest of this story, and the original charts, see Chapter Seven of *Understanding Statistical Process Control, Third Edition,* by Wheeler and Chambers.

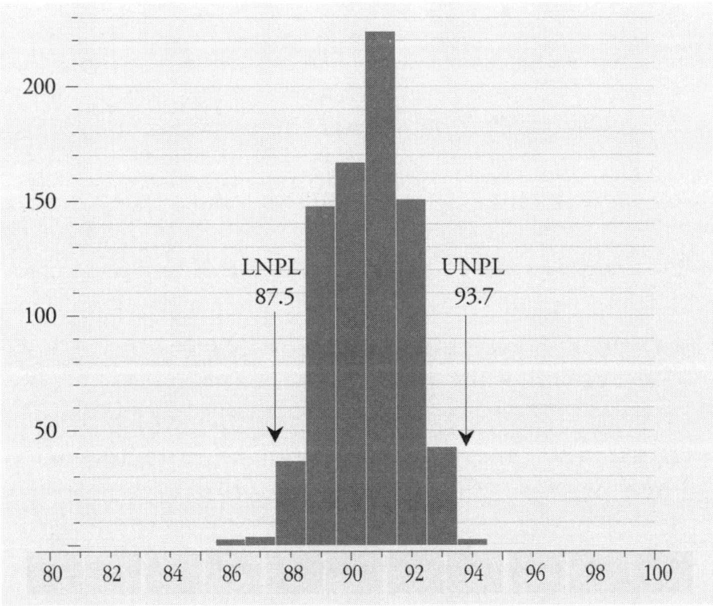

Figure 2.20: Tokai Rika Histogram for Days 186 to 380

As will be shown in Chapter Five, World Class Quality requires that you operate On Target with Minimum Variance. As we saw in Chapter One, this is equivalent to operating on target and predictably. And the only way to operate a process predictably is to use process behavior charts.

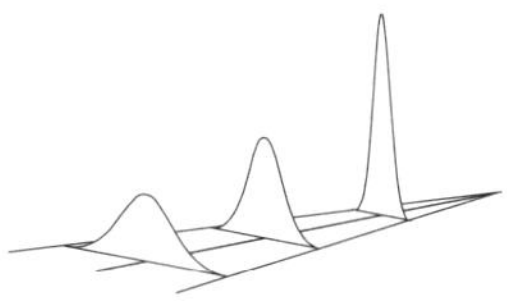

World Class Quality Requires Process Behavior Charts

Chapter Three

The Opportunities

The question of whether a process is meeting requirements is well known to all who work in production. While the observed scrap and rework rates would seem to provide a simple answer to this question, the fact that there is a fork in the road to process improvement complicates things. Can we get more from our current process, or do we need to change the process in some major way in order to meet the requirements? A proper understanding of the roles played by the various Capability and Performance Indexes will help to identify the nature of the opportunities that exist for each situation.

3.1 Performance and Potential

When a process is operated predictably, the Natural Process Limits will characterize both the process performance and the process potential—what the process has done in the past and what it is likely to continue to do in the future.

The Tokai Rika example provides examples of this. The limits shown in Figure 2.20 came from Days 126 to 140. They characterize the process potential. The histogram shows the data from Days 186 to 380. It displays the subsequent process performance. The message of Figure 2.20 is that when a process is operated predictably the performance and the potential are one and the same.

To have another illustration of a predictable process we will use 500 values generated using a Bead Board (Table 3.1). These data were arranged in time order and divided into 100 subgroups of size five. The limits were based on the first ten subgroups, which have a grand average of 10.08 and an average range of 4.00. Using the formulas from Table A.3 we find:

$$\text{Limits for Averages} = \bar{\bar{X}} \pm A_2 \bar{R} = 10.08 \pm 0.577\,(\,4.00\,) = 7.77 \text{ to } 12.39$$

$$\text{Limits for Ranges} = D_3 \bar{R} \text{ to } D_4 \bar{R} = 0 \text{ to } 2.114\,(4) = 0 \text{ to } 8.46$$

$$\text{Natural Process Limits} = \bar{\bar{X}} \pm E_2 \bar{R} = 10.08 \pm 1.290\,(\,4.00\,) = 4.92 \text{ to } 15.24$$

Figure 3.1 shows both the histogram of the 500 individual values and the Average and Range Chart for the Bead Board Data from Table 3.1.

Table 3.1: The Bead Board Data Arranged into 100 Subgroups of Size 5

No.	*Values*					$\bar{X}$	*R*
-1-	12	9	9	8	9	*9.4*	*4*
-2-	11	13	11	11	11	*11.4*	*2*
-3-	14	6	11	8	9	*9.6*	*8*
-4-	10	12	13	8	10	*10.6*	*5*
-5-	11	10	12	11	8	*10.4*	*4*
-6-	7	10	7	7	9	*8.0*	*3*
-7-	12	12	14	7	10	*11.0*	*7*
-8-	11	10	9	10	10	*10.0*	*2*
-9-	9	9	11	11	9	*9.8*	*2*
-10-	11	12	10	9	11	*10.6*	*3*
-11-	9	10	11	11	10	*10.2*	*2*
-12-	9	9	10	11	7	*9.2*	*4*
-13-	7	10	12	11	10	*10.0*	*5*
-14-	8	10	11	9	10	*9.6*	*3*
-15-	10	8	11	12	11	*10.4*	*4*
-16-	11	8	9	11	7	*9.2*	*4*
-17-	9	9	11	11	8	*9.6*	*3*
-18-	12	9	11	10	6	*9.6*	*6*
-19-	12	11	10	15	12	*12.0*	*5*
-20-	9	12	11	9	9	*10.0*	*3*
-21-	13	11	12	7	8	*10.2*	*6*
-22-	9	9	13	8	7	*9.2*	*6*
-23-	10	12	9	11	10	*10.4*	*3*
-24-	8	8	12	10	8	*9.2*	*4*
-25-	9	13	7	10	13	*10.4*	*6*
-26-	11	10	12	10	10	*10.6*	*2*
-27-	11	9	9	8	8	*9.0*	*3*
-28-	11	12	8	12	10	*10.6*	*4*
-29-	11	14	8	13	8	*10.8*	*6*
-30-	14	12	9	9	10	*10.8*	*5*
-31-	9	11	13	10	7	*10.0*	*6*
-32-	10	11	10	12	11	*10.8*	*2*
-33-	9	10	9	13	14	*11.0*	*5*
-34-	12	10	9	8	8	*9.4*	*4*
-35-	9	7	14	12	9	*10.2*	*7*
-36-	9	10	10	11	10	*10.0*	*2*
-37-	11	8	11	9	11	*10.0*	*3*
-38-	12	11	13	8	10	*10.8*	*5*
-39-	9	11	11	11	11	*10.6*	*2*
-40-	13	11	10	7	7	*9.6*	*6*
-41-	10	9	13	10	12	*10.8*	*4*
-42-	8	9	7	10	11	*9.0*	*4*
-43-	7	15	7	8	11	*9.6*	*8*
-44-	11	9	10	11	13	*10.8*	*4*
-45-	10	13	10	9	11	*10.6*	*4*
-46-	12	9	7	13	13	*10.8*	*6*
-47-	9	11	9	10	9	*9.6*	*2*
-48-	8	15	11	10	9	*10.6*	*7*
-49-	7	9	5	12	13	*9.2*	*8*
-50-	10	9	10	11	10	*10.0*	*2*
-51-	12	12	10	11	11	*11.2*	*2*
-52-	11	11	12	10	14	*11.6*	*4*
-53-	12	9	12	9	9	*10.2*	*3*
-53-	8	10	14	13	9	*10.8*	*6*
-55-	10	10	10	14	11	*11.0*	*4*
-56-	11	11	8	11	7	*9.6*	*4*
-57-	10	9	11	8	10	*9.6*	*3*
-58-	11	11	10	10	13	*11.0*	*3*
-59-	12	11	5	9	7	*8.8*	*7*
-60-	11	9	9	11	12	*10.4*	*3*
-61-	14	11	14	10	8	*11.4*	*6*
-62-	9	10	10	15	11	*11.0*	*6*
-63-	10	13	7	12	10	*10.4*	*6*
-64-	11	9	12	10	13	*11.0*	*4*
-65-	6	10	11	10	10	*9.4*	*5*
-66-	10	10	11	7	10	*9.6*	*4*
-67-	10	12	12	8	10	*10.4*	*4*
-68-	13	7	11	12	11	*10.8*	*6*
-69-	12	10	7	10	9	*9.6*	*5*
-70-	9	9	9	10	10	*9.4*	*1*
-71-	10	12	11	9	13	*11.0*	*4*
-72-	10	7	11	8	13	*9.8*	*6*
-73-	12	10	9	8	13	*10.4*	*5*
-74-	12	10	11	10	10	*10.6*	*2*
-75-	11	10	12	10	9	*10.4*	*3*
-76-	7	8	12	10	10	*9.4*	*5*
-77-	8	9	11	9	8	*9.0*	*3*
-78-	9	14	12	12	11	*11.6*	*5*
-79-	7	8	9	9	9	*8.4*	*2*
-80-	12	9	10	10	9	*10.0*	*3*
-81-	8	9	9	11	7	*8.8*	*4*
-82-	10	8	12	9	11	*10.0*	*4*
-83-	12	10	10	10	13	*11.0*	*3*
-84-	9	9	10	8	9	*9.0*	*2*
-85-	10	9	7	9	9	*8.8*	*3*
-86-	11	7	5	9	8	*8.0*	*6*
-87-	9	11	11	10	9	*10.0*	*2*
-88-	12	10	13	12	13	*12.0*	*3*
-89-	12	10	11	8	12	*10.6*	*4*
-90-	10	9	12	9	13	*10.6*	*4*
-91-	11	12	9	11	7	*10.0*	*5*
-92-	7	8	13	10	10	*9.6*	*6*
-93-	9	9	8	7	9	*8.4*	*2*
-94-	9	10	11	9	7	*9.2*	*4*
-95-	10	11	11	9	10	*10.2*	*2*
-96-	12	9	13	9	8	*10.2*	*5*
-97-	11	13	11	11	7	*10.6*	*6*
-98-	7	8	9	10	8	*8.4*	*3*
-99-	9	11	12	11	9	*10.4*	*3*
-100-	9	10	9	10	10	*9.6*	*1*

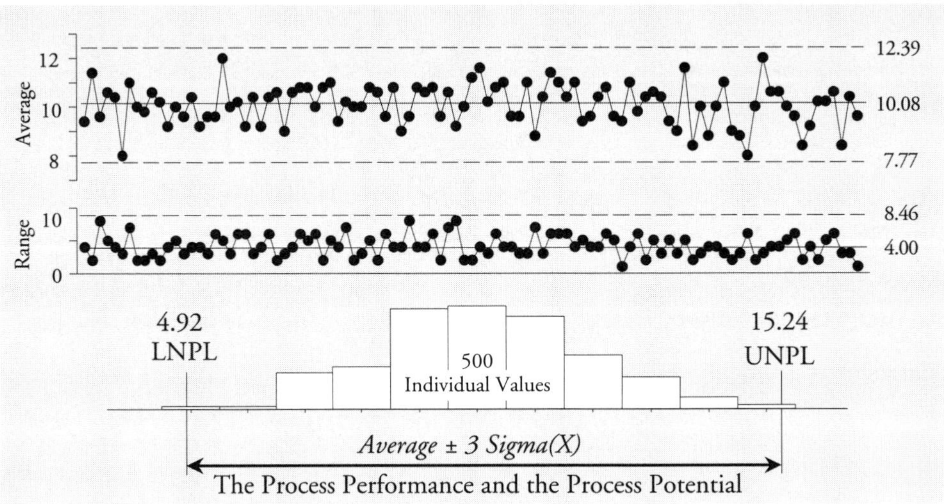

Figure 3.1: The Performance and Potential of a Predictable Process Are the Same

Even though the limits are based on the first 10 subgroups, they do a reasonable job of characterizing the process performance simply because the Bead Board is a prototype for a predictable process. Hence, for a predictable process, the three-sigma Natural Process Limits for individual values characterize both the process performance and the process potential.

WITHIN SUBGROUP VERSUS GLOBAL MEASURES OF DISPERSION

In Figure 3.1, and throughout this book, the notation *Sigma(X)* will be used when referring to any *summary within-subgroup measure of dispersion* that could be used to estimate the standard deviation parameter of the distribution of X when this parameter is well-defined (see Table A.1 for formulas). *Three-sigma limits* will always refer to limits computed using one of these summary within-subgroup measures of dispersion.

Three-standard-deviation limits will refer to limits computed using the average plus or minus three times the global standard deviation statistic, s. This descriptive statistic may be used to characterize the past outcomes and summarize past performance.

For Table 3.1 the global standard deviation statistic for the first 10 subgroups is $s = 1.828$, and the three-standard-deviation limits for X would be $10.08 \pm 3\ (1.83) = 4.60$ to 15.56. These values hardly differ from those shown in Figure 3.1.

Table 3.2: Batch Weight Data

Batch No.	*Batch Weights (kilograms of product exiting blender)*									
1-10	905	930	865	895	905	885	890	930	915	910
11-20	920	915	925	860	905	925	925	905	915	930
21-30	890	940	860	875	985	970	940	975	1000	1035
31-40	1020	985	960	945	965	940	900	920	980	950
41-50	955	970	970	1035	1040	1000	1000	990	1000	950
51-60	940	965	920	920	925	900	905	900	925	885
61-70	1005	1005	950	920	875	865	880	960	925	925
71-80	875	900	905	990	970	910	980	900	970	900
81-90	895	885	925	870	875	910	915	900	950	880
91-100	910	965	910	880	900	920	940	985	965	925
101-110	925	975	905	890	950	975	935	940	900	915
111-120	980	880	905	915	960	900	915	920	865	980
121-130	935	840	900	965	890	875	1020	780	900	900
131-140	800	960	845	820	910	885	940	930	925	850
141-150	965	1010	1030	980	1010	950	940	1005	880	930
151-160	845	935	905	965	975	985	975	950	905	965
161-170	905	950	905	995	900	840	1050	935	940	920
171-180	985	970	915	935	950	1030	875	880	955	910
181-190	1050	890	1005	915	1070	970	1040	770	940	950
191-200	1040	1035	1110	845	900	905	910	860	1045	820
201-210	900	860	875	1005	880	750	900	835	930	860
211-220	960	950	1020	975	950	960	950	880	1000	1005
221-230	990	1020	980	1020	920	960	1000	1000	860	1130
231-240	830	965	930	950	945	900	990	865	945	970
241-250	915	975	940	870	890	915	935	1060	1015	1100
251-259	810	1010	1140	805	1020	1110	975	970	1090	

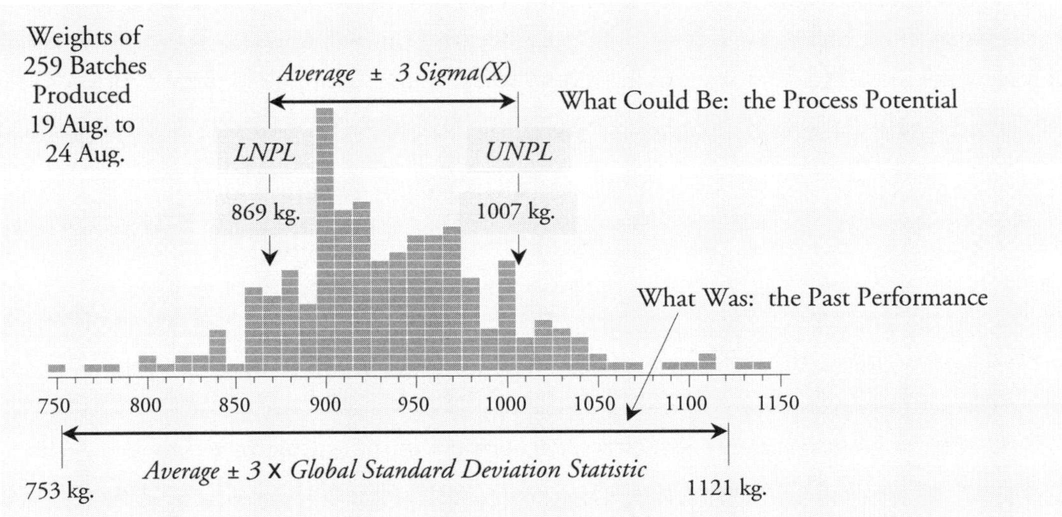

Figure 3.2: The Potential and the Performance of an Unpredictable Process Are Different

The Batch Weight Data of Table 3.2 are the weights, in kilograms, of each batch of a compound as it came out of the blender. For these 259 values the average is 937 kg and the global standard deviation statistic is $s = 61.3$ kg. The histogram is shown in Figure 3.2, where we see that the three-standard-deviation limits of 937 ± 3 (61.3) = 753 kg to 1121 kg do a reasonable job of describing this week's production—only 3 of the 259 batches have weights that are slightly outside this interval.

The X Chart in Figure 3.3 has limits based on the first 60 values. The average for this baseline period is 938.1, and the average moving range is 25.8. With 54 of the 259 individual

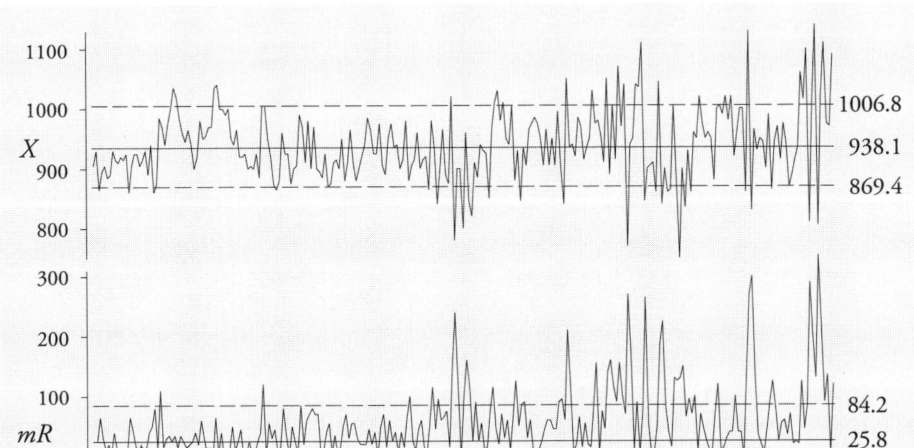

Figure 3.3: *XmR* Chart for the Batch Weight Data

values outside these limits, this process is clearly not operating up to its full potential. The Natural Process Limits of 938.1 ± 2.66 (25.8) = 869 to 1007 on the X Chart approximate what this process has the potential to do. They define the hypothetical process potential.

As may be seen in Figure 3.2, whenever a process is operated in an unpredictable manner there will be a discrepancy between the process performance and the process potential. *This discrepancy represents the extent to which the current process can be improved without making any fundamental process changes.*

An unpredictable process speaks with two voices—its hypothetical potential is completely different from its actual performance. But a predictable process may be said to operate with a single voice—the process performance and the process potential are one and the same. This is why we will need to distinguish between computations done using within-subgroup measures of dispersion and those done using global measures of dispersion. This distinction is fundamental to clear thinking about process potential and process performance, and it will be the fulcrum that provides the leverage for everything we do in this book.

3.2 Capability Indexes

The Capability Ratio, C_p, compares the specifications with the process potential by dividing the [space available within the specifications] by the [space required by the process]:

$$C_p = \frac{\text{space available}}{\text{space required}} = \frac{USL - LSL}{UNPL - LNPL} = \frac{USL - LSL}{6\ Sigma(X)}$$

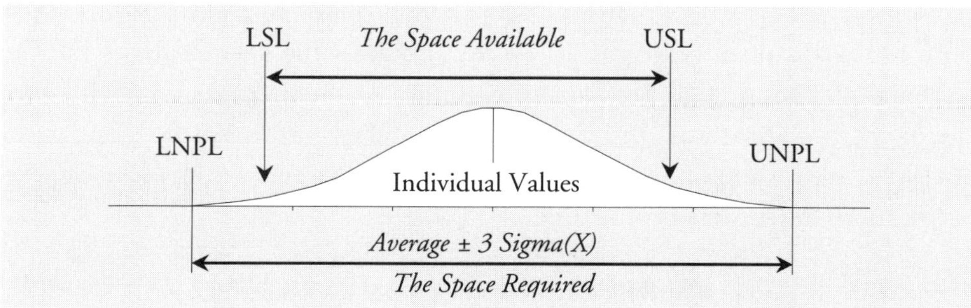

Figure 3.4: The Capability Ratio Is the Space Available Divided by the Space Required

Table 3.3: Capability Ratios

	Sigma(X)	*LSL*	*USL*	C_p
Tokai Rika, Days 1-24	1.476	80	100	2.26
Tokai Rika, Days 126-140	1.034	80	100	3.22
Bead Board, Subgroups 1-10	1.720	8.5	15.5	0.68
Batch Weights, Batches 1-60	22.87	900	1100	1.46

Thus, the Capability Ratio is an index number that characterizes the elbow room available within the specifications. It does not take into account how the process is centered within those specifications. To characterize the location of the process we use the Distance to the Nearer Specification, *DNS*, which may be defined as:

$$DNS = \text{Minimum of } [\ (Average - LSL) \text{ and } (USL - Average)\ ]$$

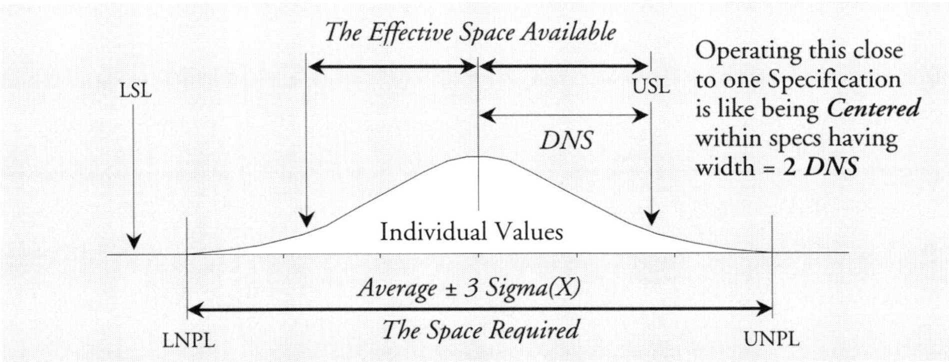

Figure 3.5: The Centered Capability Ratio Is the Effective Space Available Divided by the Space Required by the Process

Operating this close to one specification is like being centered within specifications that have a width equal to twice the *DNS* value. Therefore, the Centered Capability Ratio, C_{pk}, is defined as:

$$C_{pk} = \frac{\text{effective space available}}{\text{space required}} = \frac{2\ DNS}{UNPL - LNPL} = \frac{2\ DNS}{6\ Sigma(X)} = \frac{DNS}{3\ Sigma(X)}$$

When a process is operated right at the mid-point of the specifications the Centered Capability Ratio will be equal to the Capability Ratio. As the process aim deviates from the mid-point of the specifications, the Centered Capability Ratio will be smaller than the Capability Ratio. Hence, the size of C_{pk} relative to C_p will characterize how close to the center of the specifications the process is operating.

Table 3.4: Capability Ratios and Centered Capability Ratios

	Sigma(X)	*LSL*	*USL*	*Average*	*DNS*	C_p	C_{pk}
Tokai Rika 1-24	1.476	80	100	89.80	9.80	2.26	2.21
Tokai Rika 126-140	1.034	80	100	90.57	9.43	3.22	3.04
Bead Board 1-10	1.720	8.5	15.5	10.08	1.58	0.68	0.31
Batch Weights 1-60	22.87	900	1100	938.1	38.1	1.46	0.56

Comparing Cp and C_{pk} values in Table 3.4 we can see that the first two processes are reasonably well centered, while the last two are not centered. When a process is operated predictably these two Capability Indexes will characterize *both* the process performance and the process potential.

However, we cannot distinguish which processes are operated predictably using just these two Capability Indexes. When a process is operated unpredictably the process performance will differ from the process potential. Here the Capability Indexes will characterize the *hypothetical* process potential, while the Performance Indexes will summarize past performance.

3.3 Performance Indexes

The Performance Ratio, P_p, compares the past process performance with the specifications by dividing the [space available within the specifications] by the [space used by the process outcomes in the past]. Since the global standard deviation statistic, s, provides a descriptive summary for the past data we generally define the Performance Ratio as:

$$P_p = \frac{\textit{space available}}{\textit{space used in past}} = \frac{USL - LSL}{6\,s}$$

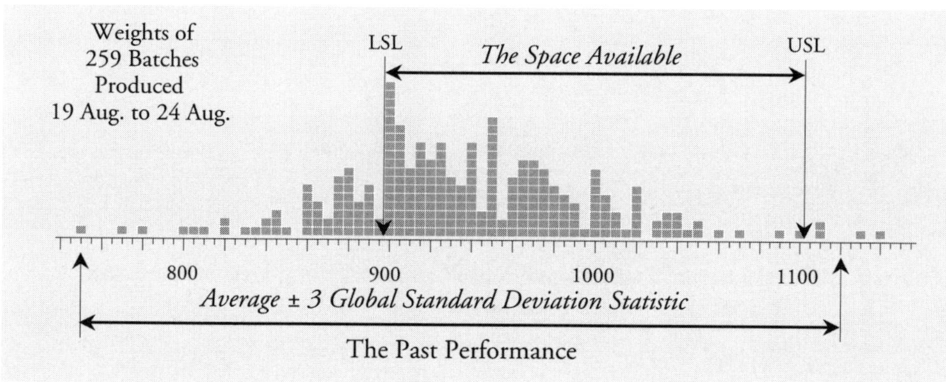

Figure 3.6: The Performance Ratio Compares the Space Available with the Past Performance

Inspection of the formulas will show that the only difference between the Capability Ratio, C_p, and the Performance Ratio, P_p, is the way the dispersion for the denominator is computed. The Capability Ratio uses one of the within-subgroup *Sigma(X)* formulas from Table A.1, while the Performance Ratio uses the global standard deviation statistic, s.

In a similar manner, the Centered Performance Ratio, P_{pk}, is defined to be:

$$P_{pk} = \frac{\textit{effective space available}}{\textit{space used in past}} = \frac{2\,DNS}{6\,s} = \frac{DNS}{3\,s}$$

When the historical average has been near the mid-point of the specifications P_{pk} will be close to P_p. Thus, the discrepancy between P_p and P_{pk} will characterize how centered the process has been. At the same time, substantial discrepancies between each of these Performance Indexes and their corresponding Capability Indexes will characterize how unpredictable the process has been.

Table 3.5: Performance Ratios and Centered Performance Ratios

	s	LSL	USL	*Average*	DNS	P_p	P_{pk}
Tokai Rika 1-24	1.389	80	100	89.80	9.80	2.40	2.35
Tokai Rika 126-140	1.212	80	100	90.57	9.43	2.75	2.59
Bead Board 1-10	1.828	8.5	15.5	10.08	1.58	0.64	0.29
Batch Weights 1-259	61.3	900	1100	937.0	37.0	0.54	0.20

We would interpret the P_p value of 0.54 for the Batch Weight Data to mean that the space available within the specifications is only 54% of the space used by this process in the past. This agrees with Figure 3.6. The P_{pk} value of 0.20 suggests that the effective space available in Figure 3.7 is only 20% of the space used in the past.

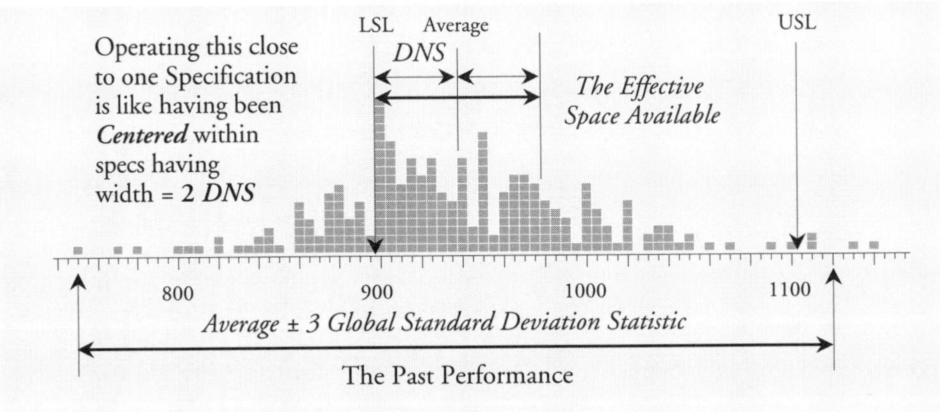

Figure 3.7: The Centered Performance Ratio Compares the Effective Space Available with the Space Used by the Process in the Past

When Capability Indexes are computed for an unpredictable process they will approximate the *hypothetical* process potential, while the Performance Indexes will characterize the *actual* past performance. Thus, these four ratios will characterize the gap between performance and potential, as seen in Table 3.6.

Table 3.6: Capability and Performance Indexes

Data Set	C_p	C_{pk}	P_p	P_{pk}	*Interpretation*
Tokai Rika 1-24	2.26	2.21	2.40	2.35	Predictable, On target, Capable
Bead Board	0.68	0.30	0.64	0.29	Predictable, Off target, Not Capable
Batch Weights	1.46	0.56	0.54	0.20	Unpredictable, Off target, Could be Capable

For the Batch Weight Data, the dissimilarities between the Performance Indexes and the Capability Indexes tell us the process has been unpredictable. The discrepancy between the two Performance Indexes tells us the process has been off target. The Capability Ratio tells us this process has the potential to operate capably, while the other values tell us that it has not been operated within the specifications in the past.

3.4 The Gaps

The Capability and Performance Indexes are related as shown in Figure 3.8. There the top tier represents the actual capability of a process that is operated predictably, or the hypothetical capability of a process that is operated unpredictably.

The bottom tier represents the actual performance of a process that is operated unpredictably.

The left side represents what happens when the process is centered at the mid-point of the specifications, while the right side represents what happens when the process is not centered at the mid-point of the specifications.

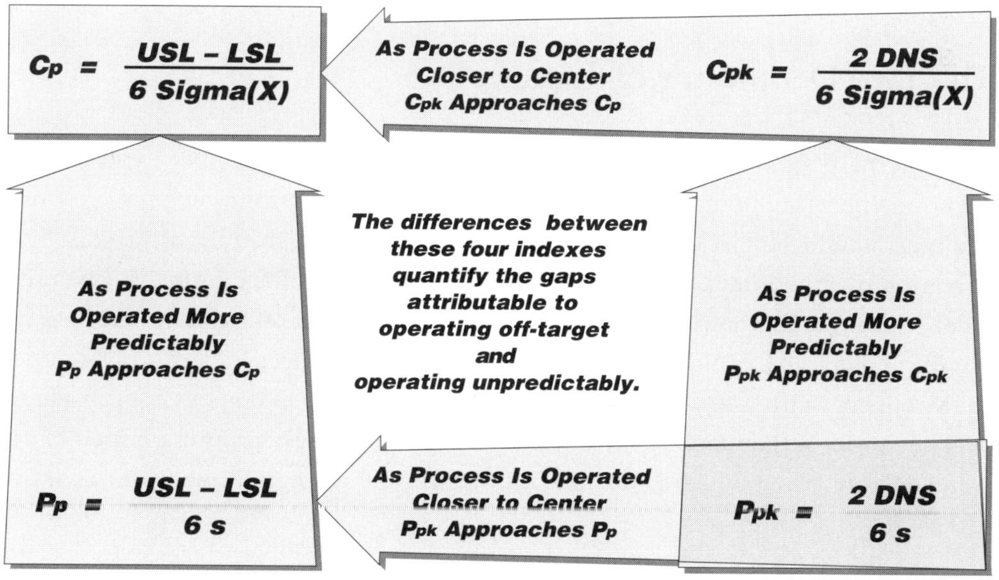

Figure 3.8: How the Capability and Performance Indexes Define the Gaps Between Performance and Potential

- When a process is operated predictably and on target, these four indexes will be four estimates of the same thing. (See the first row of Table 3.6)
- When a process is operated predictably but not centered within the specifications, the discrepancy between the right and left sides of Figure 3.8 will quantify the effects of being off center. With a predictable process, the two indexes on the right side of Figure 3.8 will both estimate the same thing while the two indexes on the left side will be two estimates of another quantity. (See the second row of Table 3.6)
- When a process is operated unpredictably, the indexes in the bottom row of Figure 3.8

will be smaller than those in the top row, and these discrepancies will quantify the gap due to unpredictable operation. (See the last row of Table 3.6)

- When a process is operated unpredictably and off target, the four indexes will represent four different quantities. (See the last row of Table 3.6.)

Thus, while the Capability Ratio, C_p, is the best-case value, the Centered Performance Ratio, P_{pk}, is the worst-case value. The gap between these two values is the opportunity that exists for improving the current process by operating it up to its full potential.

At the same time the Capability Ratio, C_p, approximates what can be done without reengineering the process. If this best-case value is not good enough, then it may be necessary to upgrade the process in order to meet requirements.

To see how this works in practice we will need to consider how these four indexes fit in with the two definitions of trouble.

3.5 The Two Definitions of Trouble

In the past there was only one criterion required to be a good supplier: you had to ship very few nonconforming items. If your proportion of nonconforming items took a turn for the worse you would be "in trouble," and you would stay in trouble until your fraction nonconforming dropped back down to "an acceptable level." Based on this one criterion most suppliers would alternate between being in trouble and operating okay. As long as you were operating okay you could have an attitude of benign neglect toward your operations. But when you were in trouble you would have to bring in the problem-solving team to fix the process. The world of the manufacturer was characterized by alternating periods of benign neglect and intense panic which could be summarized by a single dimension, as shown in Figure 3.9.

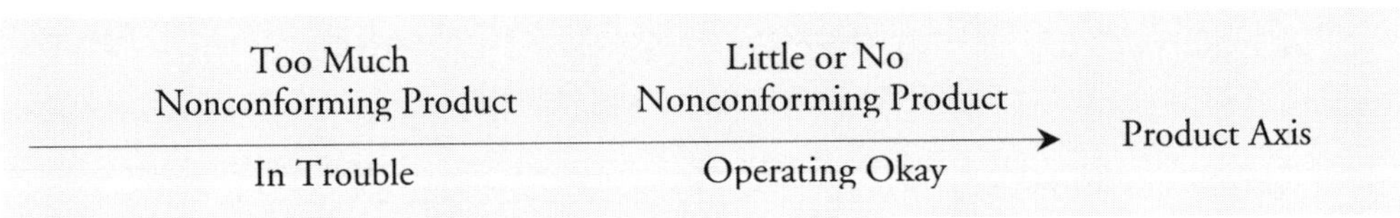

Figure 3.9: The Old Definition of Trouble

Walter Shewhart gave us a new definition of trouble. This second definition of trouble focuses on the production process rather than the product. If your production process is operating up to its full potential then you are operating okay, but if your production process is not operating at full potential then you are in trouble. And the only methodology for determining if a process is operating at full potential is the process behavior chart. Thus, the

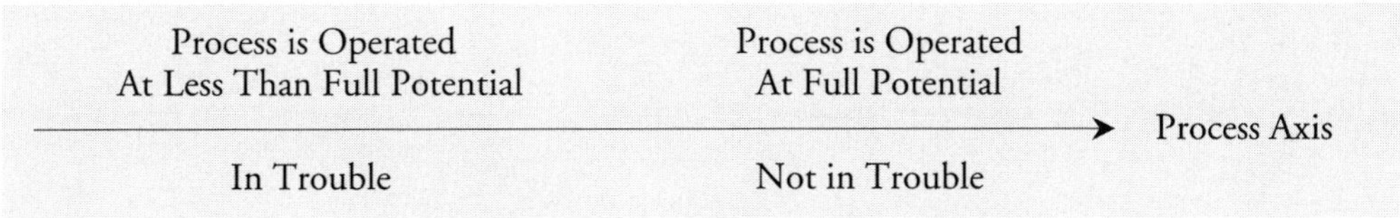

Figure 3.10: The New Definition of Trouble

new definition of trouble would look like Figure 3.10.

While both definitions are valid, neither one is sufficient to fully characterize a production process. Hence we need to combine these two definitions to get the full picture. When we do this we end up with the four possibilities listed below and shown in Figure 3.11.

1. Conforming Product and a Predictable Process (*no trouble*),
2. Nonconforming Product and a Predictable Process (*product trouble*),
3. Conforming Product and an Unpredictable Process (*process trouble*), and
4. Nonconforming Product and an Unpredictable Process (*double trouble*).

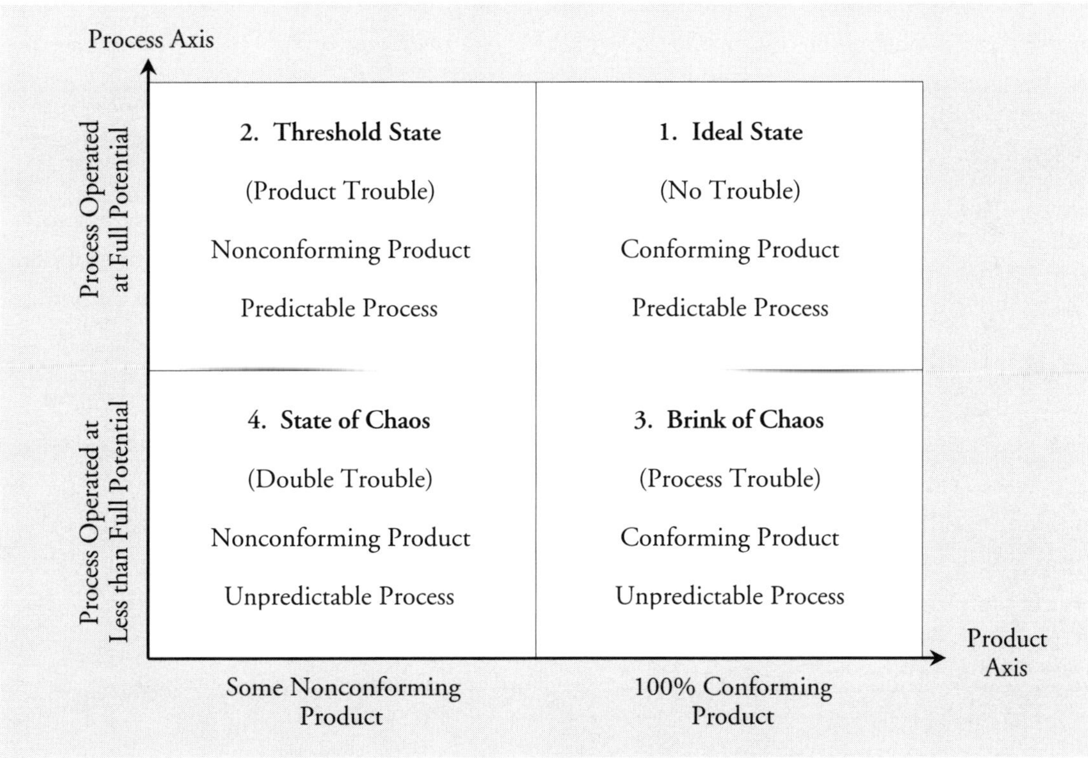

Figure 3.11: The Four Possibilities for Any Process

3.6 The Four Possibilities

THE IDEAL STATE (NO TROUBLE)

The first of these four categories is the "Ideal State." A process in this state is predictable and is producing 100 percent conforming product. The predictability of the process will be the result of deliberate efforts on the part of the manufacturer. A predictable process is an achievement, requiring constancy of purpose and the effective use of process behavior charts. The conformity of the product will be the result of having Natural Process Limits that fall inside the specification limits.

When your process is operating in the Ideal State you will usually find the Centered Capability Ratio, C_{pk}, to be greater than 1.00. (Be careful to avoid reversing this sentence. Having a C_{pk} that exceeds 1.00 does not mean that you are in the Ideal State.)

When your process is in the Ideal State you, and your customer, can expect the conformity of the product to continue as long as the process behavior remains predictable. Since the product stream for a predictable process can be thought of as being homogeneous, the measurements taken to maintain the process behavior chart will also serve to characterize the product produced by the predictable process.

How does a process get to be in this Ideal State? Only by satisfying four conditions:

1. The process must be inherently predictable over time.
2. The manufacturer must operate the process in a predictable and consistent manner; the operating conditions cannot be selected or changed arbitrarily.
3. The process average must be set at the proper level.
4. The Natural Process Limits must fall inside the specification limits for the product.

Whenever one of these conditions is not satisfied, the possibility of shipping nonconforming product exists. When a process satisfies these four conditions, the manufacturer can be confident that nothing but conforming product is being shipped. The only way that a manufacturer can know that these four conditions apply to a process, and the only way to establish and maintain these conditions day after day, is by the use of process behavior charts.

For reasons that will be explained later, getting your process in the Ideal State is not the end of your journey of continual improvement, but it certainly is better than being in one of the other three states.

The Tokai Rika process is being operated in the Ideal State. All four of the Capability and Performance Indexes are comfortably greater than 1.00. In spite of the occasional excursions, this process ships 100% conforming product, and it does so day after day.

THE THRESHOLD STATE (PRODUCT TROUBLE)

The second of these four categories is the "Threshold State." A process in this state will be predictable, but it will be producing some nonconforming product. When your process is operating in the Threshold State the Centered Capability Ratio, C_{pk}, will be less than 1.00.

As before, the predictability of the process will be the result of deliberate and persistent efforts on the part of the producer—a predictable process does not occur by accident. Moreover, because the process is predictable, it must be thought of as operating as consistently as it currently can operate. Nevertheless, the existence of some nonconforming product will be the result of one or both of the Natural Process Limits falling outside the specification limits.

The fact that the process is predictable puts a new twist on product trouble. First of all, as long as the process remains predictable, the nonconforming product will persist. Therefore you cannot wait for things to spontaneously improve. Second, the ultimate solution to the problem of nonconforming product will require you to move this process up to the Ideal State. This will only happen when you either change the process or change the specifications.

If the Capability Ratio, C_p, is greater than 1.00, then the process has enough elbow room to operate in the Ideal State and the nonconforming product is likely to be due to a faulty process aim. In this case the manufacturer will need to tweak the Control Factors to adjust the process aim. Here the process behavior chart can help to determine how various adjustments affect the process average.

If the Capability Ratio, C_p, is less than 1.00, then the process is not likely to have enough elbow room to meet the specifications even if it is operated on target. Here there will be a need to reduce the process variation. Since a predictable process is already operating as consistently as it currently can operate, the reduction of the process variation will usually require a major change in the process itself. As you experiment with major process changes the process behavior chart will allow you to evaluate the effects of your changes. Thus, process behavior charts will not only help you achieve a predictable process, but they also help in moving the process from the Threshold State to the Ideal State.

As always, a short-term solution to the existence of nonconforming product is to use 100% nondestructive testing. Unfortunately, as has been proven over and over again, 100% screening of product is imperfect and expensive. The only way to guarantee that you will not ship any nonconforming product is to avoid making any in the first place. Sorting should be nothing more than a stop-gap measure, not a way of life.

THE BRINK OF CHAOS (PROCESS TROUBLE)

The third state is the "Brink of Chaos." Processes in this state are unpredictable even though they are currently producing 100% conforming product. With the traditional view the existence of 100% conforming product is considered to be evidence that the process is "operating okay." Unfortunately, this view inevitably leads to benign neglect. Processes in the Brink of Chaos will usually have a Centered Performance Ratio, P_{pk}, value that is greater than 1.00.

The characterization of process behavior emphasizes that processes on the Brink of Chaos are unpredictable—in spite of the current 100% conformity, the process is changing unpredictably, and the 100% conformity can disappear at any time.

The reason that any process is unpredictable is that it is subject to the effects of Assignable Causes. So while the conformity to specifications may lull the producer into thinking all is well, the Assignable Causes will continue to change the process until it will eventually produce some nonconforming product. The change from 100% conforming product to some nonconforming product can come at any time, without the slightest warning. When this change occurs the process will be in the "State of Chaos." The producer will suddenly discover that the process is in trouble, yet will have no idea of how it got there, nor any idea of how to get it out of trouble.

There is no way to predict what such a process will yield tomorrow, or next week, or even in the next hour. Since the unpredictability of such a process is due to Assignable Causes, and since Assignable Causes are dominant causes that are not being controlled by the manufacturer, the only way to move out of the Brink of Chaos is to first eliminate the Assignable Causes. This requires the use of process behavior charts.

THE STATE OF CHAOS (DOUBLE TROUBLE)

The State of Chaos exists when an unpredictable process is producing some nonconforming product. The unpredictable process means that the producer is confronted with a changing level of nonconformity in the product stream. Here the Centered Performance Ratio, P_{pk}, will be less than 1.00.

A manufacturer whose process is in the State of Chaos knows that he has a problem, but he usually does not know what to do to correct it. Moreover, his efforts to correct the problem are ultimately frustrated by the random changes in the process which result from the presence of the Assignable Causes. When he makes a needed modification to the process, the effect will be short-lived because the Assignable Causes continue to change the process. When he makes an unnecessary modification, a fortuitous shift by the Assignable Causes may mislead him. No matter what he tries, nothing works for long because the process is always changing. As a result, he finally despairs of ever operating his process rationally, and he begins to speak in terms of "magic" and "art."

The only way to make any progress in moving a process out of the State of Chaos is to first eliminate the Assignable Causes. This will require the use of process behavior charts.

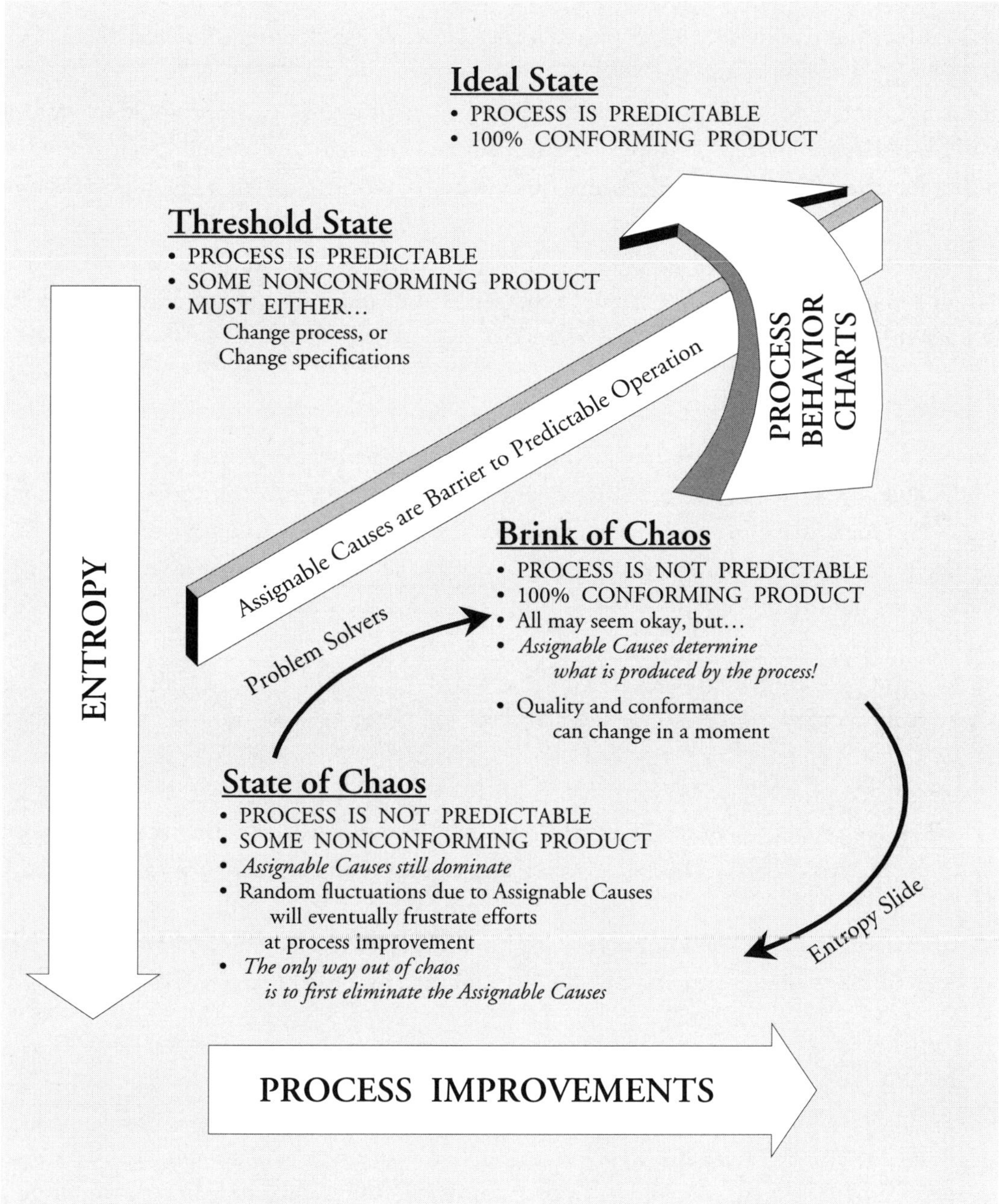

Figure 3.12: The Four Possibilities and the Cycle of Despair

THE EFFECT OF ENTROPY

All processes belong to one of these four states. But processes do not always remain in one state. It is possible for a process to move from one state to another. In fact there is a universal force acting on every process that will cause it to move in a certain direction. That force is entropy. It continually acts upon all processes to cause deterioration and decay, wear and tear, breakdowns and failures.

Entropy is relentless. Because of it every process will naturally and inevitably migrate toward the State of Chaos. The only way this migration can be overcome is by continually repairing the effects of entropy. Of course this means that the effects for a given process must be known before they can be repaired. With such knowledge, the repairs are generally fairly easy to make. On the other hand, it is very difficult to repair something when one is unaware of it. Yet if the effects of entropy are not repaired, it will come to dominate the process and force it inexorably toward the State of Chaos.

THE CYCLE OF DESPAIR

Since everybody knows that they are in trouble when their process is in the State of Chaos it is inevitable that problem-solvers will be appointed to drag the process up from the State of Chaos. With luck, these problem-solvers can get the process back to the Brink of Chaos—a state which is erroneously considered to be "out of trouble" in most operations.

Once they get the process back to the Brink of Chaos the problem-solvers are sent off to work on another problem. As soon as their backs are turned, the process begins to move back down the entropy slide toward the State of Chaos.

New technologies, process upgrades, and all the other "magic bullets" which may be tried can never overcome this Cycle of Despair. You may change technologies—often a case of jumping out of the frying pan and into the fire—but the benign neglect, which inevitably occurs when the process is on the Brink of Chaos, will allow entropy to drag the process back down to the State of Chaos. This is why focusing solely upon conformance to specifications will condemn you to forever cycle between the State of Chaos and the Brink of Chaos. In the words of one engineer, "I spend my working life on project teams whose average half-life is two weeks, implementing solutions that have an average half-life of two weeks."

3.7 Summary

It is important to operate a process to meet requirements. But the fact that some processes are operated up to their full potential, while others are not, makes the issue of operating to meet requirements more complex than simply upgrading the bad processes and ignoring the good processes.

Traditional process upgrades are focused on moving from the left side to the right side of Figure 3.12. Operating a process up to its full potential is focused on moving from the bottom to the top of Figure 3.12. Economic operation requires that we do both.

Chapter Four

The Costs of Using Conforming Product

The excess costs associated with using conforming product will be used throughout the remainder of this book. Since these costs have generally been ignored, this chapter is provided for those who need a concrete example of such costs.

For the past two hundred years quality has been characterized as conformance to specifications. Mass production techniques were used to make stuff, and the specifications were used to sort the good stuff from the bad stuff—make enough stuff and some of it is bound to be good! As long as the economies of mass production were greater than the losses due to scrap and rework this approach made sense. However, today the emphasis is upon lean production. Your customers no longer want to pay for your scrap and rework, so they are interested in your process capabilities. But there is more to the costs of variation than simple scrap and rework—there are also costs associated with using conforming product. This chapter will illustrate these costs using a simple example.

4.1 The Problem of Production

Consider the task of assembling an axle and bearing. How do you make these two parts so that you end up with an assembly that works? Where do you set the target value for the axle diameter? Where do you set the target value for the bearing diameter? And where do you set the specification limits for these two dimensions?

Say the target diameter for the axle is 3.000 mm with specifications of ± 0.015 mm. (The exact values are not really critical to the following discussion; the basic results will remain the same no matter how we might change the values.) Also, in the interest of simplicity, assume that all of our axles are perfect cylinders—these axles show no out-of-roundness and they show no taper.

Assume that the nominal diameter for the bearing is 3.008 mm, 8 microns larger than the target axle diameter, with specifications of 3.008 ± 0.015 mm. Again, for the sake of simplicity, assume that we are making perfectly cylindrical bearings with no taper and no out-of-roundness.

Since the process for making the axles is different from that for making the bearings, we can only represent the combined set of specifications as a rectangle or a square on a two-dimensional plot.

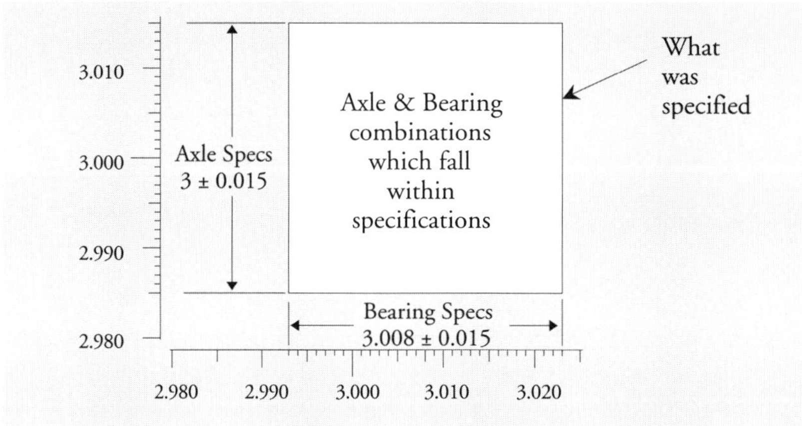

Figure 4.1: Specifications for Axles and Bearings

The square in Figure 4.1 defines those combinations of axle diameters and bearing diameters that will be in conformance to the specifications. However, the functionality of the assembly will not depend upon either the axle diameter or the bearing diameter. It will instead depend upon the difference between the axle diameter and the bearing diameter, which we shall call the gap:

$$Gap = Bearing\ Diameter - Axle\ Diameter$$

The actual clearance between the axle and the bearing will be one-half of this total gap value. If the gap is less than 1 micron the operator will have trouble in assembling the two parts together and the assembly will fail early due to inadequate lubrication. When the axle diameter exceeds the bearing diameter the two parts will not assemble—a situation known as an interference fit.

On the other hand, if the gap exceeds 15 microns the assembly will be too loose. Experience has shown that virtually all of the assemblies with gaps greater than 15 microns will fail during the warranty period and will have to be replaced.

Thus, the job of producing assemblies that will function as intended requires gaps in the range of 1 to 15 microns. This region intersects with the specification rectangle to produce the diagonal band shown in Figure 4.2.

So the square is what we specified and the diagonal band is what we want. Unfortunately, there is no way to define specifications for axles and specifications for bearings that will result in a diagonal region for conforming product. It simply cannot be done. While we may want functional *assemblies,* we have to set specifications on the *individual components.*

Now, to characterize what we are going to get, let us assume that the production processes are predictable and capable: the bearing process is centered at 3.008 millimeters with a standard deviation of 5 microns, while the axle process is centered on a value of 3 millimeters with a standard deviation of 4 microns.

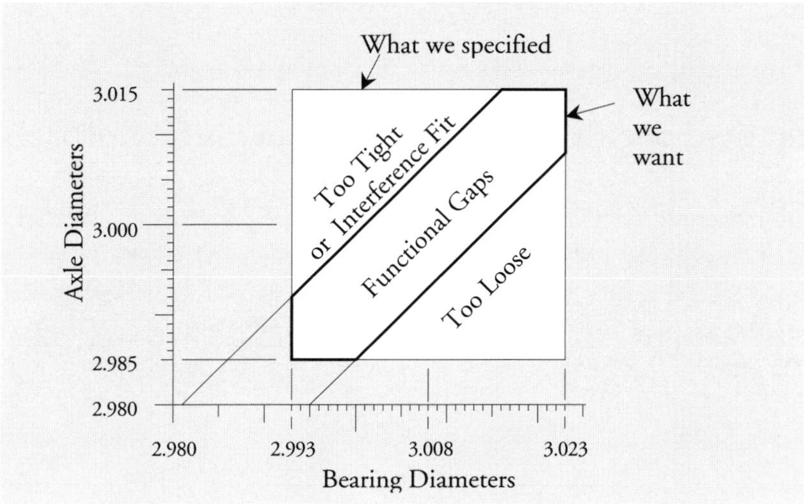

Figure 4.2: The Region of Functional Assemblies

When parts from these two independent production processes are combined we will get combinations shown by the ellipse in Figure 4.3. There the square is what we specified; the diagonal band is what we want; and the ellipse is what we are going to get.

So what can we do? Even though all of our parts are conforming, a substantial number of the pairs of axles and bearings will not assemble, and when we finally get a pair to fit together a substantial number of the resulting assemblies will be so loose that they will fail soon after being placed into service.

Figure 4.3 captures the essence of the problem of production: what we specify is different from what we want, and both differ from what we are going to get. Given this unholy trinity,

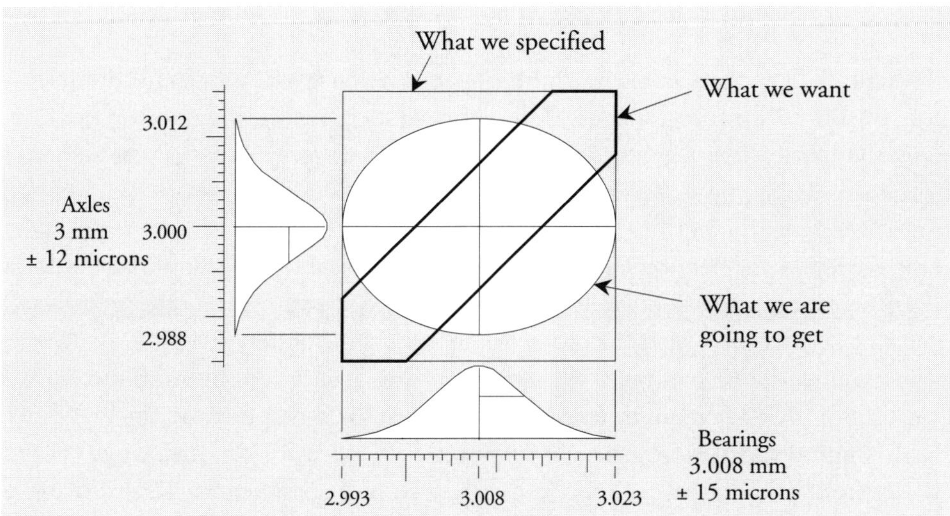

Figure 4.3: The Result of Two Predictable and Capable Processes

what can we do? The next five sections will consider different traditional approaches and their consequences.

4.2 Tighten the Specifications: The Center Cut Approach

What would happen if we simply tightened up on the specifications so that the square in Figure 4.3 would fit entirely within the diagonal band shown? Such a square would be 7 microns on a side. If the center of the square remained the same as before, we would have specifications on the axles of 3.0 ± 0.0035 mm, while the specifications for the bearings would be 3.008 ± 0.0035 mm.

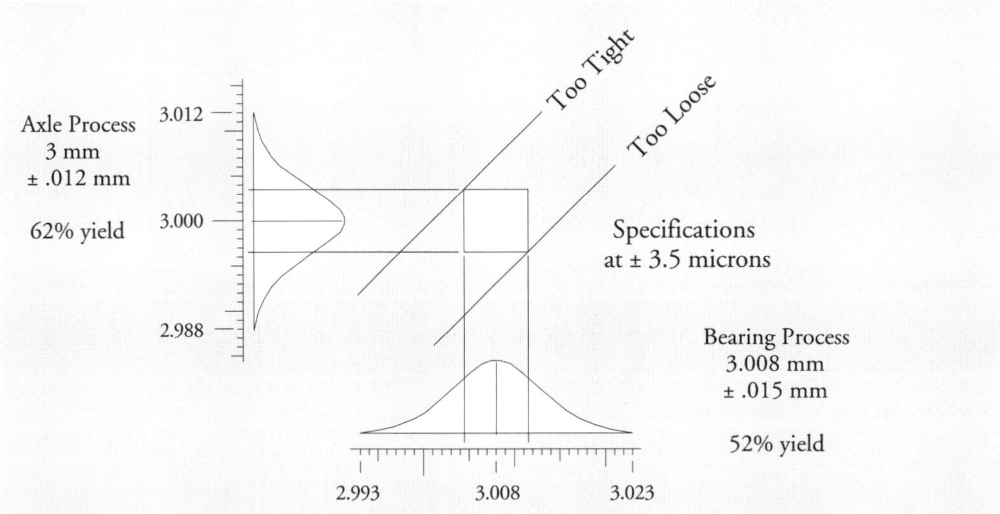

Figure 4.4: Tighten the Specifications: The Center Cut Approach

This "center-cut" approach, using tighter specifications, will eliminate the problems of interference fits and warranty failures. However, just changing the specifications will not change what the fabrication processes will give us. Since these processes were already operating as consistently as they could be operated, we will have to resort to inspection and sorting in order to meet the new, tighter specifications.

With the original specifications the axle process was capable, meaning that we had a yield of 100 percent without inspection. With these tighter specifications the axle process will have a yield of about 62 percent. Thus, in order to meet the production schedule we will need to produce the inverse of 0.62 or about 161 percent of the scheduled amount and then inspect them all to sort out the 62 percent that conform to the tightened specifications.

With the original specifications the bearing process was capable, meaning that we had a yield of 100 percent without inspection. With these tighter specifications the bearing process will have a yield of about 52 percent. In order to meet the production schedule we will need to produce the inverse of 0.52, or about 192 percent of the scheduled amount, and then inspect all of these bearings to find the 52 percent that conform to the tightened specifications.

Thus, with these tighter specifications the production process consists of (1) excess production of axles and bearings, (2) 100% inspection of all axles and bearings, (3) assembly of functional units. Assume that an axle costs $0.40 to fabricate, a bearing costs $0.35 to fabricate, the inspection of an axle costs $0.25, the inspection of a bearing costs $0.25, and the assembly of a unit costs $0.25. Then, ignoring the costs of disposing of the considerable pile of scrap, these tightened specifications will involve fabrication, inspection and assembly costs of:

$$\underset{\text{make axles}}{1.61\,(0.40)} + \underset{\text{sort axles}}{1.61\,(0.25)} + \underset{\text{make bearings}}{1.92\,(0.35)} + \underset{\text{sort bearings}}{1.92\,(0.25)} + \underset{\text{assemble}}{1.0\,(0.25)} = \$\,2.449 \text{ per assembly}$$

This cost of $2.45 per assembly is considerably greater than we would have experienced with the original specifications of ± 15 microns. There, without resorting to inspection, we had 100 percent yields at fabrication, giving a nominal cost per unit of:

$$0.40 + 0.35 + 0.25 = \$\,1.00 \text{ per assembly}$$

However, this nominal cost of $1.00 per assembly ignores the problems of interference fits and the deferred warranty costs associated with the original specifications. To incorporate these costs we will need to begin by looking at what would happen with the random assembly of axles and bearings under the original specifications.

4.3 Random Assembly with the Original Specifications

If we return to the original situation where the bearings were produced by a predictable process centered at 3.008 mm with a standard deviation of 5 microns, and the axles were produced by a predictable process centered at 3.0 mm with a standard deviation of 4 microns, then the ellipse of Figure 4.3 will characterize the pairings of axles and bearings that will occur with random assembly. Figure 4.5 shows the distribution of gaps that would occur at the initial attempts to assemble these parts together. This distribution of gaps will have a mean of 8 microns and a standard deviation of 6.40 microns. If we postulate a continuous supply of parts, so that we can use sampling with replacement, it will simplify our analysis. Under these conditions about 137 times out of a thousand the two parts will not fit together. When this happens the two parts will be disassembled and another bearing or axle will be tried. About 19 of these 137 refits will still not assemble and a third part will have to be tried. About 3 of these 19 refits will still not go together and a fourth part will be needed to get an assembly. Thus, to assemble 1000 units there will be approximately 1000 + 137 + 19 + 3 = 1159 assembly operations and 137 + 19 + 3 = 159 disassembly operations. (In practice, with discrete shipments of parts, the small axles and the large bearings will get used up and you will be left with a pile of unusable components at the end of your build cycle. This will make the excess costs larger. However, in the interest of simplicity, these excess costs were not considered here.)

Assuming that both assembly and disassembly operations are equivalent in labor content, this amounts to an average of 1318 operations to produce 1000 units. At $ 0.25 per operation, and with the other costs noted earlier, the combined labor and material costs for 1000 units will be, on the average:

$$\underbrace{1000\,(0.40)}_{\text{make axles}} + \underbrace{1000\,(0.35)}_{\text{make bearings}} + \underbrace{1318\,(0.25)}_{\text{assemble \& refit}} = \$\,1079.5$$

which is a cost of $ 1.0795 per unit. This is eight percent more than the nominal cost of $ 1.00 per unit. This eight percent is entirely due to the additional labor to handle the problems of interference fits, and interference fits are solely due to variation. If there was no variation, we would have no interference fits.

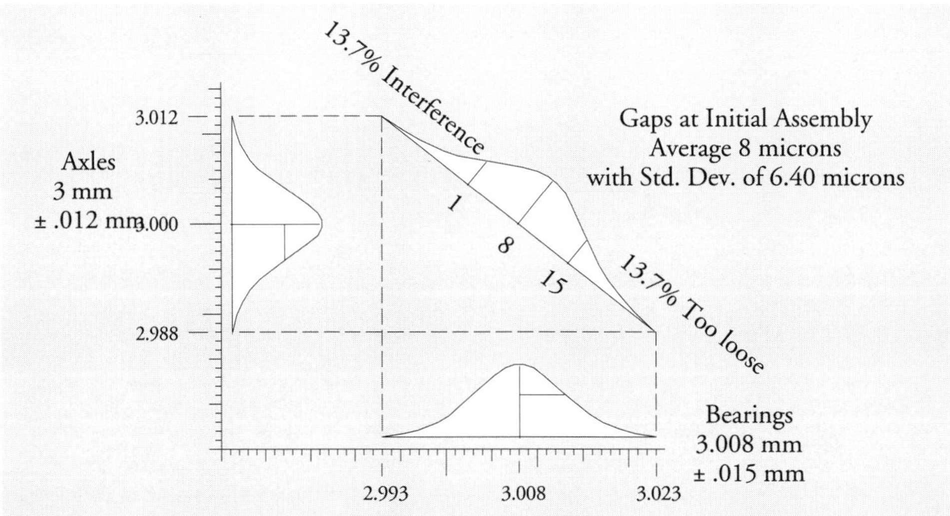

Figure 4.5: The Gap Distribution

Variation also causes a substantial percentage of the assemblies to end up in the category labeled as "too loose." These are units that have a gap in excess of 15 microns. As shown in Figure 4.5 we will have 13.7 percent of our initial assemblies in this category. However, since the interference fits will be reworked, the final percentage that are too loose will be greater than the percentage shown in Figure 4.5. We compute the percentage of shipped units that are too loose in the following manner:

$$\frac{\textit{initial fraction too loose}}{(1 - \textit{initial fraction interference fits})} = \frac{0.1371}{(1 - 0.1371)} = 0.1589$$

This means that, on the average, 159 out of 1000 units shipped will have a gap that exceeds 15 microns, and 841 will have a gap between 1 and 15 microns. Since past experience has shown that virtually 100 percent of the units with a gap in excess of 15 microns will fail during the warranty period, this average of 159 units per 1000 represents a deferred cost of production. If we use a conservative figure of $ 6.00 per unit as the cost of a warranty failure we end up with an eventual cost per 1000 units of:

make axles		make bearings		assemble & refit		warranty costs		
1000 (0.40)	+	1000 (0.35)	+	1318 (0.25)	+	159 (6.00)	=	\$ 2033.5

for an effective unit cost of \$ 2.0335. While this is less than the \$ 2.45 found using the tightened specifications, it is still over twice the nominal per unit cost of \$ 1.00. Random assembly with the original specifications is cheaper than using the tightened specifications, but it is not as economical as we would like. In looking for a place to further reduce costs it will be helpful to consider the equation above. There we see that a large chunk of the excess costs are warranty costs. Since warranty costs are 24 times greater than refitting costs, a reasonable way to try to further reduce the effective unit cost of \$ 2.0335 found above might be to reduce the bearing sizes to minimize the number of units that are too loose.

4.4 Tweaking the Target Value for the Bearing Diameter

If the bearing process was changed to have a smaller average it would increase the interference fits at initial assembly while reducing the number of units that end up being too loose. Figure 4.6 shows the situation for a bearing process centered on 3.003 millimeters. In this and the following scenarios we will assume that the specifications have been changed to coincide with the outcomes from the bearing process, so that we will use 100 percent of the bearings produced.

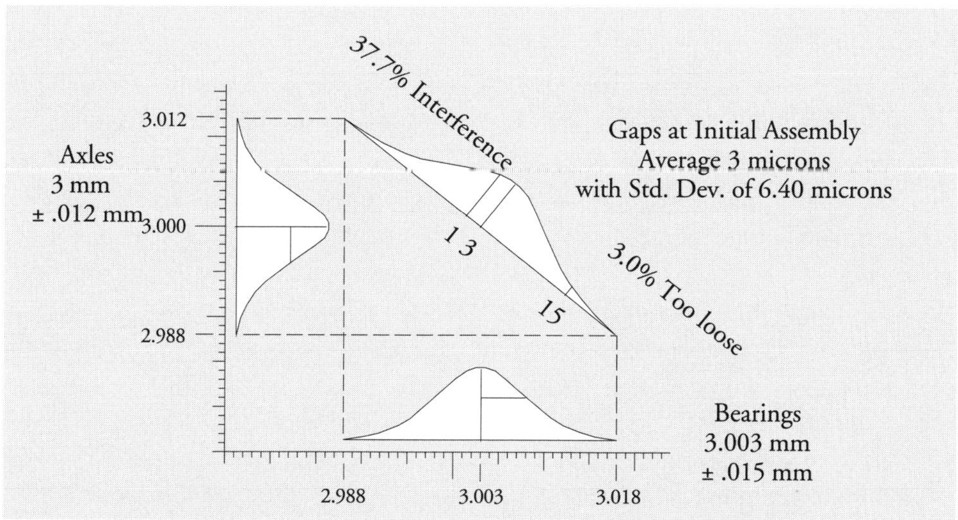

Figure 4.6: Average Gap for Initial Assembly at 3 Microns

With random assembly of axles and bearings we will have, on the average, about 378 interference fits out of a thousand attempts.

When we disassemble and refit these 378 units we will get another 143 interference fits.

When we disassemble and refit these 143 units we will get another 54 interference fits.
When we disassemble and refit these 54 units we will get another 20 interference fits.
When we disassemble and refit these 20 units we will get another 8 interference fits.
When we disassemble and refit these 8 units we will get another 3 interference fits.
When we disassemble and refit these 3 units we will get yet 1 more interference fit.

Thus, in order to get 1000 units to ship we will have to assemble 1000 units, disassemble 606 units, and refit 606 units for a total of 2212 operations. Using the costs defined earlier this results in a total initial assembly cost of:

$$\underset{\text{make axles}}{1000\,(\,0.40\,)} + \underset{\text{make bearings}}{1000\,(\,0.35\,)} + \underset{\text{assemble \& refit}}{2212\,(\,0.25\,)} = \$\,1303.0$$

While this is higher than the \$ 1079.5 found earlier, the objective was not to lower the initial cost figure, but rather to lower the deferred costs of warranty failures. With the bearing diameters averaging 3.003 mm and with 0.3774 of the initial assemblies showing an interference fit, the random assembly approach will result in:

$$\frac{\textit{initial fraction too loose}}{(1-\textit{initial fraction interference fits})} = \frac{0.0305}{(1-0.3774)} = 0.0490$$

that are too loose. Thus, out of 1000 shipped units about 49 will be likely to fail under warranty while the remaining 951 are likely to function as intended. When we include the deferred costs of warranty failures the total cost of 1000 shipped units becomes:

$$\underset{\text{make axles}}{1000\,(\,0.40\,)} + \underset{\text{make bearings}}{1000\,(\,0.35\,)} + \underset{\text{assemble \& refit}}{2212\,(\,0.25\,)} + \underset{\text{warranty costs}}{49\,(\,6.00)} = \$\,1597.0$$

which is substantially less than the corresponding values of \$ 2033.5 and \$ 2450.0 found earlier. By shifting the aim point for the bearing process we have increased the costs of refitting at assembly, but we have reduced the warranty costs by a greater amount. Now the refitting and warranty failures (both of which are costs of using conforming product) only increase the nominal cost of production by 60 percent.

But is 3.003 mm the best aim point for the Bearing Process? By carrying out the preceding analysis at different average gap values we obtain the results shown in Table 4.1.

Table 4.1: Tweaking the Target Value for the Bearing Diameter

Average Gap microns	*proportion interference fits*	*assembly operations per 1000*	*functional units per 1000*	*loose units per 1000*	*initial assembly costs*	*potential warranty costs*	*total production costs*
10	0.0799	1174	764	236	$ 1.043	$ 1.418	$ 2.461
9	0.1058	1237	805	195	$ 1.059	$ 1.170	$ 2.229
8	0.1371	1318	841	159	$ 1.079	$ 0.953	$ 2.033
7	0.1744	1422	872	128	$ 1.106	$ 0.769	$ 1.874
6	0.2174	1556	898	102	$ 1.139	$ 0.613	$ 1.752
5	0.2661	1725	919	81	$ 1.181	$ 0.484	$ 1.665
4	0.3197	1940	937	63	$ 1.235	$ 0.379	$ 1.614
3	0.3774	2212	951	49	$ 1.303	$ 0.293	$ 1.596
2	0.4379	2558	962	38	$ 1.390	$ 0.226	$ 1.616
1	0.5000	3000	971	29	$ 1.500	$ 0.173	$ 1.673
0	0.5621	3567	978	22	$ 1.642	$ 0.131	$ 1.773

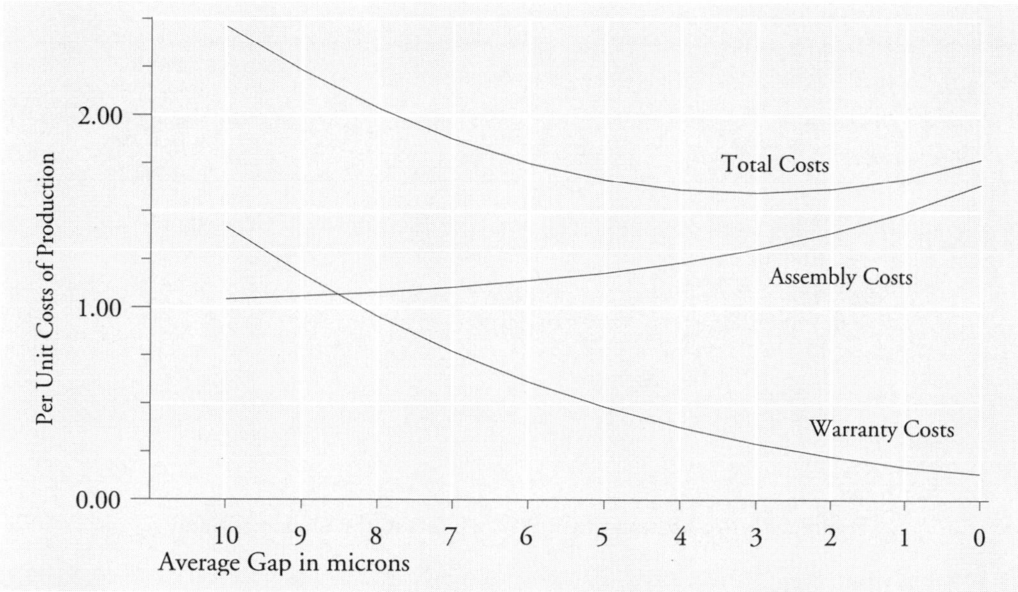

Figure 4.7: The Costs for Different Average Gaps

As we reduce the average gap the proportion of interference fits will increase, as will the number of operations per 1000 units shipped. However, at the same time, the numbers of "loose units" per 1000 units shipped will decrease. So while the costs of initial assembly increase, the potential warranty costs drop, and the minimum total cost occurs at an average gap of 3 microns, which corresponds to an average bearing diameter of 3.003 mm.

Thus, by considering a set of alternative average gaps for use with random assembly we can see that the minimum total cost per unit will occur when the average gap is 3 microns. Between the extra assembly costs and the warranty costs this total cost is 60 percent above the nominal cost of producing an assembly. Since this cost is still high, we might want to consider alternatives to random assembly. And the most common of these is the pre-sorting of the axles and bearings.

4.5 Pre-Sorting the Parts

For this approach we return to the original specifications and process characterizations. The processes are predictable and capable. The axle process is centered at 3.000 mm with a standard deviation of 4 microns while the bearing process is centered at 3.008 mm with a standard deviation of 5 microns. In order to have comparable numbers of axles and bearings in each of the paired categories we will define the three regions for axles and bearings by dividing the region between the natural process limits for each into three zones as shown in Figure 4.8.

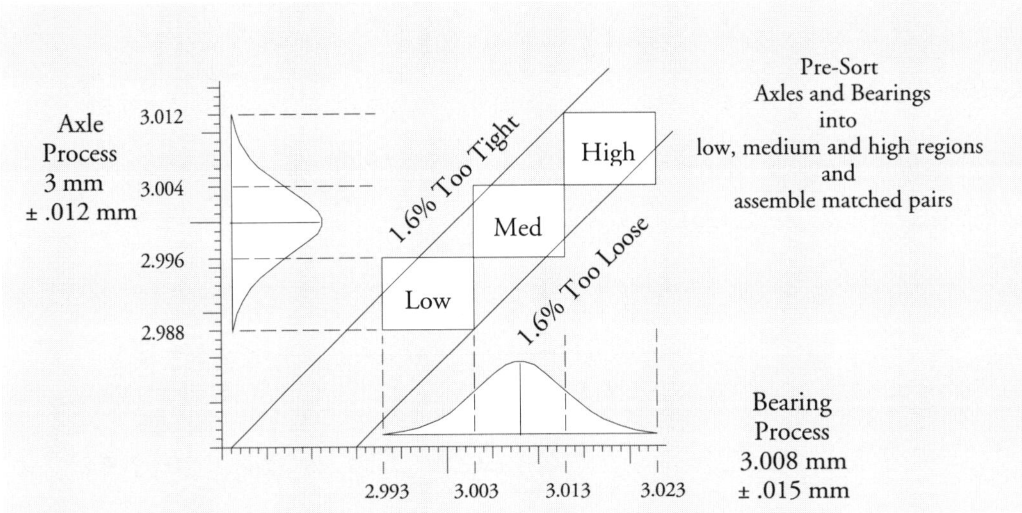

Figure 4.8: Pre-Sorting to Get More Units in the Diagonal Band

To sort bearings into groups only 10 microns wide and axles into zones only 8 microns wide will require careful work and a reasonably sophisticated measurement system. Because of the requirements of using such measurement systems, and the time involved, it is possible that the measurement of an axle or a bearing could be more labor intensive than the assembly of a unit. But, for the sake of simplicity, let us assume that the cost of sorting an axle or a bearing is equal to the cost of assembly. Then, with an average of 16 interference fits per 1000, our costs for fabricating, sorting, and assembling 1000 units to ship will be:

$$\underbrace{1000\,(0.40)}_{\text{make axles}} + \underbrace{1000\,(0.35)}_{\text{make bearings}} + \underbrace{1000\,(0.25)}_{\text{sort axles}} + \underbrace{1000\,(0.25)}_{\text{sort bearings}} + \underbrace{1032\,(0.25)}_{\text{assemble and refit}} = \$\,1508.0$$

These 1000 units will contain an average of 16 probable warranty failures. When we add in these deferred warranty costs we get a total cost of $ 1604.0 or $ 1.604 per unit. So, in the end, we still end up paying about 60 percent more than the nominal cost per unit. In this case, pre-sorting ends up being slightly more expensive than the best of the random assembly scenarios.

4.6 The Traditional Approaches to Production

The options outlined so far encompass the traditional approaches to production. These approaches are summarized in Table 4.2 and Figure 4.9. In Table 4.2 these approaches are arranged in order by the number of functional units per 1000 shipped. In every approach, for each assembly shipped, we end up having to pay a substantial premium in excess costs of one sort or another. Here these premiums range from 60 percent to 146 percent beyond the nominal cost of production. As we will see, all of these excess costs are costs of variation.

Table 4.2: The Traditional Approaches to Production Shift the Excess Costs Around

Approach Used	*Functional Units per 1000 Shipped*	*Actual Costs of Production & Use*		*Excess Costs of Scrap at Fabrication*		*Excess Costs of Inspection*		*Excess Costs of Assembly*		*Excess Costs of Warranty*
Random Assembly:										
Average Gap of 10	764	2.46	⇨	—	⇨	—	⇨	0.04	⇨	1.42
Average Gap of 9	805	2.23	⇨	—	⇨	—	⇨	0.06	⇨	1.17
Average Gap of 8	841	2.03	⇨	—	⇨	—	⇨	0.08	⇨	0.95
Average Gap of 7	872	1.87	⇨	—	⇨	—	⇨	0.11	⇨	0.77
Average Gap of 6	898	1.75	⇨	—	⇨	—	⇨	0.14	⇨	0.61
Average Gap of 5	919	1.67	⇨	—	⇨	—	⇨	0.18	⇨	0.48
Average Gap of 4	937	1.61	⇨	—	⇨	—	⇨	0.24	⇨	0.38
Average Gap of 3	951	1.60	⇨	—	⇨	—	⇨	0.30	⇦	0.29
Average Gap of 2	962	1.62	⇨	—	⇨	—	⇨	0.39	⇦	0.23
Average Gap of 1	971	1.67	⇨	—	⇨	—	⇨	0.50	⇦	0.17
Average Gap of 0	978	1.77	⇨	—	⇨	—	⇨	0.64	⇦	0.13
Pre-Sort Parts	984	1.60	⇨	—	⇨	0.50	⇦	0.01	⇨	0.10
Center Cut	1000	2.45	⇨	0.57	⇨	0.88	⇦	—	⇦	—

With the random assembly scenarios it is possible to reduce the excess costs, but only if we choose the right value for the average gap, and then hold our fabrication processes right at their aim points. All of the excess costs for random assembly come from refitting those parts that will not assemble and from warranty failures. These excess costs are all costs due to variation. If there was no variation at fabrication, none of these costs would exist.

With the pre-sort approach the excess costs are costs of inspection. Of course, the reason for this inspection is variation in the fabrication process.

The center-cut approach used tightened specifications which placed the excess costs in fabrication, inspection, and scrap. Since these fabrication processes were not capable of operating within the tightened specifications of the center cut approach, these excess costs were inevitable, and are completely attributable to variation.

Variation at fabrication is the source of the excess costs of production and use.

Except for the center cut approach, all of these approaches had 100 percent conforming product. And the only reason that the center cut approach had nonconforming product was due to a deliberate tightening of the specification limits—a decision, rather than a fact of life. Hence, not only does variation always create excess costs, but it does so even when everything is within the specifications.

Moreover, these excess costs attributable to variation are not trivial. They resulted in substantial increases in the *Actual Cost of Production and Use.*

In Figure 4.9 the values from the first two columns of Table 4.2 are plotted. In this plot the lower right hand corner is the desirable place to be. This point corresponds to a 100 percent yield of functional units and an *Actual Cost of Production and Use* of $1.00, which is the same as the nominal cost of production. The reason that none of the approaches shown in Figure 4.9 were successful in coming close to this desirable region is that none of them sought to reduce the variation at the source. While each approach attempted to deal with variation

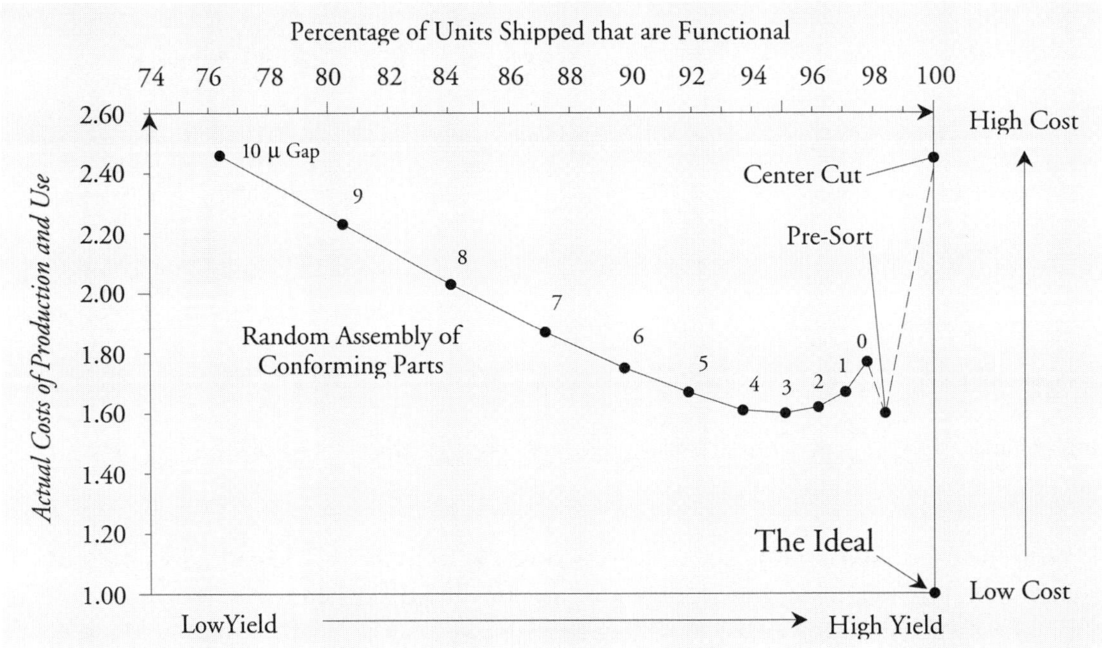

Figure 4.9: The Actual Costs of Production and Use vs. Percentage of Functional Units Shipped

after the fact, by shifting it around, none sought out ways to eliminate the variation.

In most organizations this shifting of the excess costs from one place to another will allow the costs to be hidden in the accounting system, or at least to make them someone else's responsibility. Once the costs are out of sight they are effectively forgotten. We go to great lengths to hide these costs. (In one plant I found that they had physically hidden the rework line behind a set of screens so that it could not be seen by anyone taking a plant tour. In other plants I have found people simply distorting the data to hide the excess costs in overhead.) Yet no matter how successfully we hide these excess costs, they continue to detract from the organization's bottom line. There is something about the shifting of the excess costs in Table 4.2 that brings to mind the thought of arranging deck chairs on the Titanic.

Efforts to deal with variation after the fact will invariably add costs.
Efforts to reduce variation at the source will reduce subsequent costs.

But how can we reduce variation at the source? A proven methodology for reducing variation is the use of process behavior charts to operate a process up to its full potential.

4.7 The Effects of Reducing Variation at the Source

The process behavior chart is essential to operate a process up to its full potential. By displaying the effects of any Assignable Causes that affect your process, these charts enable you to both learn about your process and take action to improve it. Experience has repeatedly shown that this simple technique for learning from the data can be used to substantially reduce the variation of a process. Reports of cutting the variation in half are common. Many report cutting the variation to one-third or one-fourth, and occasionally, even to one-fifth of previous levels. And this type of improvement is not limited to unpredictable processes. Even predictable processes can be improved by taking appropriate actions based on signals from a process behavior chart. In Section 2.7 we saw how Tokai Rika reduced the standard deviation of their predictable process by 30 percent. Therefore, the improvements postulated in this section are based on real experiences, not wishful thinking.

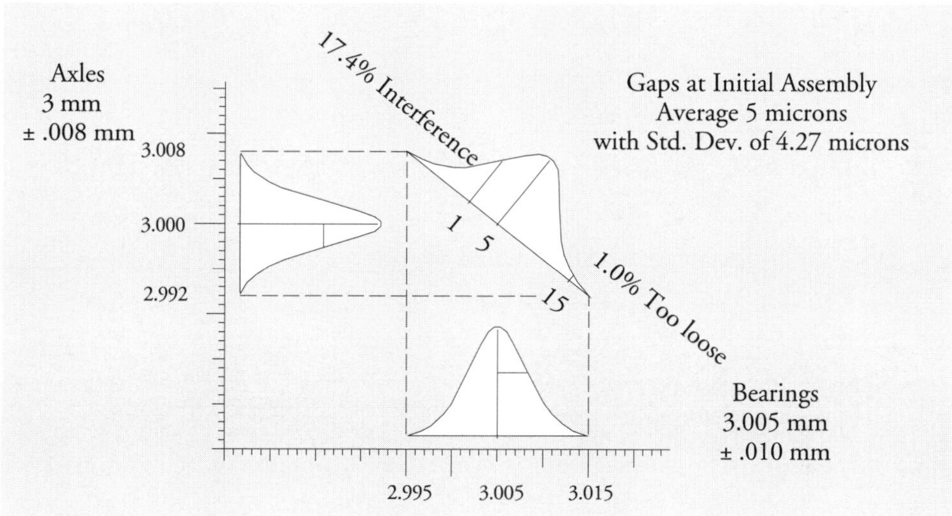

Figure 4.10: Random Assembly when the Standard Deviations are Reduced by One-Third

Consider what could be done by using process behavior charts to improve both the axle process and the bearing process. Assume that the standard deviation of each of these processes was reduced by one-third. The axle process would have a standard deviation of 2.67 microns, and the bearing process would have a standard deviation of 3.33 microns. If we set the average of the bearing process at 3.005 mm, while leaving the axle process centered on 3.0 mm, we would have a gap distribution centered on 5 microns with a standard deviation of 4.27 microns. About 17.4 percent of the initial pairings would not assemble, resulting in refitting such that it would take 1422 operations to ship 1000 units. On the average, 12 of these units would be too loose, while 988 would have a functional gap. Thus, the total cost of shipping 1000 units and taking care of the warranty costs would be:

$$\underset{\text{make axles}}{1000\,(\,0.40\,)} + \underset{\text{make bearings}}{1000\,(\,0.35\,)} + \underset{\text{assemble \& refit}}{1422\,(\,0.25\,)} + \underset{\text{warranty costs}}{12\,(\,6.00)} = \$1177.5$$

which is 18 percent above the nominal cost. This is a substantial savings over the earlier approaches.

Table 4.3: The Effects of Reducing the Standard Deviation by One-Third upon the Different Approaches

Approach Used	*Functional Units per 1000 Shipped*	*Actual Costs of Production & Use*		*Excess Costs of Scrap at Fabrication*		*Excess Costs of Inspection*		*Excess Costs of Assembly*		*Excess Costs of Warranty*
Random Assembly:										
Average Gap of 10	877	1.75	⇨	—	⇨	—	⇨	0.01	⇨	0.74
Average Gap of 9	918	1.51	⇨	—	⇨	—	⇨	0.02	⇨	0.49
Average Gap of 8	947	1.35	⇨	—	⇨	—	⇨	0.03	⇨	0.32
Average Gap of 7	967	1.24	⇨	—	⇨	—	⇨	0.04	⇨	0.20
Average Gap of 6	980	1.19	⇨	—	⇨	—	⇨	0.07	⇨	0.12
Average Gap of 5	988	1.18	⇨	—	⇨	—	⇨	0.11	⇦	0.07
Average Gap of 4	993	1.20	⇨	—	⇨	—	⇨	0.16	⇦	0.04
Average Gap of 3	996	1.26	⇨	—	⇨	—	⇨	0.24	⇦	0.02
Average Gap of 2	998	1.36	⇨	—	⇨	—	⇨	0.34	⇦	0.01
Average Gap of 1	999	1.51	⇨	—	⇨	—	⇨	0.50	⇦	0.01
Average Gap of 0	999	1.73	⇨	—	⇨	—	⇨	0.73	⇦	0.00
Pre-Sort Parts	999	1.51	⇨	—	⇨	0.50	⇦	0.01	⇦	—
Center Cut	1000	1.90	⇨	0.24	⇨	0.66	⇦	—	⇦	—

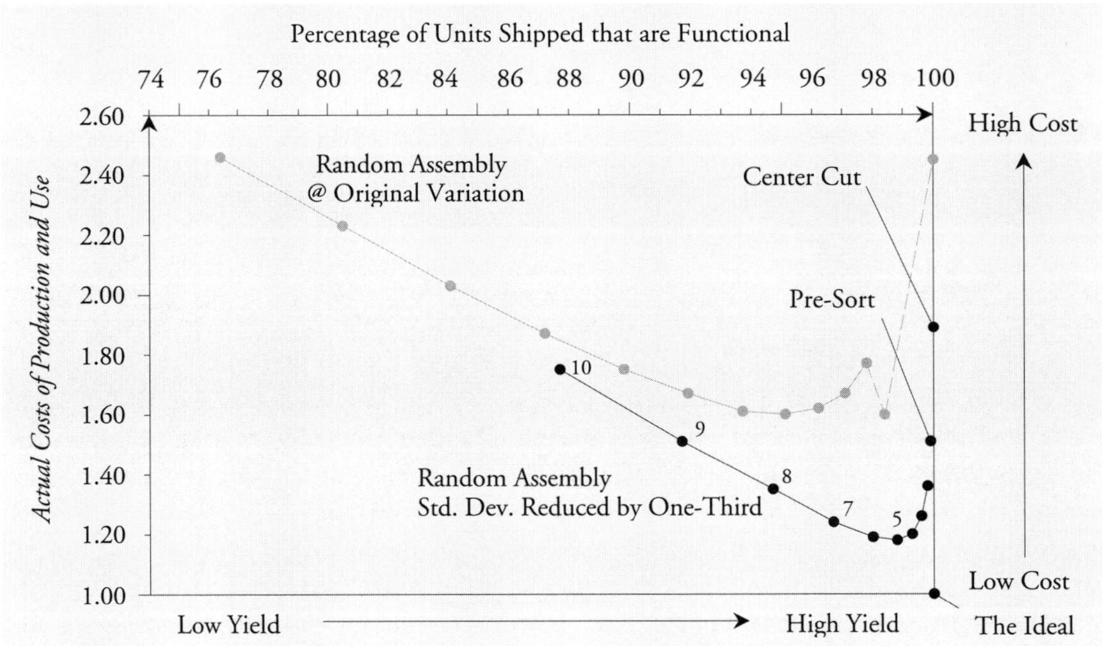

Figure 4.11: The Actual Costs of Production and Use vs. Percentage of Functional Units Shipped

Table 4.3 shows the results for random assembly for different average gaps when the standard deviation of the fabrication processes have been reduced by one-third, as well as what would happen with pre-sorting and the center cut approach. The lowest total cost with random assembly will occur when the average gap is 5 microns. At $ 1.18 per unit this cost is considerably less than the previous best of $ 1.60 per unit.

Figure 4.11 shows a plot of the values in the first two columns of Table 4.3, along with those from Table 4.2 shown in gray. Here we can see how the reduction of variation can move us toward the desirable region of high yield and low cost production.

Notice how, with reduced variation, even the center cut approach becomes cheaper than before, although it remains the highest cost alternative.

Continuing to use process behavior charts will, over time, present you with further opportunities to reduce the variation of the fabrication processes without having to resort to expensive process modifications. If you are able to eventually reduce the standard deviation to one-half of the original level you would have a situation like the one shown in Figure 4.12.

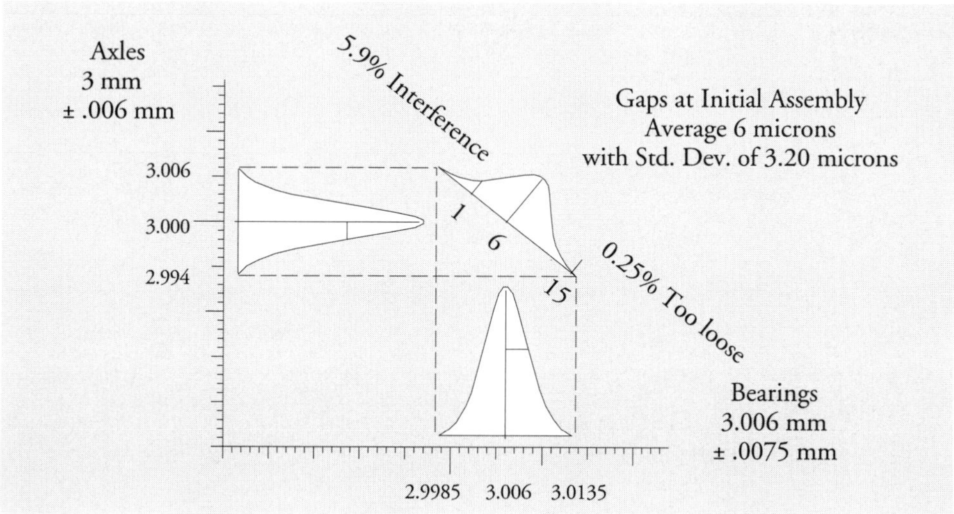

Figure 4.12: Random Assembly when Standard Deviations Are Reduced by One-Half

Here the axle process has a standard deviation of 2 microns, and the bearing process has a standard deviation of 2.5 microns. If we set the average of the bearing process at 3.006 mm while leaving the axle process centered on 3.0 mm we would have a gap distribution centered on 6 microns with a standard deviation of 3.20 microns. About 5.9% of the initial pairings would not assemble, resulting in refitting such that it would take 1126 operations to ship 1000 units. On the average about 2.6 of these units will be too loose, while 997.4 will have a functional gap. Thus, the cost of shipping 1000 units would be:

make axles		make bearings		assemble & refit		warranty costs		
1000 (0.40)	+	1000 (0.35)	+	1126 (0.25)	+	2.6 (6.00)	=	$ 1047.1

Thus, by reducing the standard deviation of the fabrication processes by one-half, the *Actual Cost of Production and Use* can be reduced to $ 1.047 per unit. This is almost 13 percent less than the best *Actual Cost of Production and Use* value found with the one-third reduction, and 55 percent less than the best *Actual Cost of Production and Use* value found with the original variation. Table 4.4 shows the effect of operating with standard deviations that are one-half of the original values.

Table 4.4: The Effects of Reducing Standard Deviation by One-Half upon the Different Approaches

Approach Used	*Functional Units per 1000 Shipped*	*Actual Costs of Production & Use*		*Excess Costs of Scrap at Fabrication*		*Excess Costs of Inspection*		*Excess Costs of Assembly*		*Excess Costs of Warranty*
Random Assembly:										
Average Gap of 10	941	1.36	⇨	—	⇨	—	⇨	0.00	⇨	0.36
Average Gap of 9	969	1.19	⇨	—	⇨	—	⇨	0.00	⇨	0.18
Average Gap of 8	985	1.10	⇨	—	⇨	—	⇨	0.01	⇨	0.09
Average Gap of 7	994	1.05	⇨	—	⇨	—	⇨	0.02	⇨	0.04
Average Gap of 6	997	1.05	⇨	—	⇨	—	⇨	0.03	⇨	0.02
Average Gap of 5	999	1.07	⇨	—	⇨	—	⇨	0.06	⇦	0.01
Average Gap of 4	1000	1.11	⇨	—	⇨	—	⇨	0.11	⇦	0.00
Average Gap of 3	1000	1.18	⇨	—	⇨	—	⇨	0.18	⇦	0.00
Average Gap of 2	1000	1.30	⇨	—	⇨	—	⇨	0.30	⇦	0.00
Average Gap of 1	1000	1.50	⇨	—	⇨	—	⇨	0.50	⇦	0.00
Average Gap of 0	1000	1.82	⇨	—	⇨	—	⇨	0.82	⇦	0.00
Pre-Sort Parts	1000	1.50	⇨	—	⇨	0.50	⇦	0.00	⇦	—
Center Cut	1000	1.67	⇨	0.10	⇨	0.57	⇦	—	⇦	—

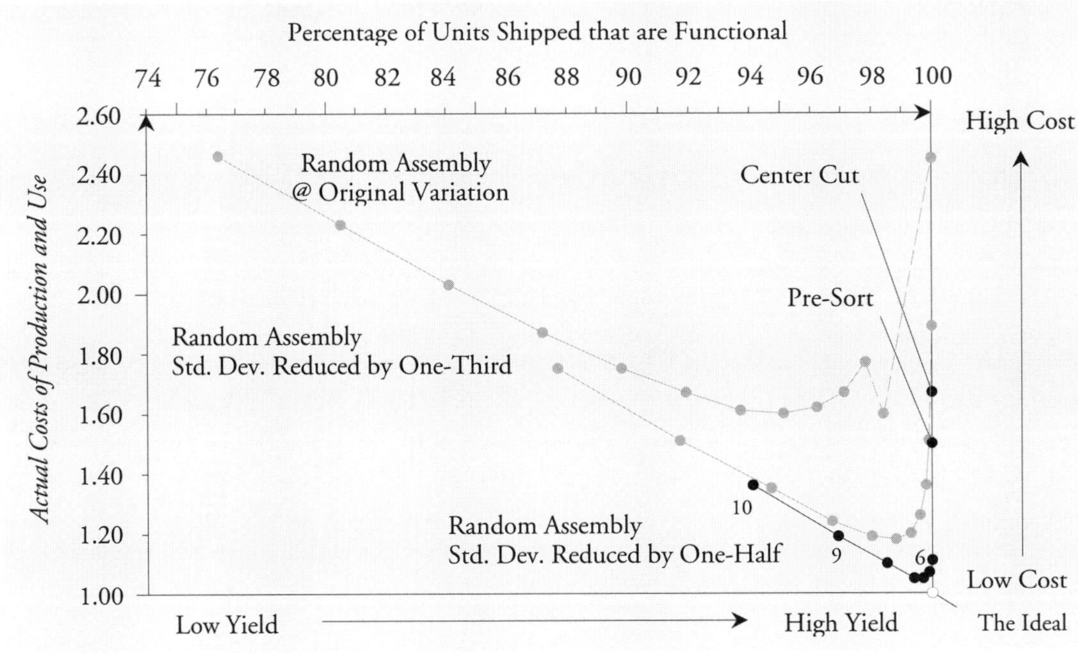

Figure 4.13: The Impact of Variation Reduction upon Actual Costs

Figure 4.13 shows a plot of the first two columns of Table 4.4, along with the corresponding points from Tables 4.2 and 4.3 shown in gray. By reducing variation at the source all of the approaches to production except the pre-sort approach come closer to the desirable state of high yield and low cost production. If you could reduce the *Effective Cost of Production and Use* by 40 or 50 percent, while simultaneously improving the lot quality, and if you could do this without making any capital expenditures, wouldn't you want to do so?

Clearly, the only way to begin to get an *Actual Cost of Production and Use* that approaches the $ 1.00 nominal cost of production is by reducing the variation at fabrication.

The secret of achieving the full benefits of random assembly is the reduction of variation at fabrication. Variation has kept you from realizing the full potential of your production processes in the past, and it will continue to be a barrier in the future. No approach that merely shifts the excess costs around will remove this barrier. There is a definite cost associated with using conforming parts, and the only reliable way to reduce that cost is to reduce the variation at fabrication.

4.8 Summary

The random assembly of interchangeable parts was the idea behind mass production. Unfortunately, this approach does not always work as intended. Problems of assembly and excessive warranty costs have plagued this approach.

Pre-Sorting of parts to facilitate assembly has long been used as an alternative to random assembly. Depending upon the costs of inspection, this is sometimes better and sometimes worse than random assembly.

The center cut approach has long been known to be a high-quality approach. Unfortunately, it has also proven to be a high-cost approach.

As was shown, all of the problems and all of the excess costs can be directly attributed to variation. *They are all costs of variation.* Therefore, since variation is the source of these excess costs, the obvious solution is to reduce variation. Process behavior charts are a powerful and effective tool for doing just that. Using process behavior charts to keep a process operating up to its full potential is the only proven way of operating with minimum variation. And this is the key to the low-cost, high-quality approach.

Figure 4.13 summarizes both the percentage of functional units and the *Actual Costs of Production and Use* for the 39 different approaches considered in this chapter. There we can see that the reduction of variation is the key to producing a high-quality product stream with high yields at low cost.

The conformance to specifications concept of quality is focused on achieving a minimum level of performance. Operating up to full potential is focused on getting all that your process can deliver. The variation reduction that is a common consequence of operating at full potential does more to reduce the *Actual Costs of Production and Use* than any of the traditional approaches. Rather than trying to shift the excess costs off onto some other aspect of the overall process, it seeks to eliminate these excess costs at the source.

The methodology of getting any process to perform up to its full potential is to use a process behavior charts to track the process behavior over time. As we saw in the Tokai Rika example of Section 2.7, all processes will be subject to upsets and changes in the natural course of events. By using process behavior charts to detect these unknown changes, and by investigating the causes of these changes, you can discover how to improve your process with little or no capital expenditures. Typically, as you find and remove Assignable Causes of excessive variation, your process variation will drop to one-half, one-third, one-fourth, or even one-fifth of its previous value. This reduced variation then simplifies all subsequent operations, resulting in additional savings, improved quality, and improved productivity.

The conformance to specifications concept of quality is obsolete. Specifications may be used to scrape the toast, but they do not create quality. All specification-based definitions of quality completely miss the point. Whenever you focus on specifications you will always end up arguing about how good the parts have to be. When you focus on getting the most out of your processes, you will get more uniform product, with less effort and greater productivity.

World Class Quality is no longer a matter of conformance to specifications. Instead, it is now defined to be On Target with Minimum Variance. Those who do not learn the difference between the specification-based approaches to quality and the approach of operating processes at full potential will only become increasingly noncompetitive.

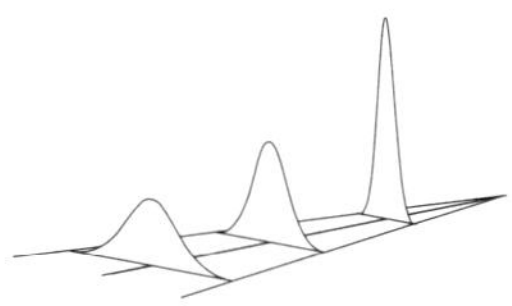

Full Potential = High Yield & Low Cost

Chapter Five

From Capabilities to Excess Costs

In Chapter Three we saw how the Capability and Performance Indexes defined the gap between performance and potential. In Chapter Four we saw how variation creates excess costs even when the components are all conforming. In this chapter we shall illustrate the linkage between the Capability Indexes and both the *Excess Costs of Production* and the *Excess Costs of Use.* This linkage will allow us to skip the detailed computations and simply use the Capability Indexes to determine the excess costs directly.

5.1 The Effective Cost of Production and Use

Let use define the *Effective Cost of Production and Use* as the actual cost of producing and using an item divided by the nominal cost of producing that item.

$$\textit{Effective Cost of Production and Use} \;=\; \frac{\textit{Actual Cost of Production \& Use}}{\textit{Nominal Cost of Production}}$$

The numerator of the ratio above will consist of the nominal cost of production plus the average excess costs per unit associated with producing and using such items.

$$\begin{aligned}\textit{Actual Cost of Production \& Use} \;=\; & \textit{Nominal Cost of Production} \\ & + \textit{Average Excess Cost of Production} \\ & + \textit{Average Excess Cost of Use}\end{aligned}$$

Thus, in order to determine the *Effective Cost of Production and Use* we will need to characterize these Average Excess Costs. These excess costs can be broken down into three components: the costs of scrap; the costs of rework; and the excess costs associated with the use of conforming product. To create our Excess Cost Function we shall need to look at each of these three excess costs.

DEFINING AN EXCESS COST FUNCTION

Since scrapping an item is a cut-your-losses action, ideally taken only when it is cheaper to scrap the item than to use it, we can think of the cost of scrap as a fixed cost. That is, let C_S denote the cost of scrapping an item. This cost will have to equal the cost of the raw materials plus the cost of all processing done to the item, the cost of inspection, and perhaps the cost of disposal as well. In Figure 5.1 this fixed cost of scrapping an item is denoted by a horizontal line beyond the scrap-side specification limit, SL_S.

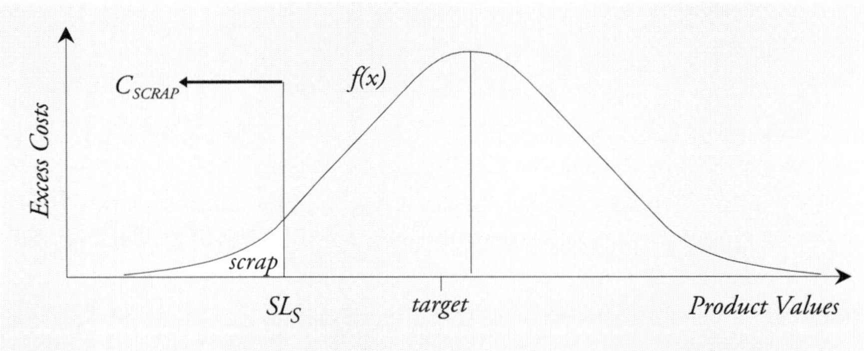

Figure 5.1: Scrap and the Cost of Scrap

Let C_R denote the average cost of reworking an item as a fixed cost of rework. This cost will usually be less than or equal to the cost of scrapping the item. In Figure 5.2 this average cost of reworking an item is denoted by a horizontal line beyond the rework-side specification limit, SL_R.

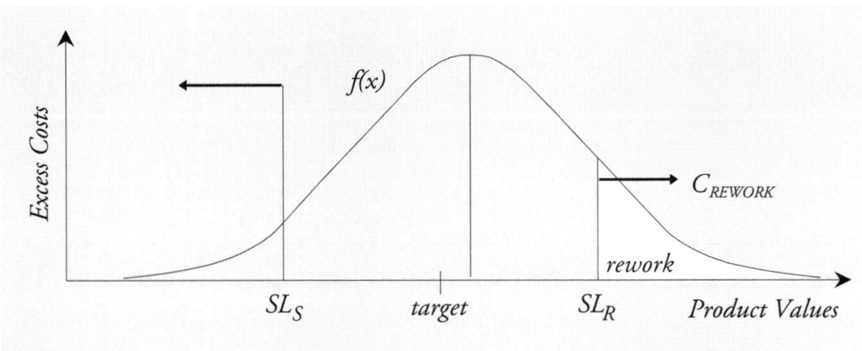

Figure 5.2: Rework and the Cost of Rework

Thus, we know the Excess Cost Function in two regions—scrap and rework. To fill in the rest of the curve we need to think about the nature of excess costs. Excess costs are the result of variation. They are costs incurred because everything is not right at the target value. If we could make everything right at the target value, there would be no excess costs of production. Therefore, we define the Excess Cost Function to be zero at the target value, as shown in Figure 5.3. Thus, in most cases we can find values for the three portions of the Excess Cost Function shown in Figure 5.3.

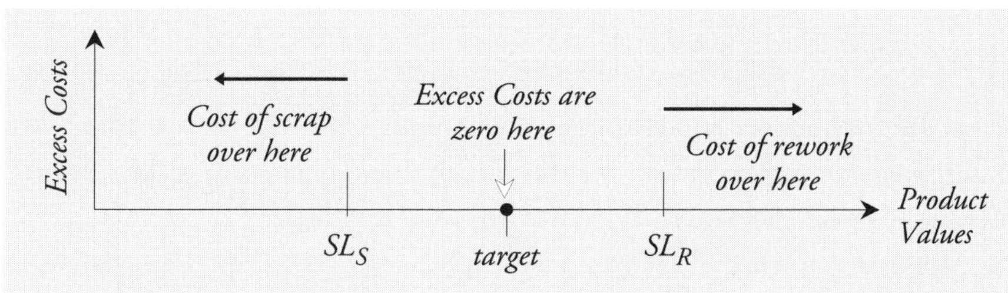

Figure 5.3: The Three Known Portions of the Excess Cost Function

But how do we fill in the gaps? Since we know that small deviations from the target can be tolerated, we prefer to use quadratic curves to connect the three known portions of the Excess Cost Function. These quadratic curves will have their minimum at the target value, and will connect with the fixed portions of the Excess Cost Function to give the piecewise continuous curve shown as $\mathcal{L}(x)$ in Figure 5.4.

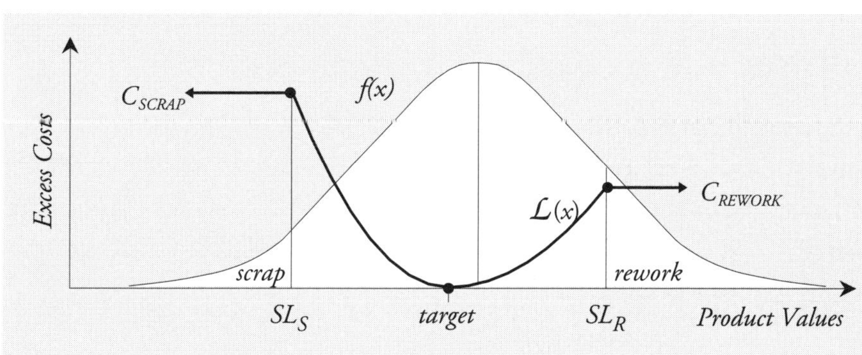

Figure 5.4: A Realistic Excess Cost Function

This Excess Cost Function suggests that when an item falls near the target there will be few excess costs associated with its use. However, as an item falls further and further away from the target the costs of using the item will increase. Moreover, these costs will increase up to that point where it will be cheaper to scrap or rework the item than it will be to try to use it. Since this is consistent with reality, the curve in Figure 5.4 is a realistic Excess Cost

Function. This Excess Cost Function will be symmetric when all nonconforming product is treated alike (either scrapped or reworked). But when nonconforming product on one side is scrapped, while that on the other side is reworked, we may have an asymmetric Excess Cost Function as shown in Figure 5.4 .

It is important to note that the realistic Excess Cost Function in Figure 5.4 is completely determined by the three points shown as black dots. These three points have *(x,y)* coordinates of:

$$(SL_S , C_S) \qquad (target, 0) \qquad (SL_R , C_R)$$

THE AVERAGE EXCESS COSTS

In Figure 5.4 the probability model *f(x)* represents the product distribution. Whether it is a description of the past or a prediction for the future does not matter at this point. When we multiply the Excess Cost Function $\mathcal{L}(x)$ by the probability model *f(x)* and then integrate over all values of *x*, we end up with the weighted average excess cost of production and use:

$$\textit{Average Excess Cost of Production and Use} \quad = \quad \int_{-\infty}^{\infty} \mathcal{L}(x)\, f(x)\, dx$$

The portion of this integral where *x* is between the specifications will define the *Average Excess Costs of Use*. The sum of the two tail portions of this integral will define the *Average Excess Costs of Production*. For more details and the specific formulas obtained for different situations see Chapter Eleven. For illustrative purposes, the following example will look at how we might evaluate the equation above in a simple situation where the process is capable.

TOKAI RIKA: DIRECT COMPUTATION FOR DAYS 1 TO 24

Table 2.7 gives the data for the Tokai Rika Lighter Detent Dimension for Days 1 to 24. This process was found to be predictable and capable. The grand average is 89.80, the target value is 90.0, the specifications are 80 to 100, and *Sigma(X)* is 1.48. Assuming that any nonconforming items are scrapped we have a symmetric Excess Cost Function and we end up with the situation shown in Figure 5.5.

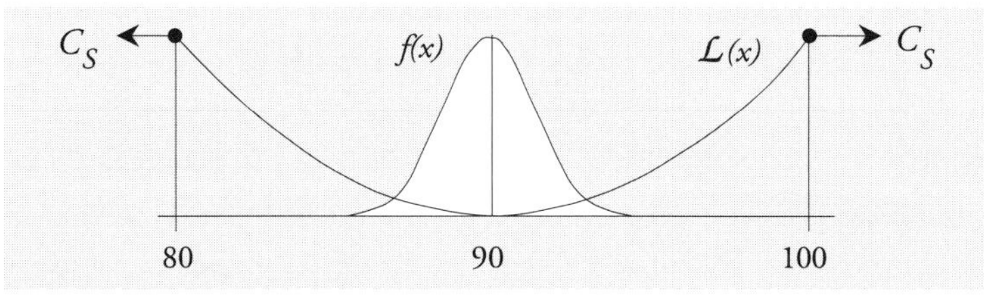

Figure 5.5: The Excess Cost Function for the Tokai Rika Lighter Detent Data

Here the reduced range of x values for $f(x)$ will turn the generalized Excess Cost Function of Figure 5.4 into a quadratic equation of the form:

$$\mathcal{L}(x) \quad = \quad K\,(\,x - \tau\,)^2 \quad = \quad \frac{C_S\,(\,x - 90\,)^2}{(\,100 - 90\,)^2}$$

Whenever the Excess Cost Function is a quadratic equation, the Average Excess Cost will reduce down to the following expression *regardless of which probability model f(x) is used*:

$$\textit{Average Excess Cost of Production and Use} \quad = \quad \int_{-\infty}^{\infty} K\,(\,x - \tau\,)^2\, f(x)\, dx \quad = \quad K\,[\,\sigma^2 \,+\, (\,\mu - \tau\,)^2\,]$$

where μ is the mean of $f(x)$, σ is the standard deviation of $f(x)$, τ is the target value, and K is a constant that defines the quadratic curve and converts everything into monetary units. Inspection of this perfectly general result will show that **the Average Excess Cost will always be minimized by operating On Target with Minimum Variance.**

We may evaluate the expression for the Average Excess Cost using:

$$\textit{Average Excess Cost} \quad = \quad \frac{C_S}{(SL_S - \tau)^2}\,[\,Sigma(X)^2 \,+\, (\,\bar{\bar{X}} - \tau\,)^2\,]$$

If we assume that the cost of scrapping a lighter shell is the same as the nominal cost of production, and that this cost is ¥ 20 then we have:

$$\begin{aligned} \textit{Average Excess Cost} \quad &= K\,[\,Sigma(X)^2 \,+\, (\,\bar{\bar{X}} - \tau\,)^2\,] \\ &= \frac{20\ ¥}{10^2}\,[\,1.48^2 + (\,89.80 - 90.0\,)^2\,] \\ &= 0.45\ ¥ \text{ per piece} \end{aligned}$$

Thus, the *Effective Cost of Production and Use* for this process is estimated to be:

$$\textit{Effective Cost of Production and Use} \quad = \quad \frac{¥\,20 \,+\, ¥\,0.45}{¥\,20} \quad = \quad 1.023$$

Tokai Rika has excess costs that amount to 2.3% of the nominal cost for this process. All of these excess costs are *Excess Costs of Use* since Figure 5.5 shows that there are no nonconforming items being produced by this predictable process.

TOKAI RIKA: USING CAPABILITY INDEXES FOR DAYS 1 TO 24

Since the Excess Cost Function is completely defined by the three points shown as dots in Figure 5.4, the Average Excess Cost will depend upon the relationship between the product distribution, $f(x)$, and the specification limits. Since the relationship between the product distribution and the specification limits is fully and completely characterized by the Capability Indexes, it is possible to skip the computations above and jump directly from the Capability Indexes to the *Effective Costs of Production and Use.*

In Table 3.4 we found the Tokai Rika Detent Data for Days 1 to 24 to have a Capability Ratio of 2.26 and a Centered Capability Ratio of 2.21. From page 174, when $C_p = 2.30$ and $C_{pk} = 2.20$ we find that:

$$\textit{Excess Costs of Production} = 0.000$$
$$\textit{Excess Costs of Use} = 0.023$$
$$\textit{Effective Cost of Production and Use} = 1.023$$

which agrees with the direct computations above.

Tables One to Ten, on pages 169 to 240, allow you to convert Capability and Performance Indexes directly into *Effective Costs of Production and Use* for various situations. Chapter Eleven gives the specific formulas used to create Tables One to Ten. When we combine this ability to convert Capability and Performance Indexes into excess costs with the fact that these four indexes define the gap between performance and potential, we gain a powerful tool for estimating the payback for various improvement projects.

5.2 The Gaps between Performance and Potential

Figure 3.8 is reproduced below as Figure 5.6. There we see how the Capability and Performance Indexes identify the gaps between performance and potential. By converting these Capability and Performance Indexes into *Effective Costs of Production and Use* we can quantify the costs associated with these gaps.

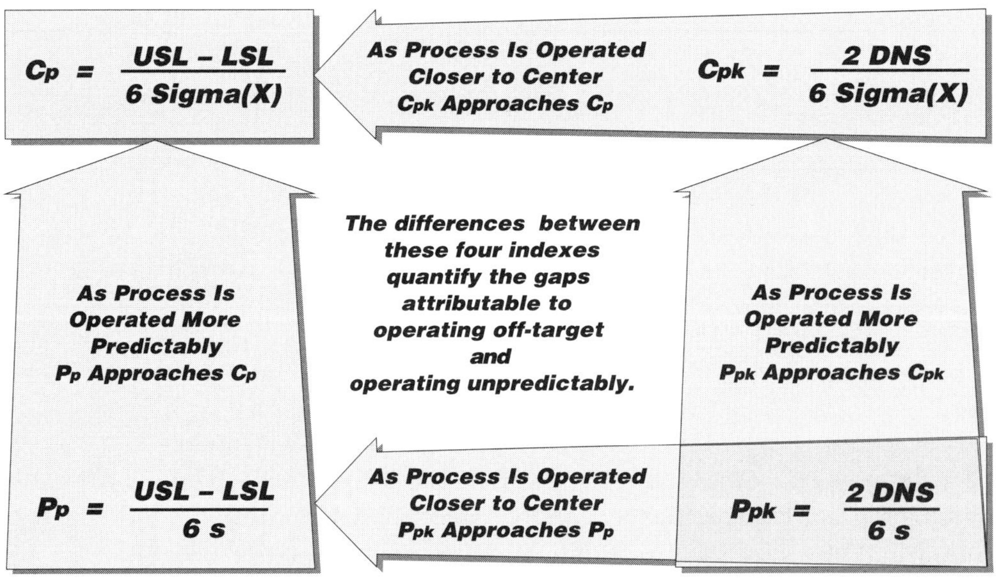

Figure 5.6: The Gaps between Performance and Potential

USING THE PERFORMANCE INDEXES

The Performance Indexes, P_p and P_{pk}, will describe the past performance of the process. When these values are based upon a reasonable amount of data, spanning a reasonable period of time, they can be used to establish a baseline. As always, what is a reasonable baseline will be a judgment call depending upon process knowledge.

Step One: Find the Baseline Costs

Use P_p and P_{pk} to find a Baseline value for the *Effective Cost of Production and Use*. While it may be difficult to predict what an unpredictable process is going to do in the future, we can be reasonably sure that spontaneous improvement is highly unlikely. If we do not do anything, the future is unlikely to be any better than the past, and it could be worse. Therefore, by characterizing the Baseline Costs for the past we have a reasonable value for purposes of comparison.

Step Two: Find the Centered Costs and Compare to the Baseline Costs

By increasing the value of P_{pk} to be the same as P_p, we can estimate what the excess costs would be if we managed to shift the process aim to coincide with the mid-point of the specifications. By comparing these Centered Costs with the Baseline Costs we can estimate the potential payback from efforts at improving the process aim.

USING THE CAPABILITY INDEXES

The Capability Indexes, C_p and C_{pk}, characterize what happens when the process is operated predictably. If the process was operated predictably in the past, these two values will be essentially the same as the Performance Indexes, and the following steps will be moot. But if the process is not being operated predictably, then the Performance Ratios will be substantially smaller than the Capability Indexes. Thus, the last two steps in our general approach to using the *Effective Cost of Production and Use* will depend upon the Capability Indexes to define the hypothetical process potential—what the process can be made to do when it is operated up to its full potential.

Step Three: Find the Predictable Costs and Compare to the Baseline Costs

This step is optional. If the process is already being operated predictably, then C_p and P_p will be two estimates of the same quantity, and C_{pk} and P_{pk} will also be two estimates of the same quantity. Thus, for a process that is already being operated predictably, this step will be redundant with Step One.

For a process that has not been operated predictably, we can use the values for C_p and C_{pk} to characterize the excess costs that correspond to operating a process predictably at the historical average. This might be of interest for processes with averages that are hard to change. However, in many cases, learning how to operate a process predictably will also provide new leverage for adjusting the process aim, making the next step of greater interest.

Step Four: Find the Minimum Costs and Compare to the Baseline Costs

By increasing the value of C_{pk} to be the same as C_p, we can estimate the excess costs associated with operating the process predictably and on target. Since this represents operation at full potential, these excess costs approximate the minimum excess costs that are consistent with the current process configuration.

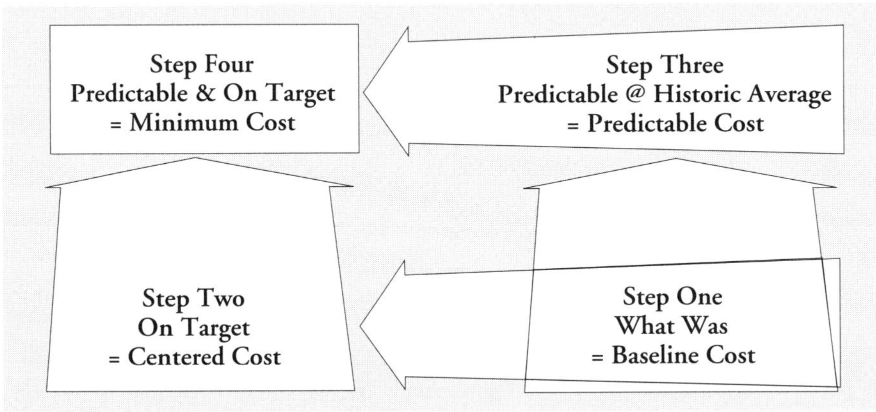

Figure 5.7: Assessing the Benefits of Process Improvements

Depending on the situation, we may be finished after Steps One and Two, or we may skip Step Three, or we might skip straight from Step One to Step Four. The path is flexible, but the point remains the same: to compare and contrast the *Effective Cost of Production and Use* for the Baseline Period with those that correspond to various improved process conditions.

Table 5.1: The Effective Costs of Production and Use under Different Conditions

Indexes Used	Effective Costs of Production and Use	Interpretation
P_p & P_{pk}	Baseline Values	What Was
P_{pk} set equal to P_p	Centered Values	On Target Operation
C_p & C_{pk}	Predictable Values	Predictable @ Historical Average
C_{pk} set equal to C_p	Minimum Values	Predictable and On Target

5.3 The Batch Weight Data

Consider the Batch Weight Data given in Table 3.2. The *XmR* Chart for these 259 data is shown in Figure 5.8. In Table 3.6 the Capability and Performance Indexes for the Batch Weight Data were found to be:

$$C_p = 1.46 \qquad C_{pk} = 0.56 \qquad P_p = 0.54 \qquad P_{pk} = 0.20$$

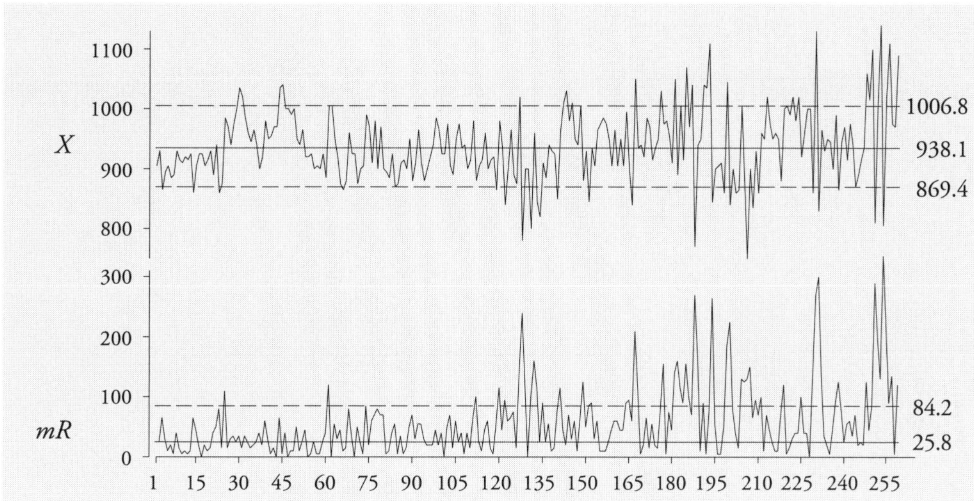

Figure 5.8: *XmR* Chart for the Batch Weight Data

THE BASELINE

A Baseline value for the *Effective Cost of Production and Use* may be obtained using the Performance Indexes. Since these indexes describe the past they will provide an *Effective Cost of Production and Use* that characterizes the past efficiency of a process. Moreover, as a historical value, this Baseline Cost will also provide a reasonable approximation to what you might expect in the future (at least until things take a turn for the worse). Certainly, you should not expect to do substantially better than the Baseline Cost until you improve the process in some way.

Since all nonconforming product is scrapped, and since the cost of scrap is close to the nominal cost of production, we can use the entries from Table One on page 171. Rounding the Performance Indexes off to 0.50 and 0.20, Table One lists an *Excess Cost of Production* of 0.394, an *Excess Cost of Use* of 0.301, and an *Effective Cost of Production and Use* of 1.694. At a Nominal Cost of $ 500 per batch, and with an annual volume of 13,000 batches per year, this works out to a Baseline annualized excess cost of:

$$\textit{Baseline Annualized Excess Cost} \; = \; 0.694 \times \$\,500 \times 13{,}000 \; = \; \$\,4{,}511{,}000$$

If they could wave a wand and make things perfect, this is the potential amount they could save over the course of a year. Since perfection is unlikely to occur, this 4.5 million dollars represents the opportunity pool that exists for improvement projects. (Such opportunity pools are sometimes known as entitlements.) In this case, because the producer and the customer are separate departments in the same company, it is appropriate to use both the *Excess Costs of Production* and the *Excess Costs of Use* in defining this opportunity pool.

THE CENTERED VALUES

Recall what happens when a process is operated on target. As the process average gets closer to the mid-point of the specifications, the value of the Centered Performance Ratio, P_{pk}, will approach the value of the Performance Ratio, P_p. Thus, if we compute the *Effective Cost of Production and Use* for the case where $P_{pk} = P_p$ we will approximate the cost that will exist when the process is operated on target. We will call these values the Centered Costs.

The difference between the Baseline Costs and the Centered Costs will be the potential savings that are likely to be realized by improving the process aim. Since efforts to improve the process aim are some of the easier types of process improvements, it is helpful to see what can be gained by this approach.

For the Batch Weight Data the Performance Indexes were:

$$P_p = 0.54 \quad \text{and} \quad P_{pk} = 0.20.$$

Once again, the Centered Costs may be obtained from Table One. With $P_{pk} = P_p = 0.50$ we find the *Excess Cost of Production* is 0.154, the *Excess Cost of Use* is 0.245, and the *Effective Cost of Production and Use* is *ECP&U* = 1.399. This represents a Centered annualized excess cost of:

$$\textit{Centered Annualized Excess Cost} \;=\; 0.399 \times \$500 \times 13{,}000 \;=\; \$2{,}593{,}500$$

This is $ 1.9 million lower than the Baseline Cost. Thus, we would estimate that we can save almost two million dollars per year by simply operating this process closer to the target value of 1000 kg. Of these savings, approximately 1.5 million will be in reduced costs of production, and the remainder will be in reduced costs of use.

THE PREDICTABLE VALUES

When a process is operated predictably the process potential and the process performance are essentially the same. This means that with a predictable process, the Baseline Costs and the Centered Costs will tell the tale. However, with an unpredictable process, the Performance Indexes will characterize the past performance, while the Capability Indexes will characterize the process potential. To estimate the payback from operating a process predictably at the historical average we begin by using C_p and C_{pk} to find the *Effective Cost of Production and Use.*

Comparing these Predictable Costs with the Baseline Costs will provide an estimate of the savings to be generated by learning how to operate a process predictably.

For the Batch Weight Data the Capability Indexes are $C_p = 1.46$ and $C_{pk} = 0.56$. Rounding these off to 1.50 and 0.60 we find from Table One an *Excess Cost of Production* of 0.037, an *Excess Cost of Use* of 0.380, and an *Effective Cost of Production and Use* is *ECP&U* = 1.418. This would translate into a Predictable annualized excess cost of:

$$\textit{Predictable Annualized Excess Cost} \;=\; 0.418 \times \$500 \times 13{,}000 \;=\; \$2{,}717{,}000$$

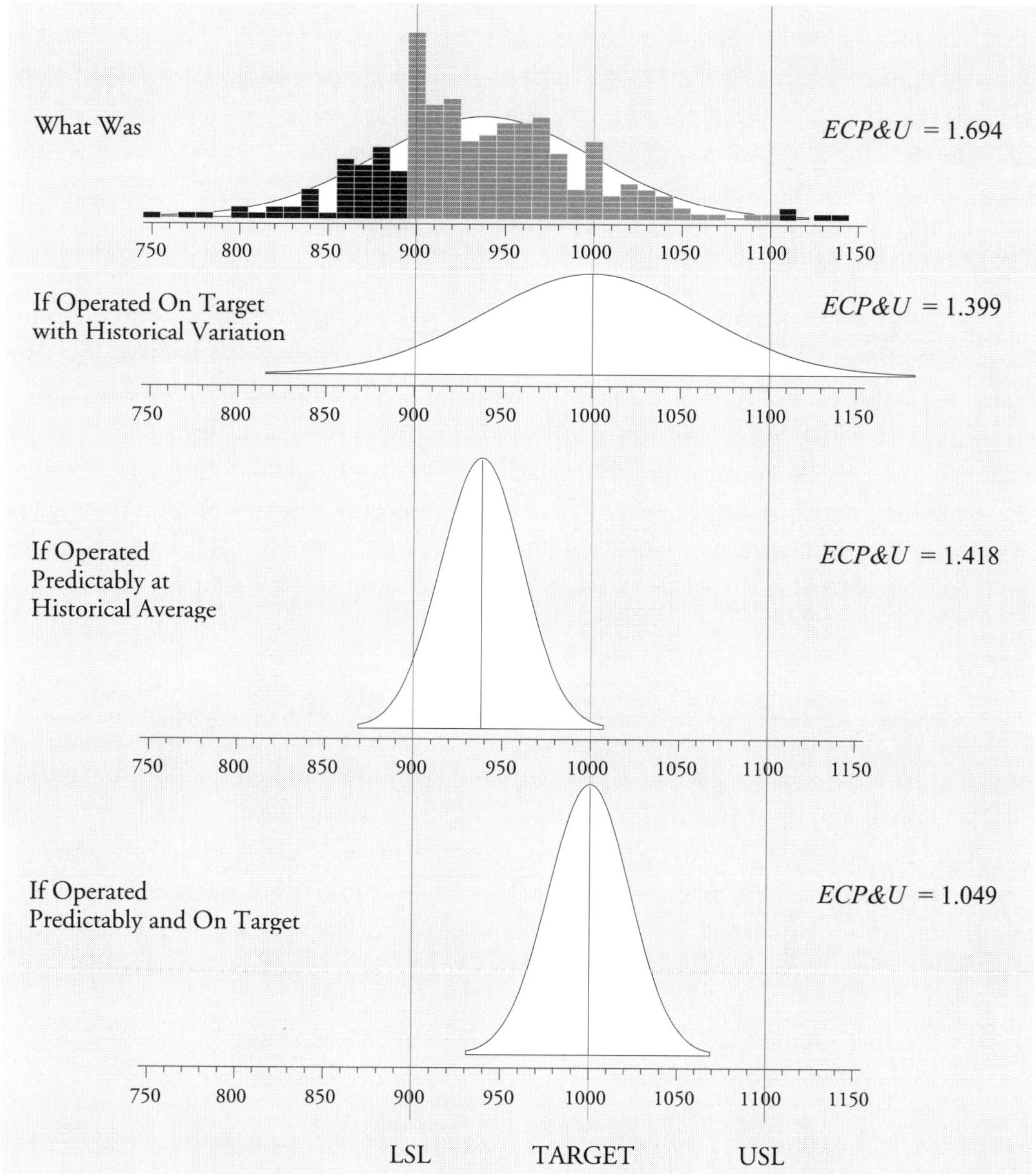

Figure 5.9: The Batch Weight Process under Different Conditions

Compared with the Baseline Cost this represents a savings of $ 1.794 million. The *Excess Cost of Production* is estimated to have gone from 0.394 to 0.037, which translates into over $ 2.3 million in savings, but the increased consistency about the historical average, which is off target, will result in an increase in the *Excess Cost of Use* from 0.301 to 0.380, which translates into a loss of over $ 500,000.

Fortunately, as we identify Assignable Causes and make them part of the set of Control Factors, we will gain additional leverage for adjusting the process aim. This means that we can usually operate closer to target as we learn how to operate a process predictably. Since these Predictable Costs postulate predictable operation at the historic average, they will generally understate the potential payback from operating predictably. For this reason, we may occasionally skip Step Three and go on to Step Four.

THE MINIMUM VALUES

Operating predictably and on target is equivalent to operating on target with minimum variance. When we do this we are operating our process up to its full potential. And this potential is characterized by the Capability Ratio, C_p. By finding the *Effective Cost of Production and Use* that corresponds to $C_{pk} = C_p$ we can describe what our process has the potential to do. By comparing these Minimum Costs with the Baseline Costs we can approximate the savings to be achieved by operating our process predictably and on target.

The Batch Weight Data has a Capability Ratio of $C_p = 1.46$. With $C_p = 1.50$ and $C_{pk} = 1.50$ we find from Table One an *Excess Cost of Production* of 0.000, an *Excess Cost of Use* of 0.049, and an *Effective Cost of Production and Use* is *ECP&U* = 1.049. This would translate into a Minimum annualized excess cost of:

$$\textit{Minimum Annualized Excess Cost} = 0.049 \times \$500 \times 13{,}000 = \$318{,}500$$

all of which would be *Excess Cost of Use*. Compared with the Baseline Cost of $4.5 million, this is a potential savings of $4.2 million per year just from learning how to operate this process up to its full potential. Finally, with excess costs that amount to less than 5% of the nominal costs (*ECP&U* = 1.049), there is not much need to think about process upgrades or reengineering here. The current process has the potential to meet the requirements for some time to come.

Table 5.2: The Excess Costs for the Batch Weight Data

	ExCP	*ExCU*	Total	
Baseline	$ 2,554,500	$ 1,956,500	$ 4,511,000	What Was
Centered	$ 1,001,000	$ 1,592,500	$ 2,593,500	If Operated On Target
Predictable	$ 247,000	$ 2,470,000	$ 2,717,000	If Operated Predictably @ Old Average
Minimum	—	$ 318,500	$ 318,000	If Operated Predictably & On Target

These various excess costs of production and use may be displayed in tabular form for emphasis. For the Batch Weight Data example, where we had annualized costs, this table might look like Table 5.2. Of course, once we have Table 5.2, it is logical to follow with Table 5.3 which shows the potential savings for different courses of action.

Table 5.3: Estimated Annual Savings for Various Process Improvements for Batch Weights

	Savings at Production	*Savings in Use*	*Total Savings*
If Operated On Target	$ 1,553,500	$ 364,000	$ 1,917,500
If Operated Predictably @ Old Average	$ 2,307,500	$ – 513,500	$ 1,794,000
If Operated Predictably & On Target	$ 2,554,500	$ 1,638,000	$ 4,193,000

By converting Capability Indexes and Performance Indexes into tables like these you will find it much easier to get the support for needed process improvements. It will also help you to avoid working on projects with little potential for payback.

5.4 The Methanol Percentages

For another example we return to the Methanol Percentages given in Table 2.6. Since this methyl alcohol is used as a precursor in subsequent chemical reactions, the methanol has specifications of 4.50% to 5.50%. These 15 values have an average of 4.927% and a global standard deviation statistic of $s = 0.359\%$. The average moving range is 0.23%, which results in a *Sigma(X)* value of 0.204%.

Table 5.4: Methanol Percentages in 15 Successive Batches of Alcohol

X	4.6	4.7	4.3	4.7	4.7	4.6	4.8	4.8	5.2	5.0	5.2	5.0	5.6	5.5	5.2
mR		0.1	0.4	0.4	0.0	0.1	0.2	0.0	0.4	0.2	0.2	0.2	0.6	0.1	0.3

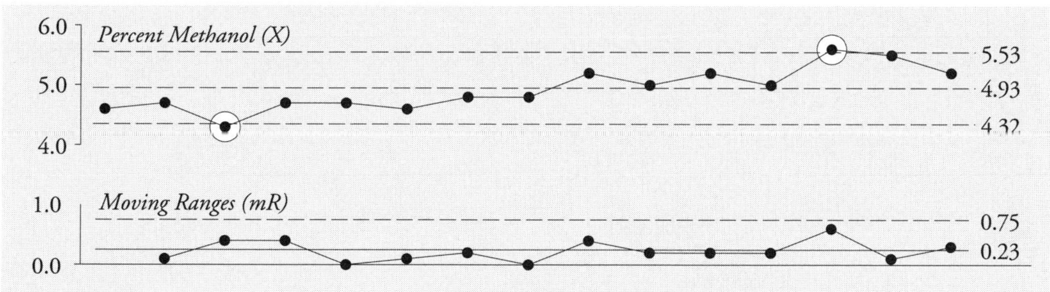

Figure 5.10: *XmR* Chart for the Methanol Data

Thus, the Capability and Performance Indexes are:

$$C_p = \frac{\textit{space available}}{\textit{space required}} = \frac{5.50 - 4.50}{6\,(0.204)} = 0.82$$

$$DNS = 4.927 - 4.500 = 0.427$$

$$C_{pk} = \frac{\textit{effective space available}}{\textit{space required}} = \frac{2\,(0.427)}{6\,(0.204)} = 0.70$$

$$P_p = \frac{\textit{space available}}{\textit{space used in past}} = \frac{5.50 - 4.50}{6\,(0.359)} = 0.46$$

$$P_{pk} = \frac{\textit{effective space available}}{\textit{space used in past}} = \frac{2\,(0.427)}{6\,(0.359)} = 0.40$$

Assuming that nonconforming batches can be blended to spec (i.e. reworked) and that the Cost of Rework is only 25% of the nominal cost of production, we can use Table Two to find the Effective Costs for this process.

BASELINE COSTS

Rounding the Performance Indexes to 0.50 and 0.40 we find the entry for the *Excess Cost of Production* to be 0.151. Deflating this by the Cost of Rework (0.25) we find our *Excess Cost of Production* to be *ExCP* = 0.038.

Likewise, the entry for the *Excess Cost of Use* is 0.214. Deflating this by the Cost of Rework we find our *Excess Cost of Use* to be *ExCU* = 0.054. Thus, our *Effective Cost of Production and Use* is *ECP&U* = 1.092. These are the Baseline Costs. The excess costs here are about 9 percent of nominal, and this is our opportunity pool.

CENTERED COSTS

Rounding both Performance Indexes to 0.50 we find the entry for the *Excess Cost of Production* to be 0.134. Deflating this by the Cost of Rework (0.25) we get an *Excess Cost of Production* of *ExCP* = 0.034.

Likewise, the entry for the *Excess Cost of Use* is 0.212. Deflating this by the Cost of Rework we get an *Excess Cost of Use* of *ExCU* = 0.053. Thus, our *Effective Cost of Production and Use* is *ECP&U* = 1.087. This is a savings of about 0.5 percent of nominal, most of which are reduced *Excess Costs of Production.*

Inspection of the *XmR* Chart shows that the essence of the unpredictable nature of this process is the wandering process aim. Hence, it is unlikely that we would operate this process predictably without also operating it on target. For this reason we jump straight to the Minimum Costs.

MINIMUM COSTS

Rounding both Capability Indexes to 0.80 we find the entry for the *Excess Cost of Production* to be 0.016. Deflating this by the Cost of Rework (0.25) we get an *Excess Cost of Production* of *ExCP* = 0.004.

Likewise, the entry for the *Excess Cost of Use* is 0.152. Deflating this by the Cost of Rework we get an *Excess Cost of Use* of *ExCU* = 0.038. Thus, our *Effective Cost of Production and Use* is *ECP&U* = 1.042. This is a savings of about 5 percent of nominal when compared to the Baseline values, and most of these savings are reductions in the *Excess Cost of Production.*

Table 5.5 summarizes these costs. Here the costs are only expressed as percentages, rather than in terms of dollars, but the table still allows for rational decisions.

Table 5.5: The *Excess Costs* for the Alcohol Batches

	ExCP	*ExCU*	Total	
Baseline	3.8%	5.4%	9.2%	What Was
Centered	3.4%	5.3%	8.7%	If Operated On Target
Minimum	0.4%	3.8%	4.2%	If Operated Predictably & On Target

Table 5.6: Estimated Savings for the Alcohol Batches

	Savings at Production	*Savings in Use*	*Total Savings*
If Operated On Target	0.4%	0.1%	0.5%
If Operated Predictably & On Target	3.4%	1.6%	5.0%

Given the nature of this process, without knowledge of the Assignable Causes that are causing this process to drift around, it is unlikely that we will be very successful in controlling the level of this process without learning how to operate it predictably. Hence, the last row is where we will need to go with this process, and we can shave about 5% off of our cost of production by operating predictably and on target. In this case, most of the potential savings lie in production rather than in use.

This example also shows that useful information is available even when we do not know the costs in terms of dollars. Do not let the difficulties of determining the nominal cost become a barrier to the use of the *Excess Costs.*

Tables like the preceding make the nature of the choices clear in terms of the potential savings for various courses of action. These tables also show that not all improvement strategies will affect both production and use in the same manner, allowing you to pick the appropriate course of action.

5.5 Tokai Rika Revisited

Now let us return to the Tokai Rika process of Section 2.7. Initially this process was being operated predictably and on target, with Performance and Capability Indexes of:

$$P_p = 2.40, \quad P_{pk} = 2.35, \quad C_p = 2.26, \quad \text{and} \quad C_{pk} = 2.21$$

Here we interpret all four of these values as estimates of the same thing. Rounding these four values off to 2.30, we find an *ExCP* value of 0.0% and an *ExCU* value of 2.1% regardless of whether we use Table One or Table Two. This gives an *ECP&U* value of 1.021. The Baseline, Centered, Predictable, and Minimum Excess Costs are all the same since this process is already operating up to its full potential. With an *Effective Cost of Production and Use* that is only 2% above the nominal cost of production, there is no real economic incentive to improve this process. As we will see in the next chapter, they were already operating in the zone of economic production.

Yet they used the opportunities that presented themselves to improve this process. By Days 126 to 140 they were operating predictably and nearly on target. From Table 3.4 the Capability Indexes for Days 126 to 140 are:

$$C_p = 3.22, \quad \text{and} \quad C_{pk} = 3.04$$

Using these two values we find an *ECP&U* value of 1.015, for a savings of about 0.6% in the *Excess Cost of Use.* Thus, there was no incentive to improve this process, and there was, at most, a marginal reduction in the *Excess Cost of Use.* So why did they bother to improve this process? We can only say that they did so because they could.

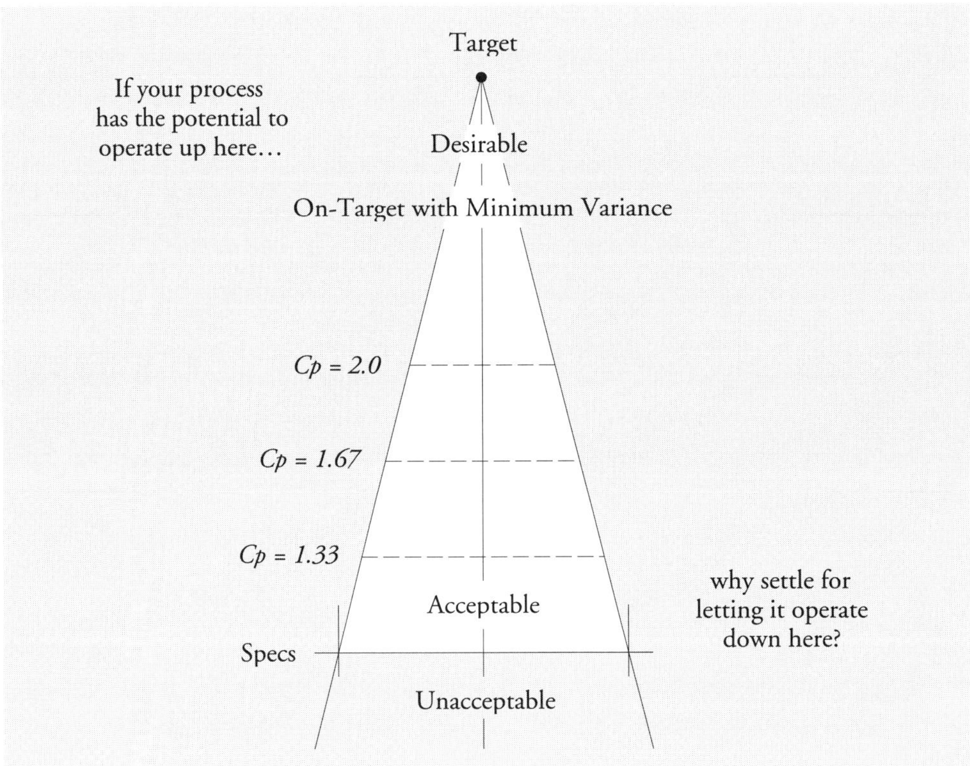

Figure 5.11: World Class Quality

In the words of Dr. Koashi Yoshida, we in the West have failed to distinguish between that which is merely acceptable and that which is desirable. We look at conformance to specifications as the finish line when it is actually the starting line. World Class Quality only happens when we operate our processes predictably and on target. When we use process behavior charts to continually squeeze our processes we find that they are capable of doing far better than we ever dreamed possible. World Class Quality is defined by On Target with Minimum Variance. Any other definition is a non-starter.

5.6 The Cover Examples

The four histograms shown on the cover come from Tables 2.2 and 2.3. The first histogram consists of the 68 data from subgroups 1 to 17 of Table 2.2. These data have an average of 4440.7 and a global standard deviation statistic of 527.1. With specifications of 3850 to 4950 these data will have Performance Indexes of P_p = 0.35 and P_{pk} = 0.32. Interpolating in Table One we find that this corresponds to an *Effective Cost of Production and Use* of 1.739.

The second histogram consists of the 68 data from subgroups 18 to 34 of Table 2.2. These data have an average of 4410.8 and a global standard deviation statistic of 402.0. With specifications of 3850 to 4950 these data will have Performance Indexes of P_p = 0.46 and P_{pk} = 0.45. Interpolating in Table One we find that this corresponds to an *Effective Cost of Production and Use* of 1.472.

The third histogram consists of the 68 data from subgroups 35 to 51 of Table 2.2. These data have an average of 4643.0 and a global standard deviation statistic of 432.6. With specifications of 3850 to 4950 these data will have Performance Indexes of P_p = 0.42 and P_{pk} = 0.24. Interpolating in Table One we find that this corresponds to an *Effective Cost of Production and Use* of 1.678.

The fourth histogram consists of the 64 data from Table 2.3 These data have an average of 4417.8 and a global standard deviation statistic of 184.4. With specifications of 3850 to 4950 these data will have Performance Indexes of P_p = 0.99 and P_{pk} = 0.96. Interpolating in Table One we find that this corresponds to an *Effective Cost of Production and Use* of 1.117.

As seen in Figures 2.5, 2.6, and 2.7, the data from Table 2.2 come from a process that is not being operated predictably, while the data from Table 2.3 come from a process that is being operated predictably. Thus, while the story told by an unpredictable process may well be changing over time, the numbers computed will generally suffice to define the size of the benefits to be obtained by operating a process predictably and on target.

5.7 The Tables for Two-Sided Specifications

Tables One to Six on pages 170 to 216 are for two-sided specifications. The specific equations used, and the assumptions regarding the computations are listed at the beginning of each table. General formulas are also provided to allow for the customization of the values to match specific situations.

Table One gives the *Excess Cost of Production*, the *Excess Cost of Use,* and the *Effective Cost of Production and Use* for those cases where all nonconforming product is scrapped. Table Two does the same for those cases where all nonconforming product is reworked.

When the nonconforming product beyond one specification is scrapped and that beyond the other specification is reworked, you will need to use Tables Three, Four, Five, or Six to find the *Excess Cost of Production*, the *Excess Cost of Use*, and the *Effective Cost of Production and Use*.

Tables Three and Four are for those cases where the process average is on the scrap side of the target value, while Tables Five and Six are for those cases where the process average is on the rework side of the target value.

Table Three assumes that the Cost of Rework is equal to the Cost of Scrap, while Table Four covers the case where the Cost of Rework is 0.1% of the Cost of Scrap. Values that correspond to other ratios between the Cost of Rework and the Cost of Scrap may be obtained by interpolating between the corresponding entries in Tables Three and Four.

Table Five assumes that the Cost of Rework is equal to the Cost of Scrap, while Table Six covers the case where the Cost of Rework is 0.1% of the Cost of Scrap. Values that correspond to other ratios between the Cost of Rework and the Cost of Scrap may be obtained by interpolating between the corresponding entries in Tables Five and Six.

Illustrative graphs and formulas are provided with each table in order to facilitate the use and customization of the values tabulated therein.

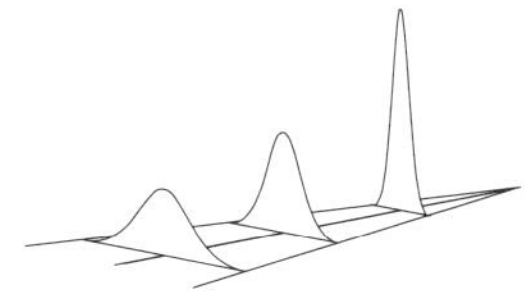

Full Potential = Minimum *ECP&U*

Chapter Six

Economic Operation

Chapter Five describes how to find the *Effective Cost of Production and Use* for two-sided specifications. Chapters Seven, Eight, and Nine will deal with one-sided specifications. This chapter will take a detour and look at what the relationship between the Capability Ratio and the Excess Costs can tell us about economic operation.

6.1 The Excess Costs of Production

Figure 6.1 shows how the *Excess Costs of Production, ExCP,* vary with the Capability Ratio, C_p. The curves in Figure 6.1 come from the values in Table One, which is for the case where all nonconforming product is scrapped. Other figures could be created based on the other tables, but those from Table One are sufficient to illustrate the essential relationships.

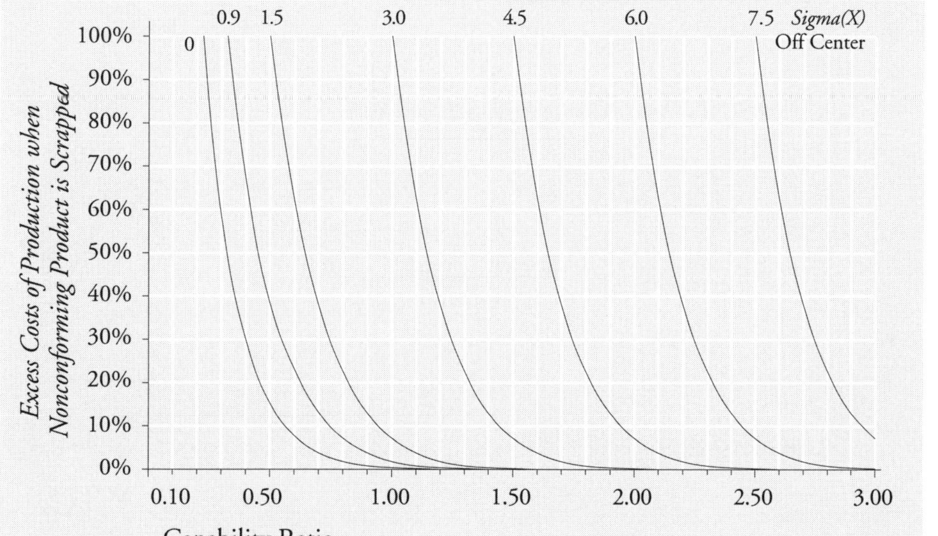

Figure 6.1: How the Excess Costs of Production of Table One Vary with the Capability Ratio

The curve labeled 0 is for the situation where the processes are centered within the specifications and $C_{pk} = C_p$. The remaining curves represent situations where the process average deviates from the midpoint of the specifications by different amounts. The curve labeled 0.9 is for those processes that are 0.9 *Sigma(X)* off center and have $C_{pk} = C_p - 0.3$. The curve labeled 1.5 is for those processes that are 1.5 *Sigma(X)* off center and have $C_{pk} = C_p - 0.5$. And so on for processes that are 3.0 *Sigma(X)*, 4.5 *Sigma(X)*, 6.0 *Sigma(X)*, and even 7.5 *Sigma(X)* off center.

Figure 6.1 shows us that while the *Excess Costs of Production* are clearly dependent upon the Capability Ratio, they are also dependent upon the extent to which the process is off center. Hence, knowledge of the Capability Ratio alone will not suffice to determine the *Excess Costs of Production*. Moreover, there is no magic value for the Capability Ratio that will guarantee that your *Excess Costs of Production* will vanish, or will even fall below some arbitrarily small level.

6.2 The Excess Costs of Use

Figure 6.2 shows how the *Excess Costs of Use, ExCU,* vary with the Capability Ratio, C_p. As before, the curves are for processes that are off center by various amounts.

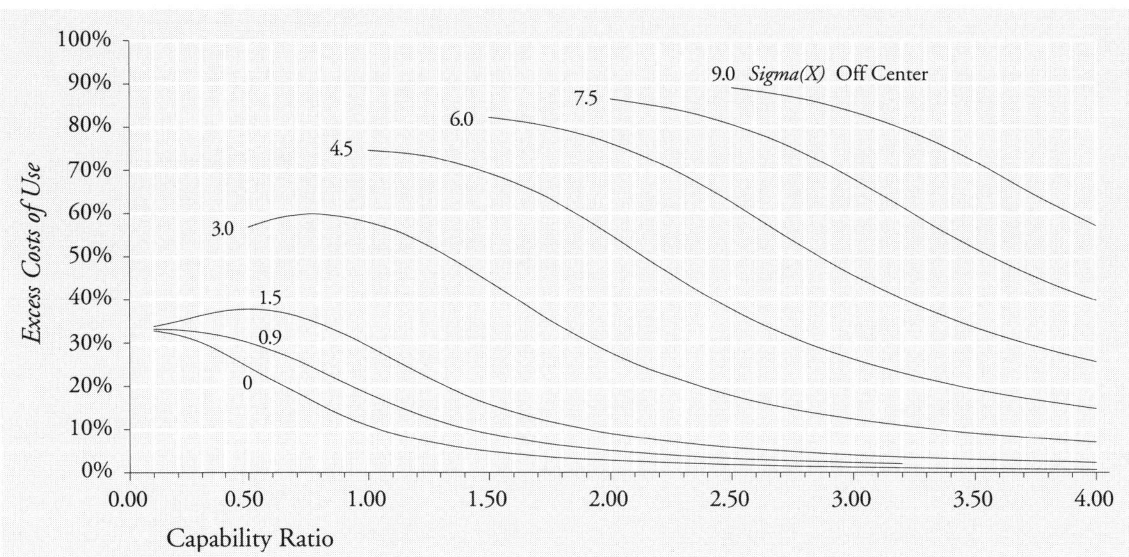

Figure 6.2 How the Excess Costs of Use of Table One Vary with the Capability Ratio

Unlike the *Excess Costs of Production* in Figure 6.1, the *Excess Costs of Use* in Figure 6.2 do not rapidly drop off with increasing values of C_p. This is because, unlike *ExCP* which is primarily a function of the tails of the product distribution, *ExCU* is concerned with the conforming product. The *Excess Costs of Use* will be small only when the bulk of the conforming product is near the mid-point of the specifications. Thus, *ExCU* is dependent upon both the width of the specifications and how well centered the predictable process is within the specifications.

So, once again we find that there is no single, magic value for the Capability Ratio that will be sufficient to guarantee that the *Excess Costs of Use* will vanish, or will even fall below some arbitrarily small value.

In the next section we shall combine the *Excess Costs of Production* with the *Excess Costs of Use* to obtain the *Effective Cost of Production and Use*. This will answer the question of why your customer is asking for Capability Indexes in excess of 1.33 and 1.50. Tables One to Ten (which follow Chapter Eleven) provide the rigorous explanation that has been lacking in so many different improvement programs. They tell us that economic production is a subset of the Ideal State.

6.3 The Effective Cost of Production and Use

In order to look at the overall economics of both production and use we will need to combine the *Excess Costs of Production* and the *Excess Costs of Use*. When we do this according to the formula given in Section 5.1 we end up with the *Effective Cost of Production and Use, ECP&U*:

$$ECP\&U = 1.0 + ExCP + ExCU$$

Using this formula we can combine the curves of Figure 6.1 with those of Figure 6.2 to get the curves shown in Figure 6.3.

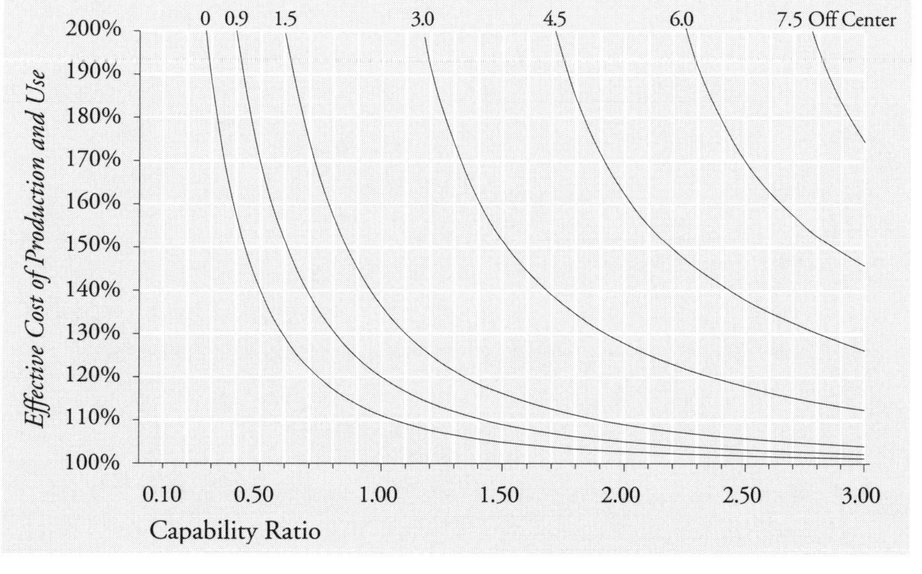

Figure 6.3: How the *Effective Costs of Production and Use* of Table One Vary with the Capability Ratio

If you have a production process that is being operated predictably and on target, then $C_p = C_{pk}$ and the bottom curve in Figure 6.3 will define the relationship between the *Effective Cost of Production and Use* and the Capability Ratio.

If you have a production process that is being operated predictably but is 0.9 *Sigma(X)* off target, so that $C_{pk} = C_p - 0.3$, then the second curve, labeled 0.9, will define the relationship between the *Effective Cost of Production and Use* and the Capability Ratio. The remaining curves are for processes that are respectively 1.5, 3.0, 4.5, 6.0, and 7.5 *Sigma(X)* off target.

As we saw in Figures 6.1 and 6.2, you might have a very large Capability Ratio, and still have a large *Effective Cost of Production and Use* simply because you are not operating on target.

In making sense of the graph in Figure 6.3 it will be helpful to recall a general property of all statistical procedures: small shifts are always harder to detect than large shifts. Process behavior charts are no exception to this rule. This means that shifts of 0.9 *Sigma(X)* or less will be hard to detect in a timely manner. Shifts of 1.5 *Sigma(X)* or less will be slow to be detected. But shifts in excess of 1.5 *Sigma(X)* will be relatively easy to detect in a timely manner.

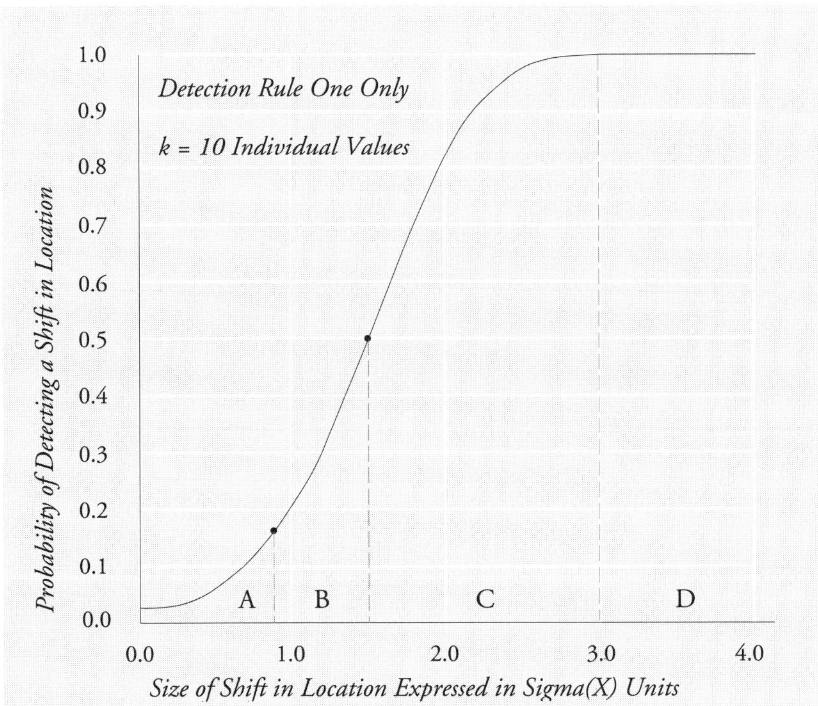

Figure 6.4: The Power Function for *X* Charts

To illustrate this, Figure 6.4 shows the theoretical power function for X Charts. The horizontal axis is given in standard deviation units, while the vertical axis represents the probability of detection. This curve is for a total of 10 observations following the shift, and shows the probability of detecting a shift using Detection Rule One (a point falling outside the three-sigma limits).

In Figure 6.4 we see that, given the same amount of data, a shift of 1.5 *Sigma(X)* is three times more likely to be detected than a shift of 0.9 *Sigma(X)*. At the same time, a shift of 3.0 *Sigma(X)* is virtually certain to be detected within 10 observations following the shift. Thus, we could redraw Figure 6.3 with zones to get Figure 6.5.

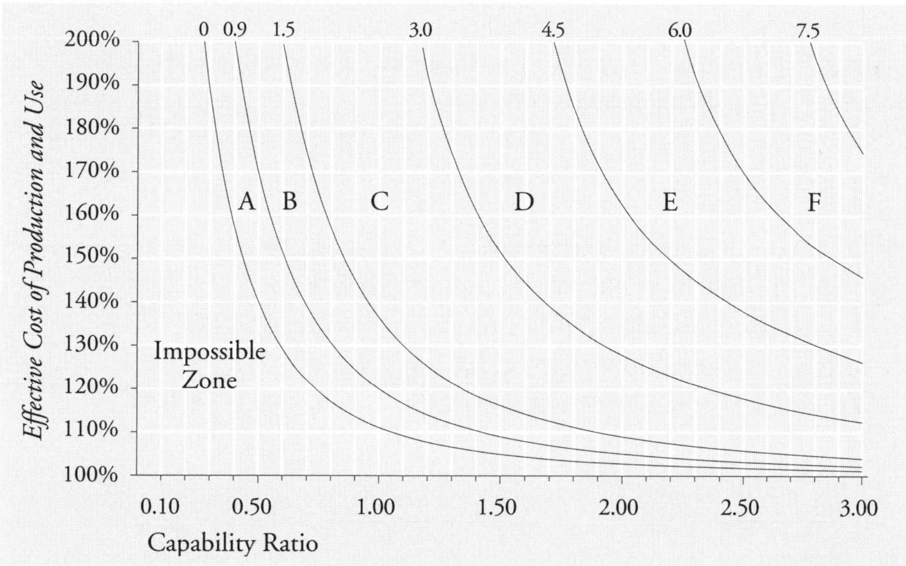

Figure 6.5: How the *Effective Costs of Production and Use* Vary with the Capability Ratio

The first zone is the impossible zone. These combinations of *ECP&U* and C_p are impossible for the situations covered by Table One.

Zone A is the zone where it is hard to detect shifts in the process. Processes in this zone are less than 0.9 *Sigma(X)* off center.

Zone B is the zone where we will be slow to detect shifts in the process. Processes in this zone are between 0.9 *Sigma(X)* and 1.5 *Sigma(X)* off center.

Zones C, D, E, F, and beyond are the zones where we will easily detect the shifts within a reasonable period of time. Processes in these zones are more than 1.5 *Sigma(X)* off center.

Of course, no real process will ever exactly match our mathematical models. There will always be some shifting and changing as the many Common Causes buffet the process. When we make allowance for this buffeting and our inability to detect small shifts in a timely manner, we have to concede that when a process is operated predictably and on target, it will generally inhabit Zone A, with occasional excursions into Zone B.

6.4 The Zone of Economic Operation

If we arbitrarily define processes with an *ECP&U* value of 1.10 or less as economical, then we can see from Figure 6.6 that we cannot *begin* to operate in the economic zone until we have a process that is operated predictably and on target, with a C_p value in excess of 1.10.

Making allowance for process excursions into Zone A, which will be hard to detect, we cannot be *assured* of economic operation until we operate our process predictably and on target and also have a C_p value in excess of 1.50.

Making allowance for occasional excursions into Zone B, we cannot *guarantee* economic operation until we operate our process predictably and on target and also have a C_p value in excess of 1.90.

Thus, with a rigorous approach to making sense of Capability Indexes, we find the minimum requirements for assured economic operation to be:

1. You will need to operate your process predictably.
2. You will need to operate your process on target.
3. You will need to have a Capability Ratio in excess of 1.5.

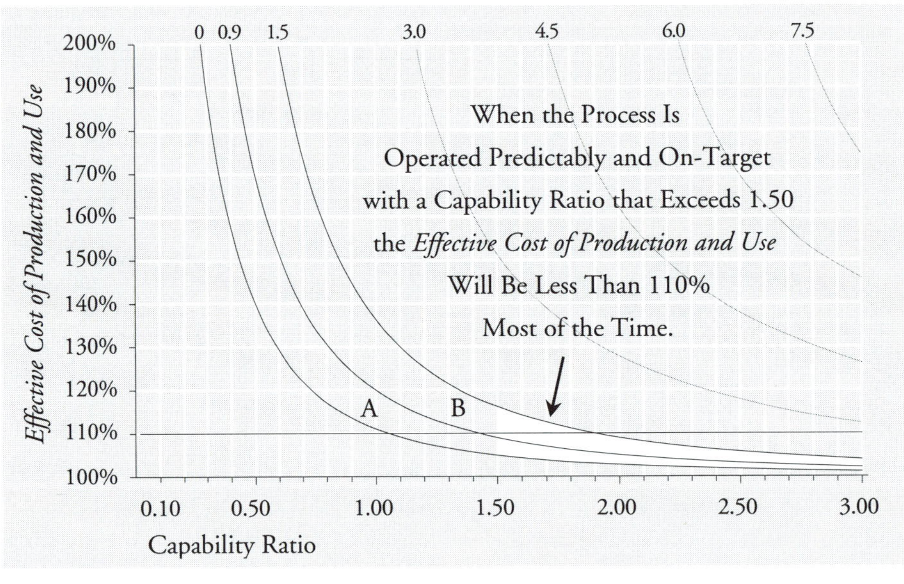

Figure 6.6: The Zone of Economic Operation

Operating a process predictably is a necessity simply because this is the only way to get your process to operate up to its full potential. To operate a process predictably is to operate that process with minimum variance. Unpredictable operation will inevitably increase the variation, which will lower the capability indexes and increase the *Effective Cost of Production and Use*.

Operating on target is a necessity simply because operating off target can take you out of the economic zone regardless of how large your Capability Ratio might be.

Therefore, any attempt to define economic operation based on Capability Indexes alone is simply not sufficient. Even if we dress our Capability Indexes up as "process sigma levels" or talk about the "parts per million defective" or the "defects per million," we still cannot define economic operation without reference to predictable and on target operation.

Figure 6.6 provides the *first* and *only* rigorous definition of what it takes to operate in the "six-sigma zone." It is what has been missing from many six-sigma programs.

6.5 The Spurious 1.5 Sigma Shift

Part of the smoke and mirror computations of various six-sigma programs has been a completely unsupported assumption that an unpredictable process will not shift around more than plus or minus 1.5 *Sigma(X)*.

Real data completely refute this assumption. Consider the Tokai Rika process from Section 2.7. This process is one of the most predictable processes you will ever see. Yet, in Figure 2.15, the average for Subgroup 36 is 2.8 *Sigma(X)* above the central line. In Figure 2.16 the average for Subgroup 60 is 6.7 *Sigma(X)* below the central line. In Figure 2.17 the average for Subgroup 105 is 2.5 *Sigma(X)* above the central line, and in Figure 2.18 the average for Subgroup 173 is 2.1 *Sigma(X)* above the central line.

These excursions serve to illustrate two points. Even a well-behaved process that is operated predictably and on target will experience occasional excursions that are substantially greater than 1.5 *Sigma(X)*. And whenever such excursions occur, shifts in excess of 1.5 *Sigma(X)* will be quickly detected by a process behavior chart.

If this reasonably predictable process can suffer repeated excursions of magnitude two-sigma, three-sigma, or even six-sigma, what hope would a truly unpredictable process have of always staying within 1.5 *Sigma(X)* of the target value? (If you should have any doubt about the answer to this question look back at the *XmR* Chart for the Batch Weight Data in Figure 5.8. There the points vary from the central line by 8 to 9 *Sigma(X)* in *both* directions.)

It is true that when a process is being *operated predictably and on target*, it will spend most of its time with a process average that is within 1.5 *Sigma(X)* of the target. However, since even the most predictable of processes will experience occasional excursions in excess of 1.5 *Sigma(X)*, there is no way to place a bound on the size of the excursions that will be seen when a process is operated unpredictably. Hence, all computations based on the assumption that an unpredictable process will not shift around more than 1.5 *Sigma(X)* are spurious.

6.6 Summary

The *Effective Cost of Production and Use* allows us to convert Capability Indexes and Performance Indexes into the language of management. Moreover, when we follow the mathematics behind this conversion to its logical conclusion we end up with a definition of those conditions needed to operate economically.

A good Capability Ratio is necessary to operate economically, but a good Capability Ratio is not sufficient to guarantee economical operation.

A predictable process is necessary to operate economically, but a predictable process is not sufficient to guarantee economical operation.

A way of operating a process on target is necessary to operate economically, but operating on target is not sufficient to guarantee economical operation.

All three of these necessary conditions must be present in order to guarantee economical operation. They only become *sufficient* when they occur together as a *group*.

Do not focus on one condition while ignoring the other two.

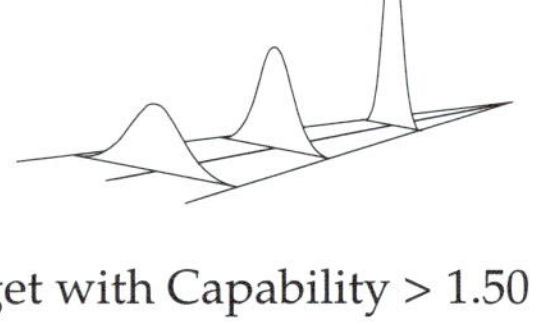

Predictable and On Target with Capability > 1.50 = Economic Operation

Chapter Seven

Excess Costs for One-Sided Specifications

The assessment of excess costs for a characteristic having a one-sided specification differs from that for a characteristic having two-sided specifications. In particular, the concept of the space available within the specs is no longer well defined and the concept of a target value changes. Therefore different tables and approaches are needed. Moreover, there are several different types of one-sided specifications which need to be considered. Among these are one-sided safety specifications and one-sided process specifications.

7.1 One-Sided Safety Specifications

Some one-sided specifications are minimums or maximums that provide a safety threshold. For example, the government has set a minimum tensile strength for the synthetic cord used in automobile brake hoses. In order to have a margin in meeting this specification, automobile manufacturers have their own, even higher, minimum tensile strength, and the automotive suppliers use a third, even higher, minimum standard.

There is a standard on the maximum concentration of hydrogen gas that can be allowed in the air space within barrels shipped by ground transport. In order to have a safety margin in meeting this specification, chemical companies set their own internal specifications on the concentration levels of given compounds.

With one-sided safety specifications you do not want to operate anywhere near the specification limit. You will commonly take action long before you get close to a safety specification.

COMPOUND M

One company making a specialty chemical produced the base fluid in one plant, and then shipped this fluid in barrels to a second plant for intermediate processing. Next this product was shipped to a third plant for three finishing operations before being shipped to the customer. To avoid hydrogen in the barrels there was a maximum specification of 1500 ppm for Compound M in the base fluid. Batches 1 to 29 had already worked satisfactorily in the customer's plant when Batch 30 failed. When Batch 31 worked the customer was convinced that the problem was in the supplier's plant. Since every batch had met every specification at every step along the way, the supplier thought that something must have gone wrong in the finishing operations, and started to troubleshoot these steps in the process. Since there was only one supplier, and only one customer, the company continued to supply the specialty chemical to the customer. Over the next 2 months, 20 batches were produced, but only 9 worked in the customer's plant. These failures produced much ill will and increased the pressure on the troubleshooting team to find the problem. Batch number 52 failed the safety specification of 1500 ppm for Compound M at Step One, so it was fed back into the distillation column in small amounts over the next few months. In the next three months an additional 33 batches were produced, but only 8 worked in the customer's plant. At this point the troubleshooting team had worked their way back upstream to Step One. When they displayed the test results for Compound M in a running record they immediately discovered the problem. This running record is shown in Figure 7.1 where the batches that worked in the customer's plant are shown with black dots.

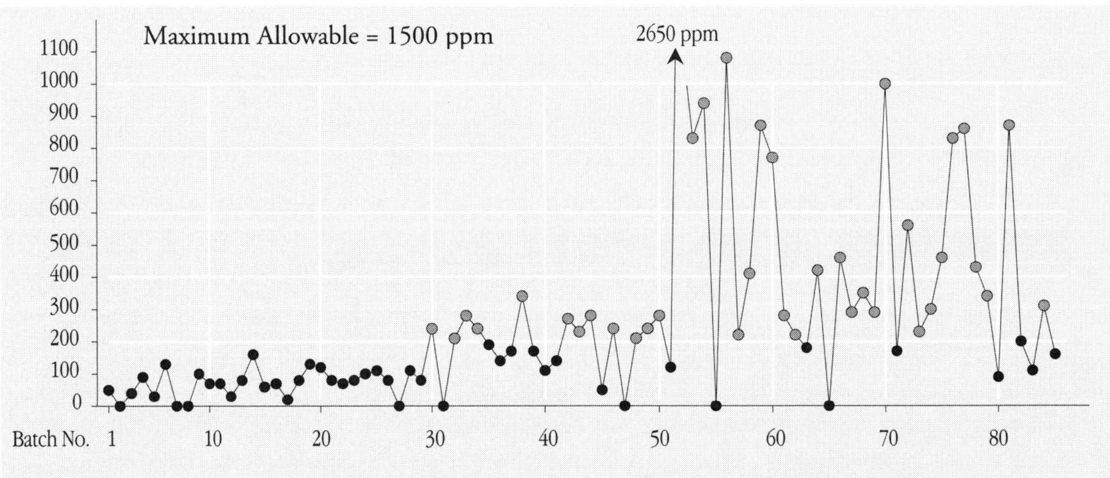

Figure 7.1: Levels of Compound M in Batches 1 to 85 of a Specialty Chemical

The first 29 batches worked because they had less than 200 ppm of Compound M. Following Batch 30, the only batches that worked were those having less than 200 ppm of Compound M. Compound M was poisoning the reaction in the customer's plant, but no one had known about this Assignable Cause. Moreover, seeing the actual levels of Compound M in the graph is completely different from knowing that Compound M is "within specification." Had they been tracking this variable with a graph they could have found the problem five months earlier and avoided all sorts of unpleasantness.

BRAKE HOSE CORD

Another example comes from an automotive supplier making brake hose who received shipments of a synthetic fiber. A sample from each of the six containers in the shipment would be obtained and tested for tensile strength. The government minimum specification for this tensile strength was 5.0 pounds. The automotive company had a more conservative internal minimum specification of 5.5 pounds. The automotive supplier had their own internal minimum specification of 6.0 pounds. The Tensile Strength Data for 11 shipments are shown in Table 7.1, and the limits for the Average and Range Chart in Figure 7.2 were based on the first five shipments.

Clearly, Shipments 1 through 10 were all part of the same Grand Lot, but Shipment 11 was not. Based on this chart, the automotive supplier rejected Shipment 11. The supplier of the brake hose cord asked why the shipment was rejected—after all, every container in Shipment 11 had a tensile strength value that exceeded the minimum specification of 6.0 lbs.

Table 7.1: Tensile Strengths for 11 Shipments of Brake Hose Cord

Shipment	-1-	-2-	-3-	-4-	-5-	-6-	-7-	-8-	-9-	-10-	-11-
Tensiles	8.7	8.8	9.0	9.4	9.1	8.9	9.5	9.3	9.3	9.1	7.2
	9.6	9.2	8.4	8.3	9.7	10.4	8.6	8.8	9.2	9.3	6.6
	9.7	9.2	8.6	8.7	9.0	8.7	9.3	8.5	8.4	8.4	7.5
	8.5	9.3	8.5	8.3	8.2	9.7	7.7	7.6	8.7	8.7	6.4
	8.7	8.5	9.8	8.7	9.5	8.9	8.6	9.1	10.0	8.8	6.6
	8.7	9.5	8.7	9.6	8.1	9.6	9.0	9.1	9.6	9.0	6.4
Averages	8.983	9.083	8.833	8.833	8.933	9.367	8.783	8.733	9.200	8.883	6.783
Ranges	1.2	1.0	1.4	1.3	1.6	1.7	1.8	1.7	1.6	0.9	1.1

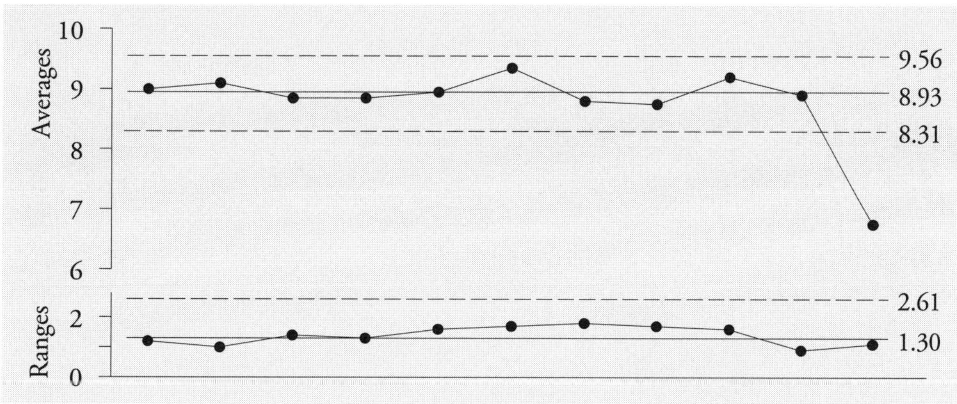

Figure 7.2: Average and Range Chart for Tensile Strengths of Brake Hose Cord

The customer responded that Shipment 11 was clearly different than what they had been using to make brake hoses, and insisted that the supplier take the shipment back. Every time the supplier protested, the customer responded by showing the chart above. Finally, after three days of negotiation with the supplier, the customer said they would accept Shipment 11 if the suppler would be willing to accept any liability that arose from brake hose produced with Shipment 11. At that point the supplier decided to take the shipment back, even though it met everyone's specifications.

The moral of these two examples of one-sided safety specifications is that knowledge that the product characteristic is "in spec" is not the same as knowing the behavior of that product characteristic over time. Predictability is the essence of meeting a safety specification. Anything else is playing roulette with safety.

7.2 One-Sided Process Specifications

Some one-sided specifications actually define where the process is to be operated. Such *one-sided process specifications* will consist of both a product value and some minimum percentage of items that are to meet or exceed the specified product value.

An example would be net weights. In filling containers you will need to have some minimum percentage with fill weights that meet or exceed the net weight on the label. Yet typically the filled containers are not screened to remove the underweight containers. If you do not have the minimum percentage required, then action is taken to increase the average fill weight. Even though the specification applies to the individual items, the action is not taken on the individual items, but rather on the process itself. This makes net weight specifications into *process specifications* rather than *product specifications* in spite of the fact that they specify a product characteristic. Whenever a specification includes a conformance percentage it becomes a process specification rather than an product specification, and the conformance percentage indirectly defines the target value for the process average.

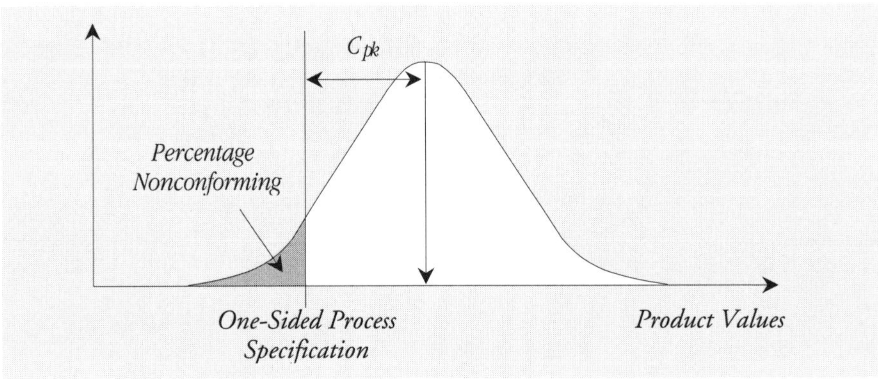

Figure 7.3: A One-Sided Process Specification

As shown in Figure 7.3, the percentage nonconforming is connected to the process average via the Centered Capability Ratio, C_{pk}. As seen in Table 7.2, the relationship between the percentage nonconforming and C_{pk} is a nonlinear relationship. These percentages of nonconforming product for the 219 different capability values given here are based on the traditional, worst-case assumption of a normal distribution for the product stream.

A distribution-free alternative to using the percentages from Table 7.2 would be to use the observed fraction nonconforming from a histogram when such data are available. Here the only limitation on the fraction nonconforming is the number examined. (You cannot get a parts per thousand number when your histogram contains less than 50 data, neither can you justify a parts per million number when you have less than one hundred thousand data.)

Table 7.2: Nonconforming Percentages Corresponding to Centered Capability Ratios

C_{pk} or P_{pk}	*0.00*	*0.01*	*0.02*	*0.03*	*0.04*	*0.05*	*0.06*	*0.07*	*0.08*	*0.09*
1.00	0.1	0.1	0.1	0.1	0.1	0.1	0.1	0.1	0.1	0.1
0.90	0.3	0.3	0.3	0.3	0.2	0.2	0.2	0.2	0.2	0.1
0.80	0.8	0.8	0.7	0.6	0.6	0.5	0.5	0.5	0.4	0.4
0.70	1.8	1.7	1.5	1.4	1.3	1.2	1.1	1.0	1.0	0.9
0.60	3.6	3.4	3.1	2.9	2.7	2.6	2.4	2.2	2.1	1.9
0.50	6.7	6.3	5.9	5.6	5.3	4.9	4.6	4.4	4.1	3.8
0.40	11.5	10.9	10.4	9.9	9.3	8.9	8.4	7.9	7.5	7.1
0.30	18.4	17.6	16.9	16.1	15.4	14.7	14.0	13.3	12.7	12.1
0.20	27.4	26.4	25.5	24.5	23.6	22.7	21.8	20.9	20.0	19.2
0.10	38.2	37.1	35.9	34.8	33.7	32.6	31.6	30.5	29.5	28.4
0.00	50.0	48.8	47.6	46.4	45.2	44.0	42.9	41.7	40.5	39.4

C_{pk} or P_{pk}	*0.00*	*-0.01*	*-0.02*	*-0.03*	*-0.04*	*-0.05*	*-0.06*	*-0.07*	*-0.08*	*-0.09*
0.00	50.0	51.2	52.4	53.6	54.8	56.0	57.1	58.3	59.5	60.6
-0.10	61.8	62.9	64.1	65.2	66.3	67.4	68.4	69.5	70.5	71.6
-0.20	72.6	73.6	74.5	75.5	76.4	77.3	78.2	79.1	80.0	80.8
-0.30	81.6	82.4	83.1	83.9	84.6	85.3	86.0	86.7	87.3	87.9
-0.40	88.5	89.1	89.6	90.1	90.7	91.1	91.6	92.1	92.5	92.9
-0.50	93.3	93.7	94.1	94.4	94.7	95.1	95.4	95.6	95.9	96.2
-0.60	96.4	96.6	96.9	97.1	97.3	97.4	97.6	97.8	97.9	98.1
-0.70	98.2	98.3	98.5	98.6	98.7	98.8	98.9	99.0	99.0	99.1
-0.80	99.2	99.2	99.3	99.4	99.4	99.5	99.5	99.5	99.6	99.6
-0.90	99.7	99.7	99.7	99.7	99.8	99.8	99.8	99.8	99.8	99.9
-1.00	99.9	99.9	99.9	99.9	99.9	99.9	99.9	99.9	99.9	99.9

In general, the process average will need to be a certain distance from the one-sided product specification in order to meet the required minimums. Commonly used percentages are given in Table 7.3 along with the corresponding target values for the Centered Capability Ratio, C_{pk}. Given a specified percentage that is to be conforming, the corresponding capability value may be converted into a target value for the process aim.

Table 7.3: C_{pk} Values Corresponding to Required Conformance Minimums

Conformance %	80%	90%	95%	98%	99%	99.5%	99.9%
Target C_{pk}	0.281	0.427	0.548	0.685	0.777	0.856	1.030

In most cases of one-sided process specifications there will be a penalty for operating with a C_{pk} value that is larger than is absolutely necessary. When this is the case the difference between the process average and the specified product value is commonly known as the "give-away." Since give-away is proportional to cost, it can be used in lieu of costs.

FILL VOLUMES

Table 7.4 contains the fill volumes for hourly samples from a bottling operation. The nominal fill volume is 750 ml, and the process specification is that 95% of the product stream must meet or exceed the nominal fill volume. The values recorded are the number of ml, in excess of 700 ml, found in each bottle tested. The eight hourly samples from each day's production are combined into daily subgroups and charted in Figure 7.4.

These data have a grand average of 52.3 ml and an average range of 7.85 ml. This process is being operated unpredictably. The *Sigma(X)* value is 2.75 ml, while the global standard deviation statistic is s = 5.81 ml. In June their give-away averaged 2.33 ml per bottle.

Table 7.4: Hourly Fill Volumes for the Month of June

June	3	4	5	6	7	10	11	12	13	14	17	18	19	20	21	24	25	26	27	28
Hourly Values	57	57	50	48	49	49	47	51	53	48	68	46	59	59	57	50	56	58	52	56
	52	59	54	45	48	41	43	53	56	51	68	53	70	53	55	46	52	50	49	51
	51	59	55	50	51	46	49	50	54	50	67	47	63	56	53	48	54	49	51	55
	50	55	53	44	55	45	46	47	56	52	66	51	60	56	52	47	54	51	51	50
	49	56	48	43	49	49	44	49	54	47	65	55	63	57	56	45	50	57	52	47
	51	50	51	49	55	44	43	45	54	51	69	45	55	58	52	49	52	55	52	50
	53	54	50	49	51	42	43	52	54	52	71	47	56	60	50	43	56	49	50	46
	51	60	54	47	54	48	50	52	59	50	64	48	65	55	55	51	53	53	47	46
Average	51.8	56.2	51.9	46.9	51.5	45.5	45.6	49.9	55.0	50.1	67.2	49.0	61.4	56.8	53.8	47.4	53.4	52.8	50.5	50.1
Range	8	10	7	7	7	8	7	8	6	5	7	10	15	7	7	8	6	9	5	10

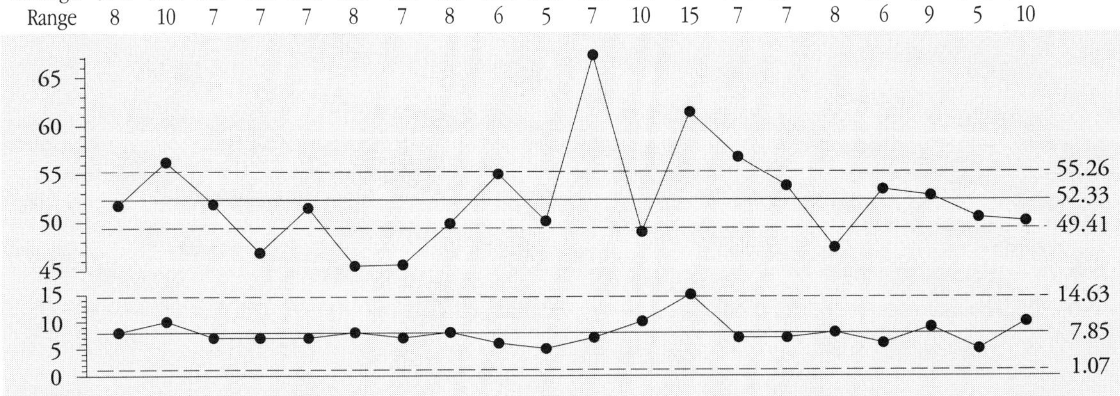

Figure 7.4: Average and Range Chart for Hourly Fill Volumes

THE BASELINE

We characterize the past performance by computing the Centered Performance Ratio:

$$P_{pk} = \frac{DNS}{3\,s} = \frac{752.3 - 750}{3\,(5.81)} = 0.13$$

In Table 7.2 this value corresponds to 34.8% nonconforming. The histogram in Figure 7.5 shows 50 out of 160, or 31%, nonconforming, which is consistent with the tabled value. Either way, this process is not meeting its specification.

PREDICTABLE OPERATION

The Average Chart suggests that fixing the problem of the day-to-day changes in the average fill volumes will be equivalent to operating this process predictably. Therefore, to determine what might be done with this process we proceed directly to consideration of what would happen with predictable operation. The Centered Capability Ratio is:

$$C_{pk} = \frac{DNS}{3\ Sigma(X)} = \frac{752.3 - 750}{3\ (2.75)} = 0.28$$

From Table 7.2 this corresponds to 20% nonconforming. Thus, operating predictably at the current average will not allow this production process to meet the process specification.

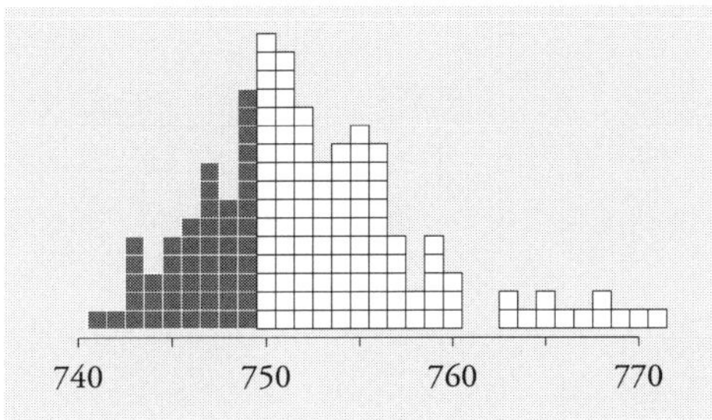

Figure 7.5: Histogram of Hourly Fill Volumes for June

PREDICTABLE AND ON TARGET OPERATION

Table 7.3 shows that to get at least 95% of the bottles to meet the 750 ml fill volume you will need a Centered Capability Ratio of 0.548 or better. With a *Sigma(X)* value of 2.75 ml we use the formula for the Centered Capability Ratio in reverse to find a target value for the process average:

$$3\ Sigma(X)\ C_{pk} = DNS = Target - 750 \text{ ml} = 3\ (2.75)\ 0.548 = 4.52 \text{ ml}$$

Thus, to meet this product specification this bottling process needs to be operated predictably at a Grand Average of 754.5 ml. Of course, if this process proves to have a smaller value for *Sigma(X)* than the one used above when it is operated predictably, then the give-away may be reduced accordingly.

Thus, while they currently have a smaller give-away than that required in order to meet the specification, they are operating in violation of the net weight specification for their product.

Whenever we have a one-sided process specification and want to minimize the give-away, the only way to do this while meeting the specification is to operate predictably and on target. The alternatives are excessive give-away, of a lack of conformance, or both.

By using the Centered Performance Ratio we can characterize the past. By using the Centered Capability Ratio we can evaluate the process potential. With one-sided process specifications we do not need *Effective Costs of Production and Use* since the give-away may be used as a surrogate for the *Excess Cost of Production.*

7.3 One-Sided Product Specifications

One-sided product specifications are specifications where action is taken to scrap or rework individual nonconforming items. Among other things this implies that each item or each batch will be evaluated relative to the specification. This situation is more directly analogous to the two-sided specification case covered earlier.

Unlike two-sided specifications, one-sided product specifications tend to creep. With two-sided specifications there will generally be a target value and a range of acceptable values around this target value. However, when setting a one-sided specification there is a tendency to purchase some insurance by setting the specification value somewhere on the acceptable side of the absolute cut-off point. In many cases this results in the specification value being placed essentially right at the target value. When this happens the concept of a target value is obscured and the definition of the Excess Cost Function for determining the *Excess Cost of Use* becomes very difficult if not completely impossible. However, we may still work with the *Excess Costs of Production*. In this section we will use the Excess Cost Function of Figure 7.6 to find these *Excess Costs of Production*. In the next section we will consider how we might incorporate the *Excess Costs of Use*.

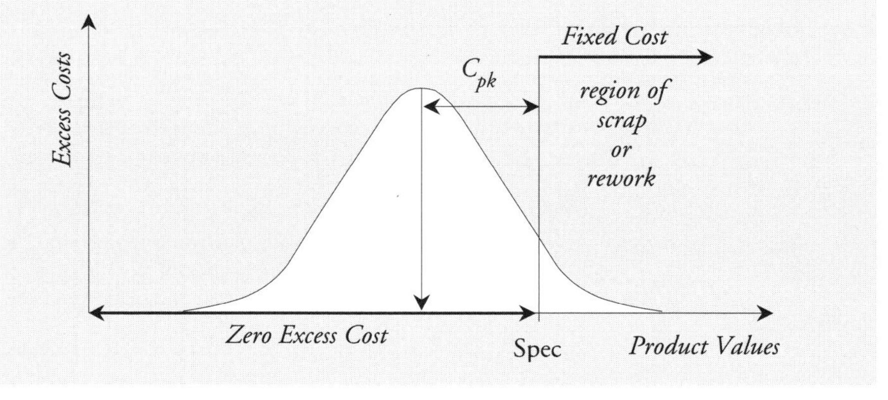

Figure 7.6: The Costs of Scrap or Rework for a One-Sided Product Specification

Table 7.5 gives the *Excess Costs of Scrap* from Table Seven and the *Excess Costs of Rework* from Table Eight which are based upon the model shown in Figure 7.6. To convert either of these values into an *Effective Cost of Production* you must add the value 1.00. To convert the resulting *Effective Cost of Production* into an *Actual Cost of Production* you would multiply by the nominal cost of production.

Table 7.5: Excess Costs of Production (*ExCP*) for One-Sided Product Specifications

Excess Costs of Scrap	C_{pk}	*Excess Costs of Rework*
14.0	*-0.50*	0.933
7.69	*-0.40*	0.885
4.43	*-0.30*	0.816
2.65	*-0.20*	0.726
1.62	*-0.10*	0.618
1.000	*0.00*	0.500
0.618	*0.10*	0.382
0.378	*0.20*	0.274
0.226	*0.30*	0.184
0.130	*0.40*	0.115
0.072	*0.50*	0.067
0.037	*0.60*	0.036
0.018	*0.70*	0.018
0.008	*0.80*	0.008
0.003	*0.90*	0.003
0.001	*1.00*	0.001

Table 7.6: Run-Out Values and Moving Ranges for Hydraulic Cylinders

X	103	68	106	56	86	84	96	73	65	73	137	59	68	102	128	270	131	81	89	99
mR		35	38	50	30	2	12	23	8	8	64	78	9	34	26	142	139	50	8	10

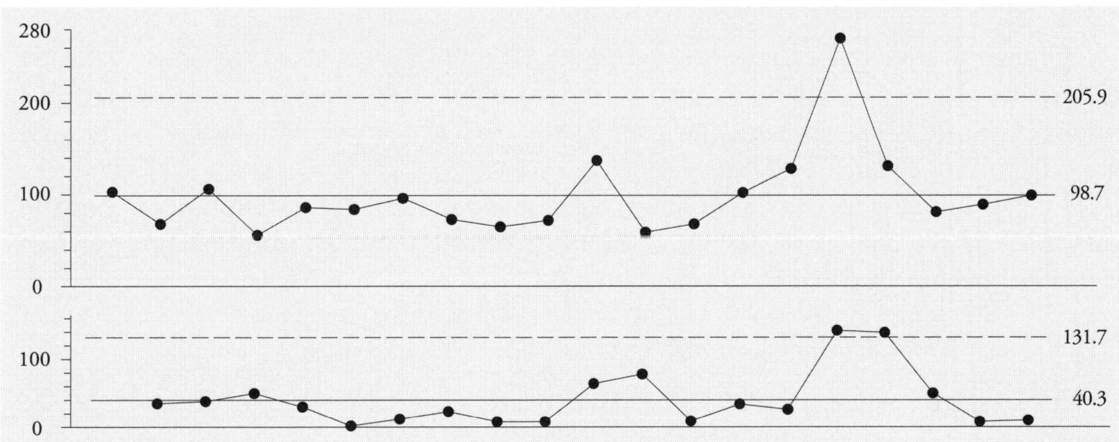

Figure 7.7: *XmR* Chart for Run-Out Data

Table 7.6 gives the Run-Out values for 20 hydraulic cylinders in millionths of an inch. The upper specification for these values is 100 millionths of an inch. The *XmR* Chart for these 20 values is shown in Figure 7.7, where we find the process to be unpredictable, and the histogram is shown in Figure 7.8, where we see that 7 of these 20 values fall above the upper specification.

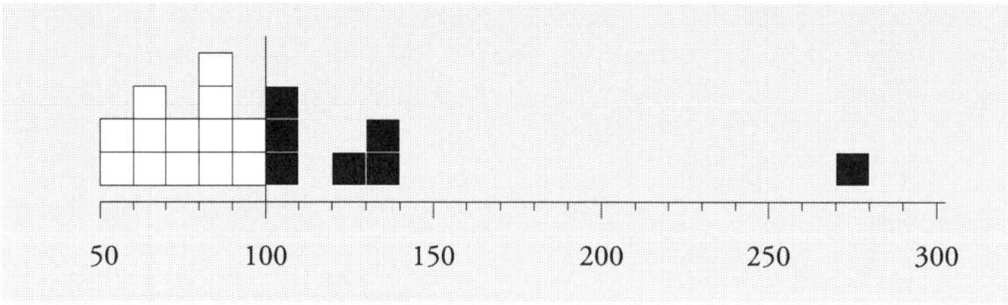

Figure 7.8: Histogram of 20 Run-Out Values

THE BASELINE

With one point outside the limits of the X Chart and 35% of the parts out of specification we have a process that is in trouble. The global standard deviation statistic is $s = 46.64$, while the average is 98.7, resulting in a Centered Performance Ratio of:

$$P_{pk} = \frac{USL - Average}{3\,s} = \frac{100 - 98.7}{3\,(46.64)} = 0.009$$

Since nonconforming cylinders are scrapped at $C_S = NC$, Table 7.5 gives an *Excess Cost of Scrap* of 1.000, which corresponds to an *Effective Cost of Production* of 2.00. Since the histogram suggests that this process needs to be operated at a different average value, we next consider the benefits of changing the aim.

OPERATION AT NEW AVERAGE

Shifting the process average from 99 down to 60 would increase the P_{pk} value to 0.28, and this would reduce this *Effective Cost of Production* to about 1.23. (The value of 60 was selected because it seemed to be both reasonable and feasible to the engineers responsible for this process.)

Since operation with an average that is right on top of the specification is never going to be satisfactory, we skip the scenario of operating predictably at the historical average and go straight to operating predictably at a new average.

PREDICTABLE OPERATION AT NEW AVERAGE

The average moving range of 40.3 corresponds to a *Sigma(X)* value of 35.7. We would estimate that learning how to operate this process predictably at an average of 60 would result in a Centered Capability Ratio of:

$$C_{pk} = \frac{USL - Average}{3\,Sigma(X)} = \frac{100 - 60}{3\,(35.7)} = 0.37$$

which corresponds to an *Effective Cost of Production* of about 1.16.

Given the opportunities shown in Table 7.7, they worked on reducing the average Run-Out and identifying the Assignable Causes of the unpredictable operation. Six weeks later they obtained the data in Table 7.8.

Table 7.7: Effective Costs of Production for Run-Out on Hydraulic Cylinders

Baseline	P_{pk} = 0.01	ECP = 2.00	Savings
Improved Aim	P_{pk} = 0.28	ECP = 1.23	0.77
Predictable @ New Aim	C_{pk} = 0.37	ECP = 1.16	0.84

Table 7.8: Run-Out Values and Moving Ranges for Hydraulic Cylinders Six Weeks Later

X	55	69	48	64	61	69	92	55	81	64	45	54	47	61	54	48	78	46	47	49
mR		14	21	16	3	8	23	37	26	17	19	9	7	14	7	6	30	32	1	2

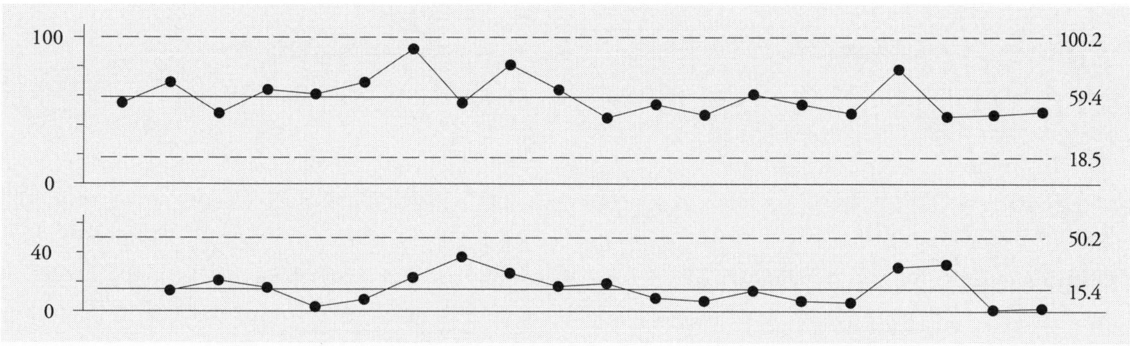

Figure 7.9: ***XmR*** **Chart for Run-Out Data Six Weeks Later**

Figure 7.9 shows a process with an average of 59.35 and an average moving range of 15.37. This gives a *Sigma(X)* value of 13.6, and a C_{pk} value of 1.00, which correspond to an *Effective Cost of Production* of 1.001. The histogram showed the need to shift the process average. The process behavior chart showed the need to find and control the Assignable Causes. By addressing both needs they improved the quality of the cylinders, virtually eliminated their scrap, and cut the *Effective Cost of Production* in half.

While the next section discusses the *Excess Costs of Use* for one-sided product specifications, the definition of a target value may prove to be difficult in some situations involving one-sided product specifications. When this happens it is usually sufficient to use the *Excess Costs of Production* as defined in this section without incorporating the *Excess Costs of Use.*

7.4 Excess Costs of Use for One-Sided Product Specifications

Inherent in having a point for taking action on the product is the concept of cutting your losses. We scrap or rework an item or a batch when it is cheaper to do so than it is to try to use the item or batch as intended. Implicit in this cut-your-losses strategy is the concept of decreasing functionality as the item approaches the action point. Therefore, in order to realistically deal with one-sided product specifications, we will need to evaluate both the excess costs associated with nonconforming items and the excess costs associated with decreased functionality on the conforming side of the limit. An Excess Cost Function that will allow us to do both will have the form shown in Figure 7.10.

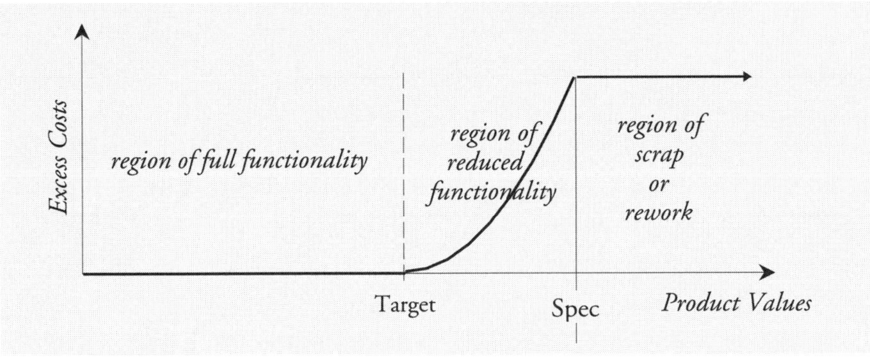

Figure 7.10: An Excess Cost Function for a One-Sided Product Specification

As an example consider the pre-load torques for ball joint assemblies. One automotive manufacturer had two suppliers for a given part number. While both were meeting the specification for the minimum pre-load torque, one supplier was operating much closer to the specification than the other. Shortly after the start of production a problem with road noise in the passenger compartment of a luxury model car caused the assembly plant to be shut down. (This was not good.) When this problem was traced to the ball joint assemblies, both suppliers were blamed for the problem and the specification for the pre-load torques was raised. However, one supplier could show, based on his process behavior charts, that his pre-load torques were predictable and had been meeting the new, revised specification from the beginning of production. Not surprisingly, this supplier soon became the sole supplier for this part.

The evaluation of the region of reduced functionality is an engineering problem, and like the initial specifications in the preceding example, may not always be set correctly. Nevertheless, any realistic appraisal of the *Effective Costs of Production and Use* requires there to be some such zone. *If the customer can define a target value for the product characteristic, then the region between this target value and the one-sided specification will define the region of reduced functionality.*

The target value for any product characteristic should define that value which has a zero excess cost of use—that point on the product value axis where, in a perfect world, the customer would like to see every item produced. The specification limit, on the other hand, should be that cut-off point where it is cheaper to scrap or rework the item that it is to use the item. (In some cases we may have to go beyond the part itself and define a specification limit by the inability of the system to meet performance expectations.) Unfortunately, as noted earlier, one-sided specifications are often set at the target value rather than at the absolute cut-off point simply as a piece of insurance on the part of the customer.

About the only way around this problem of the creeping one-sided specification is to get the customer to define a range of less-than-perfect values that are still acceptable. If this can be done, then one end-point of that range becomes the one-sided specification value and the target value may be taken as the other end-point of that range of acceptable values.

When defining a target value is a problem, you might choose to use a generic region of reduced functionality. By analogy with a Capability Ratio of 1.00 for two-sided specifications, we might take a generic region of reduced functionality to be a region defined by a C_{pk} of 1.00. While this may be arbitrary, it will usually be better than completely ignoring the *Excess Cost of Use*. However, when you feel that there is insufficient information to use the *Excess Cost of Use*, or when its use presents more difficulties that you care to work with, you can still return to the previous section and focus solely upon the *Excess Cost of Production*.

In order to avoid notational difficulties, we shall denote the width of the region of reduced functionality, which is defined by the distance from the target to the one-sided product specification, as the *Tolerance*, and we will express this distance in standard deviation units. As seen in Figure 7.11, the effective costs will depend upon both the C_{pk} value and the *Tolerance*.

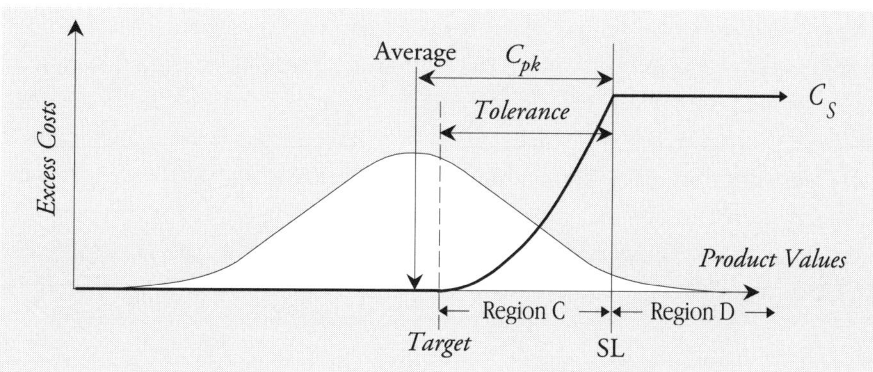

Figure 7.11: The *Effective Costs of Production and Use* Depend upon Both C_{pk} and *Tolerance*

As with C_{pk} and P_{pk}, there will be two versions of *Tolerance*. The *Descriptive Tolerance* will be:

$$\textit{Descriptive Tolerance} = \frac{\textit{Distance from Specification to Target}}{s}$$

While the *Predictable Tolerance* will be:

$$\textit{Predictable Tolerance} = \frac{\textit{Distance from Specification to Target}}{\textit{Sigma}(X)}$$

As might be expected, the *Descriptive Tolerance* will be used with P_{pk} while the *Predictable Tolerance* will be used in conjunction with C_{pk}.

Table Seven gives the *ExCP* and *ExCU* values for the case where all nonconforming product is scrapped. Table Eight does the same for the case where all nonconforming product is reworked.

In those situations where the definition of a *Tolerance* value proves difficult, you can always consider the *Excess Costs of Production* alone. However, as was the case in the example of the pre-load torques, even though the customer did not know about the region of reduced functionality, there was one, and it had a very serious effect. This is why the use of a generic region of reduced functionality is good form even when a *Tolerance* value is unavailable.

7.5 Using One-Sided Product Specifications

Tables Seven and Eight will allow the computation of *Excess Costs of Production* and *Excess Costs of Use* for one-sided product specifications. These excess costs may then be used, either singly or together, to obtain effective costs and even *Actual Costs Of Production and Use.* The process for evaluating different improvement strategies is similar to that used with two-sided specifications.

Step One: Baseline: Here we use the Centered Performance Ratio, P_{pk}, and the *Descriptive Tolerance* to characterize either the *Excess Costs of Production*, the *Excess Costs of Use*, or the *Effective Cost of Production and Use.* These Baseline values will be used with the subsequent values to assess the potential payback for different improvements.

Step Two: Improve the Aim: Unlike the two-sided case, there may not be a natural target value with a one-sided product specification. If a target value has been defined, then compute a new P_{pk} value using that target value in place of the process average. If a target value has not been defined, pick a reasonable and plausible value that represents a better process aim point than the current average, and use this aim point in place of the process average to compute a new P_{pk} value. Use this new P_{pk} value and the *Descriptive Tolerance* to determine the *Effective Cost of Production and Use* and compare it to the value found in Step One.

Steps Three and Four will be needed only when the current process has not been operated predictably.

Step Three: Predictable Operation: Here we use the Centered Capability Ratio, C_{pk}, and the *Predictable Tolerance* to characterize either the *Excess Costs of Production*, the *Excess Costs of Use*, or the *Effective Cost of Production and Use.* These values are then compared with the Baseline values to determine the potential payback from predictable operation. (You may skip this step and go directly From Step Two to Step Four at your discretion.)

Step Four: Predictable Operation with Improved Aim: Here we calculate a new C_{pk} value using either the target value or an improved aim point in place of the process average. Use this new C_{pk} value and the *Predictable Tolerance* to determine the *Effective Cost of Production and Use* and compare it to the value found in Step One.

7.6 Reassessing the Run-Out Data

As an example we shall repeat the analysis of the Run-Out Data from Table 7.6. Figure 7.7 shows these values came from an unpredictable process. Figure 7.8 shows that 7 of these 20 cylinders have a run-out value that exceeds the specified value of 100. The average for these 20 run-out values is 98.7 and the global standard deviation statistic is $s = 46.64$. The average moving range is 40.3, giving a *Sigma(X)* value of 35.7. Nonconforming cylinders are scrapped. Here we do not have a defined target value, and a generic *Tolerance* of 3.0 will correspond to a target value that is negative (and therefore meaningless). So, for purposes of this example, I am going to use a *Descriptive Tolerance* of 0.9 which corresponds to an aim point of [100 – 0.9 (46.64)] = 58, which is close to the improved aim point of 60 that was used earlier.

BASELINE

The P_{pk} value is:

$$P_{pk} = \frac{USL - Average}{3\ s} = \frac{100 - 98.7}{3\ (46.64)} = 0.009$$

From Table Seven we find an *ExCP* value of 1.000 and, with a *Descriptive Tolerance* of 0.9, we find an *ExCU* value of 0.230, giving and *Effective Cost of Production and Use* of 2.230. At the historical average, and with the present amount of variation, the cost of producing acceptable cylinders is twice the nominal cost, and the average cost of using the conforming cylinders is 23% of the nominal cost.

CENTERED ON TARGET

A *Descriptive Tolerance* of 0.9 implies a target value of 58. Recomputing P_{pk} using 58 as the target value we get:

$$P_{pk} = \frac{USL - Target}{3\ s} = \frac{100 - 58}{3\ (46.64)} = 0.30$$

From Table Seven we find an *ExCP* value of 0.226 and with a *Descriptive Tolerance* of 0.9 we find an *ExCU* value of 0.116. The *Effective Cost of Production and Use* becomes 1.342. By centering this process at 58 rather than 99 they could cut the *Excess Cost of Production* down by 77% of the nominal cost. At the same time they could also cut the *Excess Cost of Use* by 11% of nominal.

PREDICTABLE AT OLD AIM

The C_{pk} value is:

$$C_{pk} = \frac{USL - Average}{3\ Sigma(X)} = \frac{100 - 98.7}{3\ (35.7)} = 0.012$$

Since this value is essentially the same as the P_{pk} value found in Step One above, the results will turn out to be indistinguishable from those in Step One.

PREDICTABLE AND ON TARGET

Using the implied target of 58 from Step Two we get:

$$C_{pk} = \frac{100 - 58}{3\ (35.7)} = 0.39 \quad \text{and} \quad \textit{Predictable Tolerance} = \frac{42}{35.7} = 1.18$$

Rounding these values off to 0.40 and 1.20 in Table Seven we find an *ExCP* value of 0.130 and an *ExCU* value of 0.119, giving an *Effective Cost of Production and Use* of 1.249, or 125% of nominal. While the big potential savings come from improving the process aim, we estimate an additional savings of 9% of nominal can be had by operating this process predictably at the improved aim point. While the results in Steps Two, Three, and Four are all hypothetical, they are all reasonably conservative estimates of what can be accomplished.

Table 7.9: *Effective Costs of Production and Use* for Run-Out on Hydraulic Cylinders

Baseline	$P_{pk} = 0.009$	$Tol = 0.90$	$ECP\&U = 2.230$	Savings
Improved Aim	$P_{pk} = 0.300$	$Tol = 0.90$	$ECP\&U = 1.342$	0.89
Predictable @ Old Aim	$C_{pk} = 0.012$	$Tol = 0.90$	$ECP\&U = 2.230$	0.00
Predicable @ New Aim	$C_{pk} = 0.390$	$Tol = 1.18$	$ECP\&U = 1.249$	0.98

Table 7.8 contains the data from this process six weeks after that of Table 7.6. Figure 7.9 shows that this process was then being operated predictably at an average of 59.4. The average moving range of 15.4 gives a *Sigma(X)* value of 13.65. Thus, using 58 as the target value, their actual C_{pk} and *Predictable Tolerance* values are:

$$C_{pk} = \frac{100 - 59.4}{3\,(13.65)} = 0.99 \quad \text{and} \quad \textit{Predictable Tolerance} = \frac{100 - 58}{13.65} = 3.08$$

From Table Seven, the *Excess Cost of Production* is 0.001 and the *Excess Cost of Use* is 0.054. Their *Effective Cost of Production and Use* is 1.055, compared with 2.230 six weeks before. Originally the process had been unpredictable and operating at the wrong average. Now it is operating predictably at the target value. By attending to these two problems they more than doubled their productivity, eliminated their scrap, and made the excess costs of using their product virtually disappear. The Japanese customer was so pleased that this North American plant soon became their sole supplier for hydraulic cylinders.

7.7 The Contamination Data

For a further example consider the contamination values, in parts per million, for twenty batches shown in Table 7.10 and Figure 7.12. There we find the process to be unpredictable, with an average of 100.0, a median range of 16.0, a *Sigma(X)* value of 16.8, and a global standard deviation statistic of $s = 33.27$.

BASELINE

The maximum allowable level for contamination in this product is 150 ppm, and nonconforming batches are recycled through the distillation column at a cost that is 80% of the nominal cost of production. The Centered Performance Ratio is:

$$P_{pk} = \frac{USL - Average}{3\,s} = \frac{150 - 100}{3\,(33.27)} = 0.50$$

Table 7.10: Contamination Data and Moving Ranges in ppm

X	114	87	76	87	112	95	78	94	98	90	93	199	86	72	77	92	85	182	74	109
mR		27	11	11	25	17	17	16	4	8	3	106	113	14	5	15	7	97	108	35

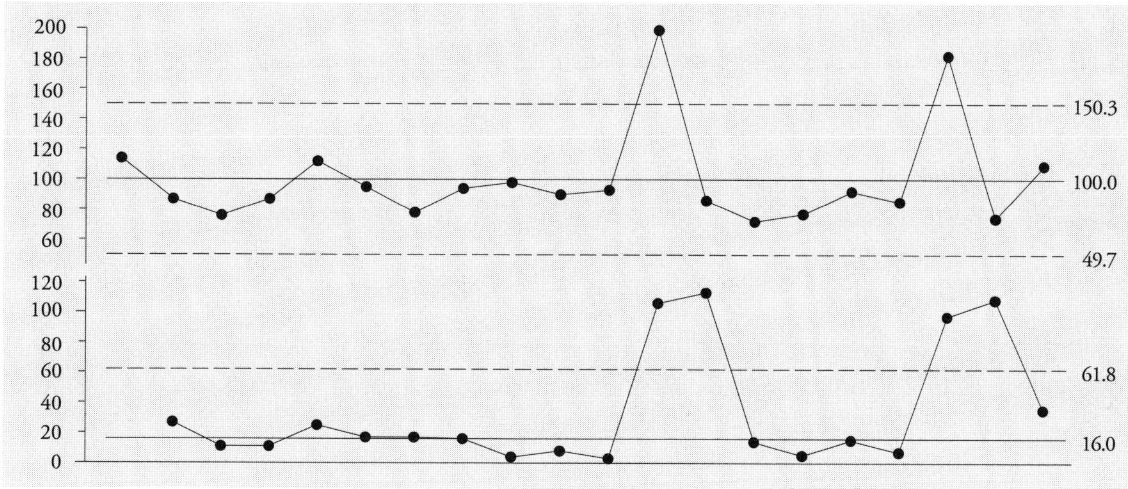

Figure 7.12: *XmR* Chart for Contamination Data

Assuming that the customer prefers a contaminant level that is below 80 ppm, the *Descriptive Tolerance* value becomes (150 – 80)/33.27 = 2.10 standard deviation units. Scaling down the values from Table Eight, we find our Baseline *ExCP* to be 0.067 (0.80) = 0.054 and our Baseline *ExCU* to be 0.169 (0.80) = 0.135. Thus, we end up with an *Effective Cost of Production and Use* of 1.189.

CENTERED ON TARGET

Shifting the aim of this predictable process to a value near 80 ppm would change the P_{pk} value to:

$$P_{pk} = \frac{USL - Target}{3\,s} = \frac{150 - 80}{3\,(33.27)} = 0.70$$

From Table Eight, with a *Descriptive Tolerance* of 2.1, we find our Centered *ExCP* to be 0.018 (0.80) = 0.014 and our Centered *ExCU* to be 0.088 (0.80) = 0.070. Thus we end up with a Centered *Effective Cost of Production and Use* of 1.084, which is 10.5% less than the Baseline value.

PREDICTABLE AT OLD AIM

Operating this process predictably at the current average will give a hypothetical Centered Capability Ratio of:

$$C_{pk} = \frac{USL - Average}{3\,Sigma(X)} = \frac{150 - 100}{3\,(16.8)} = 0.99$$

At the same time, the *Predictable Tolerance* value would become:

$$\textit{Predictable Tolerance} = \frac{150 \text{ ppm} - 80 \text{ ppm}}{16.8 \text{ ppm}} = 4.17$$

From Table Eight, we find our Predictable *ExCP* to be 0.001 (0.80) = 0.001 and our Predictable *ExCU* to be 0.134 (0.80) = 0.107. Thus, we end up with a Predictable *Effective Cost of Production and Use* of 1.108, which is 8.1% less than the Baseline value.

PREDICTABLE AND ON TARGET

If this process is operated predictably at the target value of 80 ppm the C_{pk} and *Predictable Tolerance* values would be:

$$C_{pk} = \frac{150 - 80}{3\,(16.8)} = 1.39 \quad \text{and} \quad \textit{Predictable Tolerance} = \frac{150 - 80}{16.8} = 4.17$$

From Table Eight we find our Minimum *ExCP* to be 0.000 (0.80) = 0.000 and our Minimum *ExCU* to be 0.028 (0.80) = 0.022. Thus we end up with a Minimum *Effective Cost of Production and Use* of 1.022, which is 16.7% less than the Baseline value.

Table 7.11: *Effective Costs of Production and Use* for the Contamination Data

Baseline	$P_{pk} = 0.50$	$Tol = 2.10$	*ECP&U* = 1.189	Savings
On Target	$P_{pk} = 0.70$	$Tol = 2.10$	*ECP&U* = 1.084	0.105
Predictable @ Old Aim	$C_{pk} = 0.99$	$Tol = 4.17$	*ECP&U* = 1.108	0.081
Predicable & On Target	$C_{pk} = 1.39$	$Tol = 4.17$	*ECP&U* = 1.022	0.167

But what if the customer wanted zero ppm contamination? Clearly this would be impossible. The next chapter will consider the problem of what to do when the target value is also a boundary value.

The previous examples show the importance of using the context of the problem to ask the interesting questions with your analysis. The four steps are merely guides. They should be used as a framework for organizing the interesting questions. The purpose of analysis is insight, and the only reason to collect data is to take action. If your analysis allows you to see what action to take, to understand why you need to take that action, and to communicate both the need and the reason for that action to others, then it is a successful analysis.

7.8 Summary

The only reliable way to meet one-sided safety specifications is to operate your process predictably at a point that is comfortably far from the safety specification.

One-sided process specifications specify a minimum percentage of the product stream that is supposed to meet some product specification. These percentages define a target value for the process. The only way to meet such specifications, while minimizing the give-away, is to operate the process predictably and on target. Table 7.2 gives the percentage of nonconforming product for 219 Centered Capability Ratios ranging from –1.09 to 1.09. Table 7.3 gives the target Centered Capability Ratios for selected percentages of conforming product.

One-sided product specifications are those where nonconforming items are scrapped or reworked. In order to minimize the costs of production and use with one-sided product specifications you will need to operate your process at or beyond some minimum distance from the specification value. In this effort predictable operation is desirable simply because it corresponds to operating with minimum variation and this reduces the minimum distance needed. Tables Seven and Eight give the *Excess Costs of Production* and the *Excess Costs of Use* for use with one-sided product specifications.

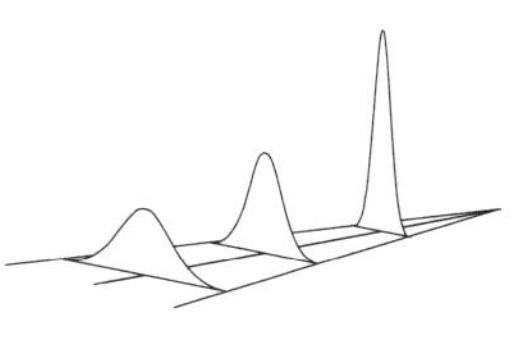

Full Potential = Minimum *ECP&U*

Chapter Eight

Excess Costs for Targets at Boundaries

In Chapter Five we looked at finding the *Effective Cost of Production and Use* for two-sided specifications. In Chapter Seven we looked at doing the same for one-sided product specifications. In this chapter we will consider one-sided product specifications that have a target value that is equal to a boundary condition.

8.1 The Excess Cost Function

When the target value coincides with a boundary condition we will have a one-sided Excess Cost Function as shown in Figure 8.1. As before, we will assume that the excess costs are zero at the target value. We will also assume that there will be a fixed cost of scrap or rework for the nonconforming items. In between the zero cost and the fixed cost we will once again use a quadratic curve to characterize the excess costs associated with conforming product that deviates from the target value.

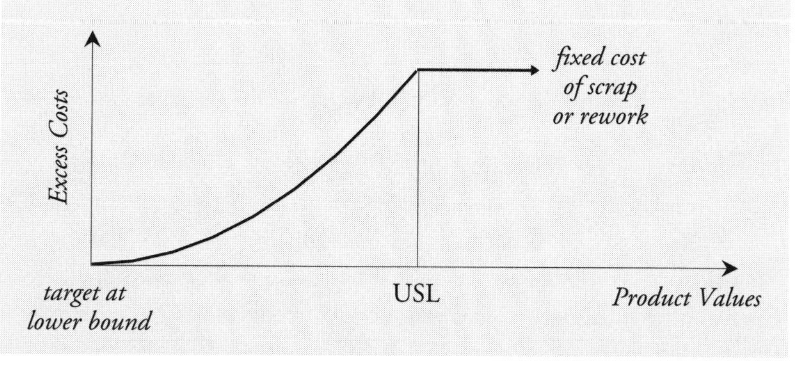

Figure 8.1: Excess Cost Function when the Target Is at a Boundary Value

Of course a boundary condition places a restriction on the possible values for the product characteristic. To incorporate this restriction into our computations we will need to use a bounded probability model, such as the one shown in Figure 8.2, to compute the excess costs.

Inspection of Figure 8.2 will reveal that this chapter may be thought of as a special case of the one-sided product specification problem addressed in Chapter Seven. As was done in Chapter Seven, and as shown in Figure 8.2, we shall characterize the relationship between the process outcomes and the specification and boundary values using the Centered Capability Ratio, C_{pk}, and the *Predictive Tolerance*; or, in the case of an unpredictable process, the Centered Performance Ratio, P_{pk}, and the *Descriptive Tolerance*. As before:

$$Predictive\ \ Tolerance = \frac{distance\ from\ specification\ to\ target}{Sigma(X)}$$

and

$$Descriptive\ Tolerance = \frac{distance\ from\ specification\ to\ target}{global\ standard\ deviation\ statistic}$$

The *Descriptive Tolerance* will characterize the past process performance. This is the logical *Tolerance* value to be used in conjunction with P_{pk} when finding Baseline values. The *Predictive Tolerance* will characterize the hypothetical process potential. This is the logical *Tolerance* value to be used in conjunction with C_{pk} when assessing what the process can be made to do.

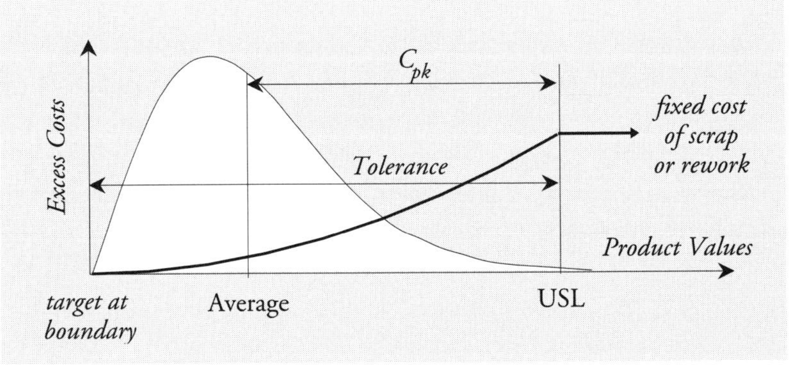

Figure 8.2: Tolerance and Centered Capability Ratio for Target at Boundary Case

Note that the Centered Capability Ratio, C_{pk}, and the Centered Performance Ratio, P_{pk}, will be defined relative to the one-sided specification *even when the average is closer to the boundary than it is to the specification value*, as shown in Figure 8.2. This is important since some boundary conditions are listed as specification values even though they are something quite different. Boundary values are inappropriate for computing C_{pk} or P_{pk}. (When using capability or performance values generated by software you will need to verify that the ratios are properly computed.)

The tables for use with Target at Boundary situations are Tables Nine and Ten. The details of their computation, and the derivation of the formulas used, are given in Chapter Eleven.

8.2 Using Tables Nine and Ten

Step One: Compute Baseline Values: To evaluate the Baseline condition we begin as in the last chapter with a computation of the Centered Performance Ratio, P_{pk}, and the *Descriptive Tolerance* (using the global standard deviation statistic, *s*). Using these two values we can obtain the Baseline *ExCP, ExCU* and *ECP&U* values from either Table Nine or Table Ten as appropriate.

When the target is at a boundary condition your process will always be off target. In general the objective will be to operate as close to the target as possible. Since the process average will generally be *at least* one standard deviation away from the boundary condition, it is likely that the historical average will have been influenced by the historical dispersion displayed by the process. In considering a value for a new aim point we shall make use of this relationship between location and dispersion. In addition, since we are unlikely to change the aim without also changing the dispersion, Step Two from the earlier procedures no longer makes much sense.

Moreover, since the usual Centered Capability Ratio, C_{pk}, is computed using the historical average, and since the average is likely to change with predictable operation in a Target at Boundary situation, the usual Centered Capability Ratio will not be of much use here. To assess what is likely to happen with predictable operation we need to adjust the average value. However, we cannot simply move the average to the target value since this cannot happen in reality. Therefore, Step Three will begin with the computation of a new aim point and the computation of a Centered Capability Ratio based on this new aim point.

Step Three: Characterize the Effects of Predictable Operation: Compute a new aim point and use this new aim point with the predictable scale factor and the specification to compute a Centered Capability Ratio, C_{pk}.

When the target is at a lower bound you will need to find the new aim point by subtracting the global standard deviation statistic, *s*, from the historical average and then adding back the predictable scale factor, *Sigma(X)*.

$$\textit{New Aim Point} = \textit{Average} - s + \textit{Sigma(X)}$$

When the target is at a upper bound you will need to find the new aim point by adding the global standard deviation statistic, *s*, to the historical average and subtracting off the predictable scale factor, *Sigma(X)*.

$$\textit{New Aim Point} = \textit{Average} + s - \textit{Sigma(X)}$$

Next, use the predictable scale factor, *Sigma(X)*, to compute a *Predictive Tolerance* and a C_{pk} value. Use the *Tolerance* and C_{pk} values to find excess costs and an effective cost for predictable operation at this new aim point.

Step Four: Characterize the Best Case Values: Use the bottom entry in the column for the *Predictive Tolerance* in Table Nine or Ten to approximate the Best Case for predictable operation with a Target at Boundary.

8.3 The Run-Out Data

Table 8.1 repeats the Run-Out values for 20 hydraulic cylinders first given in Table 7.6. The specification for these values is no more than 100 millionths of an inch. The *XmR* Chart for these 20 values is shown in Figure 8.3, and the histogram is shown in Figure 8.4.

Table 8.1: Run-Out Values and Moving Ranges for Hydraulic Cylinders

X	103	68	106	56	86	84	96	73	65	73	137	59	68	102	128	270	131	81	89	99
mR		35	38	50	30	2	12	23	8	8	64	78	9	34	26	142	139	50	8	10

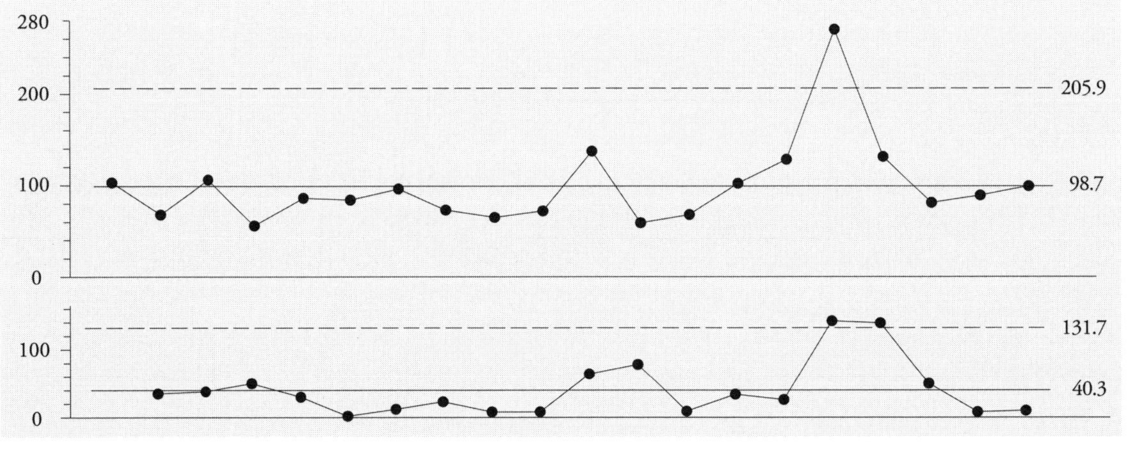

Figure 8.3: *XmR* Chart for Run-Out Data

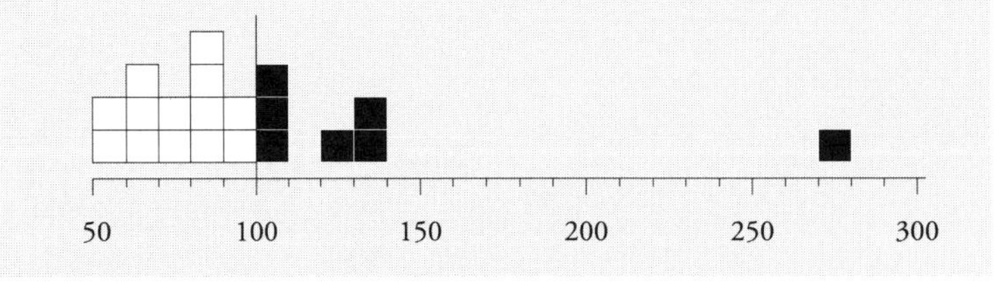

Figure 8.4: Histogram of the 20 Run-Out Values

BASELINE VALUES

As may be seen in Figures 8.3 and 8.4 we have an unpredictable process with 35% non-conforming. With an average of 98.7 millionths, an upper specification of 100 millionths, and a global standard deviation statistic of s = 46.64 millionths we have:

$$P_{pk} = \frac{USL - Average}{3\ s} = \frac{100 - 98.7}{3\ (46.64)} = 0.009 \quad \text{and} \quad Descriptive\ Tolerance = \frac{100}{46.64} = 2.14$$

Since nonconforming cylinders are scrapped we use Table Nine to find our Baseline values. These are an *Excess Cost of Production* of 0.869, an *Excess Cost of Use* of 0.466, and an *Effective Cost of Production and Use* of 2.335.

PREDICTABLE VALUES:

From the *XmR* Chart in Figure 8.3 we find the average moving range to be 40.3, which gives a *Sigma(X)* value of 35.7 millionths. With a historical average of 98.7, and $s = 46.6$, we find a new aim point of:

$$New\ Aim\ Point = Average - s + Sigma(X) = 98.7 - 46.6 + 35.7 = 87.8 \text{ millionths}$$

From this we find a Centered Capability Ratio of:

$$C_{pk} = \frac{USL - Aim\ Point}{3\ Sigma(X)} = \frac{100 - 87.8}{3\ (35.7)} = 0.11$$

and

$$Predictive\ Tolerance = \frac{100}{35.7} = 2.8$$

Rounding this down to 2.7 we find our Predictable values from Table Nine. These are an *Excess Cost of Production* of 0.581, an *Excess Cost of Use* of 0.487, and an *Effective Cost of Production and Use* of 2.067. While these are slightly better than the Baseline values, they are not as good as they could be with a lower aim point.

BEST CASE VALUES

To see what might happen with a better aim point, we start with the entry found in the preceding step and simply go down to the last entry in that column. With a Tolerance of 2.7, the last entry in Table Nine is for a C_{pk} of 0.6. A process operating here would have an *Excess Cost of Production* of 0.062, an *Excess Cost of Use* of 0.127, and an *Effective Cost of Production and Use* of 1.189. Achieving this C_{pk} value would require operating this process predictably at:

$$Average = USL - 3\ C_{pk}\ Sigma(X) = 100 - 3\ (0.6)\ 35.7 = 35.7$$

If this average is not feasible for some reason, then the Best Case situation is likely to be somewhere between the results of this step and the preceding step.

Table 8.2: Excess Costs for the Hydraulic Cylinders

	ExCP	*ExCU*	*ECP&U*
Baseline Values @ Aim of 99	0.869	0.466	2.335
Predictable @ Aim of 88	0.581	0.487	2.067
Predictable @ Best Aim of 36	0.062	0.127	1.189

It should be noted that since the process will always be off target when the target is at the boundary condition, the Best Case *Excess Costs of Use* will generally be larger than those found in other situations.

8.4 The Contamination Data

For a further example consider the contamination values, in parts per million, for twenty batches shown in Table 7.10 and reproduced here in Table 8.3 and Figure 8.5. We find the process to be unpredictable with an average of 100.0 and a median moving range of 16.0. This gives a *Sigma(X)* value of 16.8. The global standard deviation statistic is $s = 33.27$. The specification for contamination is less than 150 ppm, and nonconforming batches are reworked at a cost of 80% of nominal. However, unlike Chapter Seven, where the target value was 80 ppm, now the target is perfection, 0 ppm.

Table 8.3: Contamination Data and Moving Ranges in ppm

X	114	87	76	87	112	95	78	94	98	90	93	199	86	72	77	92	85	182	74	109
mR		27	11	11	25	17	17	16	4	8	3	106	113	14	5	15	7	97	108	35

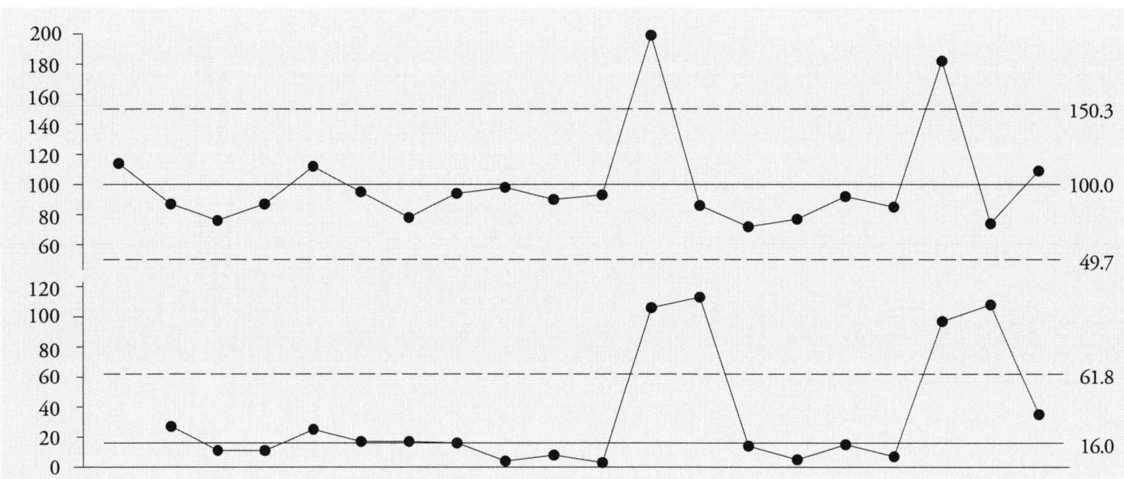

Figure 8.5: *XmR* Chart for Contamination Data

BASELINE VALUES

The Centered Performance Ratio is:

$$P_{pk} = \frac{USL - Average}{3\,s} = \frac{150 - 100}{3\,(33.27)} = 0.50$$

and the *Descriptive Tolerance* is:

$$Descriptive\ Tolerance = \frac{USL - target}{s} = \frac{150 - 0}{33.27} = 4.51$$

From Table Ten we find the Baseline *Excess Cost of Production* to be 0.070 (0.80) = 0.056, and the Baseline *Excess Cost of Use* to be 0.410 (0.80) = 0.328. (Due to the shift in the target value, this *ExCU* is considerably larger than the 0.135 found in Chapter Seven.) Combining these Excess Costs we find a Baseline *Effective Cost of Production and Use* of 1.384.

PREDICTABLE VALUES:

With a *Sigma(X)* value of 16.8 ppm, a historical average of 100, and *s* = 33.27, we find a new aim point of:

$$\textit{New Aim Point} = \textit{Average} - s + \textit{Sigma(X)} = 100 - 33.27 + 16.8 = 84 \text{ ppm}$$

From this we find a Centered Capability Ratio of:

$$C_{pk} = \frac{\textit{USL} - \textit{Aim Point}}{3\ \textit{Sigma(X)}} = \frac{150 - 84}{3\ (16.8)} = 1.31$$

and

$$\textit{Predictive Tolerance} = \frac{150}{16.8} = 8.93$$

From Table Ten, with *Tolerance* = 9.0 and C_{pk} = 1.3, we would compute an *Excess Cost of Production* of 0.000 and an *Excess Cost of Use* of 0.333 (0.80) = 0.266. Thus the *Effective Cost of Production and Use* drops to 1.266. By learning to operate this process predictably, they have the potential to operate in specification, with an overall average of 84 ppm. This would eliminate about 5.6% *Excess Costs of Production* and cut the *Excess Costs of Use* from 32.8% to 26.6%.

BEST CASE VALUES

With a *Sigma(X)* value of 16.8 ppm, and a *Predictable Tolerance* value of 9.0, about the best that can be expected from this process would correspond to a C_{pk} value of 2.70. This would correspond to operating at a process average of:

$$\textit{Average} = \textit{USL} - 3\ C_{pk}\ \textit{Sigma(X)} = 150 - 3\ (2.7)\ 16.8 = 14 \text{ ppm}$$

At this average the *Excess Cost of Use* would be 0.022 (0.80) = 0.018 and there would be no *Excess Costs of Production*.

Table 8.4: Excess Costs for the Contamination Data

	ExCP	*ExCU*	*ECP&U*
Baseline Values @ Aim of 100	0.056	0.328	1.384
Predictable @ Aim of 84	0.000	0.266	1.266
Predictable @ Best Aim of 14	0.000	0.018	1.018

Thus, predictable operation at an average of 84 would cut the *Effective Costs of Production and Use* by about 12%. Operating at lower average values could further reduce the *Excess Costs of Use* by up to another 24%.

8.5 Summary

The target at boundary condition is a special case of a one-sided product specification with a target value. As before, we use the Centered Performance Ratio and the Descriptive Tolerance to find Baseline values for the *Excess Cost of Production*, the *Excess Cost of Use*, and the *Effective Cost of Production and Use*.

Predictive values for these excess costs may be obtained using the Centered Capability Ratio and the Predictive Tolerance. Finally, using the entries on the bottom diagonal of either Table Nine or Table Ten, we find the Best Case values for the excess costs. The Centered Capability Ratio value for this Best Case scenario can be used in reverse to find the corresponding process average. This process average is the Best Aim point for your process.

World Class Quality Requires Operating Predictably and On Target

Chapter Nine

When Nominal Costs Change with Aim

In Chapters Five through Eight the nominal cost of producing an item or a batch was assumed to be constant. In practice, this assumption will generally be sufficient to provide useful estimates of the Excess Costs. However, with one-sided product specifications and with targets at boundaries there are situations where the assumption of a constant nominal cost of production will not work. This chapter illustrates an approach that works when the nominal cost of production is variable.

To this end, we return to the basic equation from Section 5.1

$$\textit{Effective Cost of Production and Use} = \frac{\textit{Actual Cost of Production \& Use}}{\textit{Nominal Cost of Production}}$$

When the nominal cost of production remains fixed, knowledge of the *Effective Cost of Production and Use* is equivalent to knowledge of the *Actual Cost of Production and Use*. However, when the nominal cost of production changes with the process average, the problem of comparing the costs at different averages will be one of having to compute the *Actual Cost of Production and Use* at each process average and with different amounts of process variation. While this is not overly complex, it does require an organized approach.

9.1 One-Sided Product Specifications without a Target

This is the situation discussed in Section 7.3, except that now the nominal cost must be shown as a function of the product values. Figure 9.1 is essentially Figure 7.6 with the addition of this Nominal Cost Function. As the process average moves to the left the fraction of nonconforming product drops, which reduces the scrap or rework costs. However, as the process average moves to the left the average cost of producing each item or batch goes up. Thus, we have conflicting forces, and the objective is to balance the dropping costs of scrap or rework against the increasing nominal costs of production.

In order to incorporate both of these conflicting forces into our calculations we will have to compute the *Actual Costs of Production* as a function of the process average. By plotting these Actual Costs against the process averages we will have a curve that we can use to find the optimum aim points under different conditions.

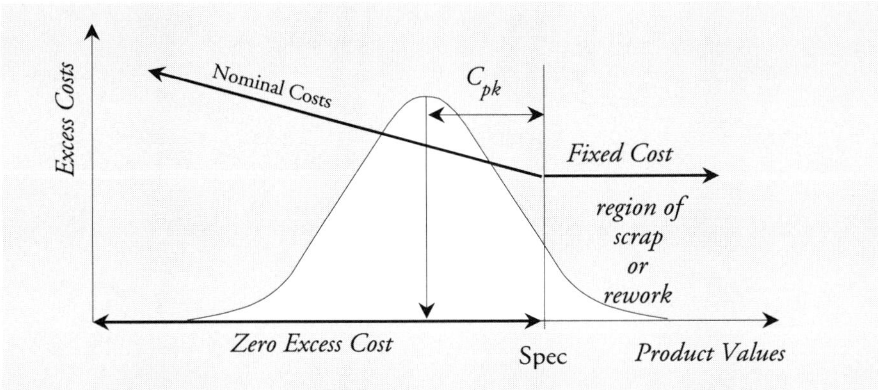

Figure 9.1: Excess Costs and Nominal Costs for a One-Sided Product Specification

The Excess Cost Function of Figure 9.1 has no target value, and therefore does not allow for the computation of any *Excess Costs of Use*. Here we are only looking at the *Actual Costs of Production*, without consideration of the costs of use. As may be seen in Table 9.1, which lists the *Excess Cost of Scrap* from Table Seven and the *Excess Cost of Rework* from Table Eight, the *Excess Costs of Production* for one-sided product specifications will depend solely upon the values of C_{pk} or P_{pk}.

Table 9.1: Excess Costs of Production (*ExCP*) for One-Sided Product Specifications

Excess Costs of Scrap	C_{pk} or P_{pk}	*Excess Costs of Rework*
14.0	***-0.50***	0.933
7.69	***-0.40***	0.885
4.43	***-0.30***	0.816
2.65	***-0.20***	0.726
1.62	***-0.10***	0.618
1.000	***0.00***	0.500
0.618	***0.10***	0.382
0.378	***0.20***	0.274
0.226	***0.30***	0.184
0.130	***0.40***	0.115
0.072	***0.50***	0.067
0.037	***0.60***	0.036
0.018	***0.70***	0.018
0.008	***0.80***	0.008
0.003	***0.90***	0.003
0.001	***1.00***	0.001

While the excess costs depend upon C_{pk}, the nominal costs will depend upon the value of the process average. In order for both the excess costs and the nominal costs to be expressed as functions of the same thing, we will need to convert the Centered Capability Ratios, or the Centered Performance Ratios, of Table 9.1 into values for the process average. For Figure 9.1 we do this by inverting the formula for C_{pk} or P_{pk} to get:

$$Process\ Average = Specification\ Value - [\,3\ s\ P_{pk}\,] \qquad \text{(Eq. 9.1)}$$

or

$$Process\ Average = Specification\ Value - [\,3\ Sigma(X)\ C_{pk}\,] \qquad \text{(Eq. 9.2)}$$

When the one-sided specification is a minimum value these equations become:

$$Process\ Average = Specification\ Value + [\,3\ s\ P_{pk}\,] \qquad \text{(Eq. 9.3)}$$

or

$$Process\ Average = Specification\ Value + [\,3\ Sigma(X)\ C_{pk}\,] \qquad \text{(Eq. 9.4)}$$

Since the tables contain only selected values for C_{pk} or P_{pk}, we start with these tabled values and convert them into values for the process average using one of the formulas above. We only need to do this out to the point where the *ExCP* values go to zero.

Table 9.2: Excess Costs of Production (*ExCP*) and Corresponding Process Averages

C_{pk}	*Excess Costs of Production*	*Process Averages*
0.00	0.500	a_0
0.10	0.382	$a_{.1}$
0.20	0.274	$a_{.2}$
0.30	0.184	$a_{.3}$
0.40	0.115	$a_{.4}$
0.50	0.067	$a_{.5}$
0.60	0.036	$a_{.6}$
0.70	0.018	$a_{.7}$
0.80	0.008	$a_{.8}$
0.90	0.003	$a_{.9}$
1.00	0.001	$a_{1.0}$

Next we will need to find the nominal cost of production at *each* of the values for the process average:

Table 9.3: Excess Costs of Production (*ExCP*), Process Averages, and Nominal Costs of Production

C_{pk}	*Excess Costs of Production*	*Process Averages*	*Nominal Costs of Production*
0.00	0.500	a_0	$NC(a_0)$
0.10	0.382	$a_{.1}$	$NC(a_{.1})$
0.20	0.274	$a_{.2}$	$NC(a_{.2})$
0.30	0.184	$a_{.3}$	$NC(a_{.3})$
0.40	0.115	$a_{.4}$	$NC(a_{.4})$
0.50	0.067	$a_{.5}$	$NC(a_{.5})$
0.60	0.036	$a_{.6}$	$NC(a_{.6})$
0.70	0.018	$a_{.7}$	$NC(a_{.7})$
0.80	0.008	$a_{.8}$	$NC(a_{.8})$
0.90	0.003	$a_{.9}$	$NC(a_{.9})$
1.00	0.001	$a_{1.0}$	$NC(a_{1.0})$

For each value of the process average, we find the *Actual Cost of Production* according to:

$$\textit{Actual Cost of Production} = [\,1.0 + \textit{Excess Cost}\,]\ \textit{Nominal Cost} \qquad \text{(Eq. 9.5)}$$

The addition of 1.0 is required to turn the *Excess Cost* into an *Effective Cost*. These *Actual Costs* are then plotted as a function of the process average.

Table 9.4: Finding the Actual Costs of Production when the Nominal Cost Changes with Aim

C_{pk}	*Excess Costs of Production*	*Process Averages*	*Nominal Costs of Production*	*Actual Costs of Production*
0.00	$ExCP_0$	a_0	$NC(a_0)$	ACP_0
0.10	$ExCP_{.1}$	$a_{.1}$	$NC(a_{.1})$	$ACP_{.1}$
0.20	$ExCP_{.2}$	$a_{.2}$	$NC(a_{.2})$	$ACP_{.2}$
0.30	$ExCP_{.3}$	$a_{.3}$	$NC(a_{.3})$	$ACP_{.3}$
0.40	$ExCP_{.4}$	$a_{.4}$	$NC(a_{.4})$	$ACP_{.4}$
0.50	$ExCP_{.5}$	$a_{.5}$	$NC(a_{.5})$	$ACP_{.5}$
0.60	$ExCP_{.6}$	$a_{.6}$	$NC(a_{.6})$	$ACP_{.6}$
0.70	$ExCP_{.7}$	$a_{.7}$	$NC(a_{.7})$	$ACP_{.7}$
0.80	$ExCP_{.8}$	$a_{.8}$	$NC(a_{.8})$	$ACP_{.8}$
0.90	$ExCP_{.9}$	$a_{.9}$	$NC(a_{.9})$	$ACP_{.9}$
1.00	$ExCP_{1.0}$	$a_{1.0}$	$NC(a_{1.0})$	$ACP_{1.0}$

Of course, if we are currently operating our process unpredictably, we will need to go through this exercise twice: once using P_{pk} and the global standard deviation statistic to obtain Baseline values for the *Actual Costs*; and then again using C_{pk} and *Sigma(X)* to obtain Predictable values for the *Actual Costs*.

Table 9.5: Run-Out Values and Moving Ranges for Hydraulic Cylinders

X	103	68	106	56	86	84	96	73	65	73	137	59	68	102	128	270	131	81	89	99
mR		35	38	50	30	2	12	23	8	8	64	78	9	34	26	142	139	50	8	10

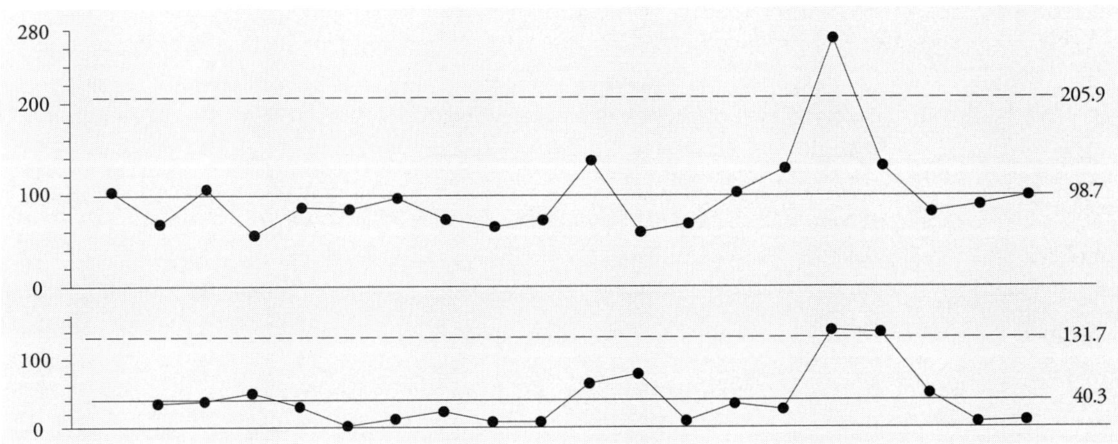

Figure 9.2: *XmR* Chart for Run-Out Data

For an example we shall use the Run-Out Data shown in Table 9.5 and charted in Figure 9.2. The upper specification for run-out is 100 millionths of an inch. The average for Table 9.5 is 98.7, while the global standard deviation statistic is $s = 46.64$. Nonconforming product is scrapped. The first step is to get the *Excess Costs of Scrap* from Table 9.1 and to use Equation 9.1 to find the corresponding process averages.

$$Process\ Average \;=\; Specification\ Value \;-\; [\,3\ s\ P_{pk}\,] \;=\; 100 - 139.92\ P_{pk}$$

Since the lower boundary is zero, negative process averages do not make sense, so we drop the last three rows from Table 9.6 before proceeding.

Table 9.6: Excess Costs of Production (*ExCP*) and Corresponding Process Averages

P_{pk}	*Excess Costs of Scrap*	*Process Averages*	
–0.50	14.0	*170.0*	
–0.40	7.69	*156.0*	
–0.30	4.43	*142.0*	
–0.10	2.65	*128.0*	
–0.10	1.62	*114.0*	
0.00	1.000	*100*	
0.10	0.618	*86.0*	
0.20	0.378	*72.0*	
0.30	0.226	*58.0*	
0.40	0.130	*44.0*	
0.50	0.072	*30.0*	
0.60	0.037	*16.1*	
0.70	0.018	*2.1*	
0.80	0.008	*–11.9*	*impossible value*
0.90	0.003	*–25.9*	*impossible value*
1.00	0.001	*–39.9*	*impossible value*

Next we need to evaluate the cost of producing cylinders with average run-out values at the various levels defined in Table 9.6. Since increased smoothness and smaller run-out values correspond to increased machine time, let us assume that the nominal costs of production in dollars per cylinder, are those given in Table 9.7.

Table 9.7: Nominal Costs of Production *(NC)* for Different Average Run-Out Values

Average Run-Out	*200*	*150*	*100*	*75*	*50*	*25*	*10*	*1*
NC	*5.00*	*6.67*	*10.00*	*13.33*	*20.00*	*40.00*	*100.00*	*1000.00*

We can either interpolate between these values to find nominal costs for the averages in Table 9.6, or we might even fit a curve to these points to obtain an equation for the Nominal Cost Function. In this case they had an equation which was:

$$Nominal\ Cost\ of\ Production \;=\; \frac{\$\,1000}{Average\ Run\text{-}Out}$$

If the Nominal Cost Function is linear, you will only need to know the costs at two points and then find the other values by interpolation or extrapolation. If you cannot figure out how the nominal costs of production vary, then perhaps you need to assume that they are fixed and use one of the approaches in Chapters Five, Seven, or Eight.

Using the equation for the Nominal Cost Function, we convert the process averages into nominal costs of production. These values are shown in Table 9.8. Next the *Actual Costs of Production* are found using Equation 9.5: for example, when P_{pk} = −0.50 we find the *Actual Costs of Production* to be [1 + 14.0] 5.88 = 88.20. These *Actual Costs* are plotted to obtain the Baseline Curve in Figure 9.3.

Table 9.8: Baseline Actual Costs of Production for the Run-Out Data

P_{pk}	*Excess Costs of Scrap*	*Process Averages*	*Nominal Cost of Production*	*Actual Cost of Production*
−0.50	14.0	170.0	5.88	88.20
−0.40	7.69	156.0	6.41	55.70
−0.30	4.43	142.0	7.04	38.23
−0.10	2.65	128.0	7.81	28.51
−0.10	1.62	114.0	8.77	22.98
0.00	1.000	100	10.00	20.00
0.10	0.618	86.0	11.63	18.82
0.20	0.378	72.0	13.89	19.14
0.30	0.226	58.0	17.24	21.14
0.40	0.130	44.0	22.73	25.68
0.50	0.072	30.0	33.33	35.73
0.60	0.037	16.1	62.11	64.41
0.70	0.018	2.1	476.19	484.76

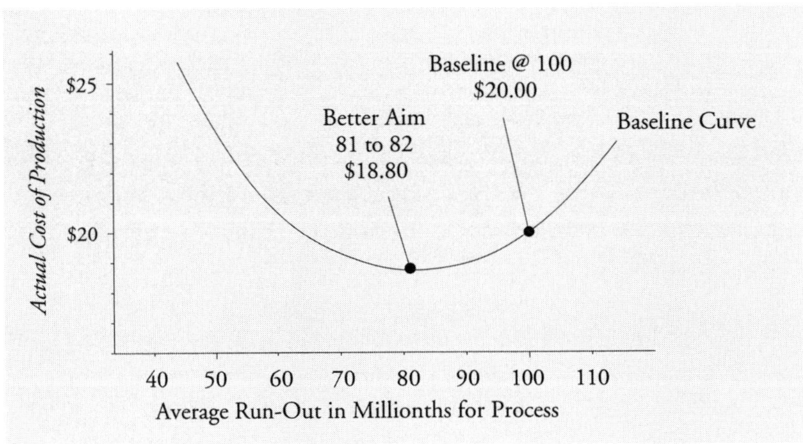

Figure 9.3: The Baseline Curve for the Actual Costs of Production for Hydraulic Cylinders

The values in Table 9.8 describe the past. They show the potential payback to be had by optimizing the process aim while operating with the historical amount of variation. Compared with their historical average of 98.7, they could save approximately $ 1.20 per cylinder by operating with an average in the neighborhood of 81 to 82, plus the customer would be getting better overall quality.

To find out what this process could do if operated predictably, we perform the whole analysis over again using *Sigma(X)* rather than *s*. This will give us a curve that shows the *Actual Costs of Production* when operating this process at full potential.

Table 9.9: Predictable Actual Costs of Production for the Run-Out Data

C_{pk}	*Excess Costs of Scrap*	*Process Averages*	*Nominal Cost of Production*	*Actual Cost of Production*
–0.50	14.0	*153.6*	*6.51*	*97.65*
–0.40	7.69	*142.8*	*7.00*	*60.83*
–0.30	4.43	*132.1*	*7.57*	*41.11*
–0.10	2.65	*121.4*	*8.24*	*30.08*
–0.10	1.62	*110.7*	*9.03*	*23.66*
0.00	1.000	*100*	*10.00*	*20.00*
0.10	0.618	*89.3*	*11.20*	*18.12*
0.20	0.378	*78.6*	*12.72*	*17.53*
0.30	0.226	*67.9*	*14.73*	*18.06*
0.40	0.130	*57.2*	*17.48*	*19.75*
0.50	0.072	*46.4*	*21.55*	*23.10*
0.60	0.037	*35.7*	*28.01*	*29.05*
0.70	0.018	*25.0*	*40,00*	*40.72*
0.80	0.008	*14.3*	*69.93*	*70.49*
0.90	0.003	*3.6*	*277.78*	*278.61*

From Figure 9.2 the average moving range is 40.3, which gives *Sigma(X)* = 35.7. Using this measure of dispersion we would recompute the process averages using Equation 9.2. Since these averages will differ from those in Table 9.8 we cannot make comparisons until we have constructed our Predicable Curve for the *Actual Costs*. The nominal costs of production at the averages of Table 9.9 are once again found by using the equation for the Nominal Cost Function. The resulting *Actual Costs of Production* characterize what this process has the potential to do. We find these values in Table 9.9, and plot the Predictable Curve in Figure 9.4.

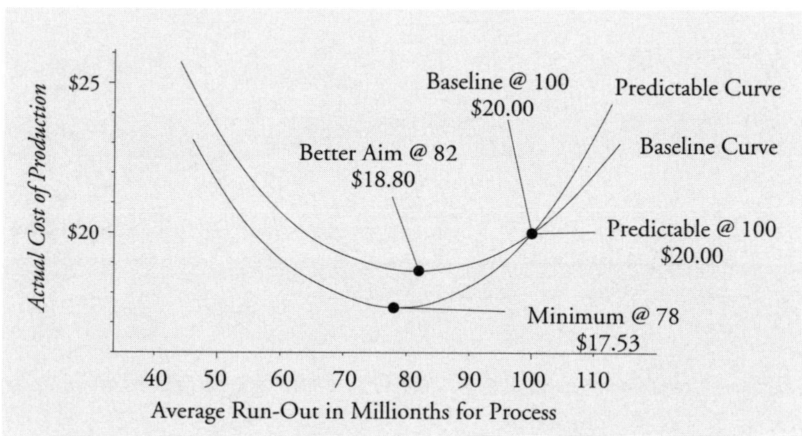

Figure 9.4: Actual Costs of Production Curves for Hydraulic Cylinders

By operating this process predictably at an average near 78 or 79 millionths they could reduce their average cost per piece by about \$2.47 compared with their Baseline cost of \$20.00 per piece. The two *Actual Cost* curves are shown in Figure 9.4. Notice how the reduced variation associated with predictable operation squeezes the *Actual Cost* curve and results in a lower minimum cost.

In this example the cost of scrapping an item was taken to be equal to the nominal cost of production at the specification. Symbolically, $C_S = NC(a_0)$. This is the situation shown in Figure 9.1. If the cost of scrapping an item exceeds this nominal cost, then all of the *Excess Cost* values in Table Seven would need to be inflated by the appropriate amount. For example, if $C_S = 1.20\ NC(a_0)$, then all of the *ExCP* and *ExCU* values in Table Seven would need to be inflated by 20%.

In a similar manner, the *Excess Cost of Rework* values in Table Eight implicitly assume that the cost of rework is equal to the nominal cost of production at the specification. If this is not approximately true, then all of the *Excess Cost* values of Table Eight may be adjusted as appropriate before they are used in constructing a table of the *Actual Costs of Production*. This adjustment will be illustrated in the next section.

9.2 One-Sided Product Specifications with a Target Value

When we have One-Sided Product Specifications with a target value we can incorporate the *Excess Costs of Use* into our computations. If, for some reason, you do not want to consider the *Excess Costs of Use*, then the procedure shown in the previous section will suffice.

Figure 9.5 is essentially Figure 7.11 with the addition of a Nominal Cost Function. As before, the *Excess Cost of Production* will depend solely upon C_{pk}, while the *Excess Cost of Use* will depend upon both C_{pk} and the *Tolerance*.

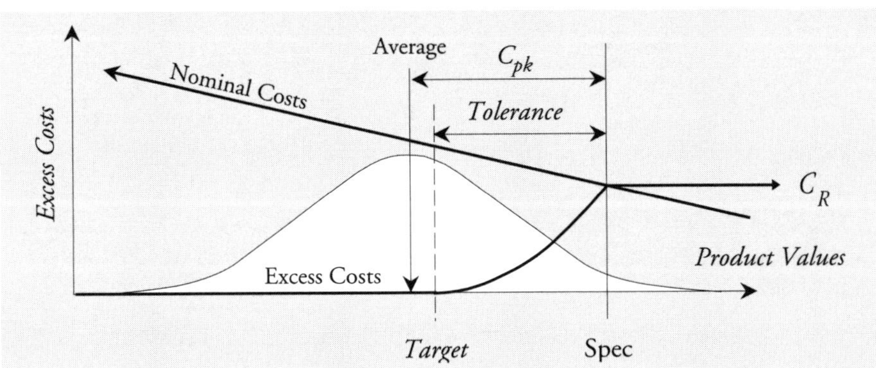

Figure 9.5: Excess Costs and Nominal Costs for One-Sided Specifications with a Target Value

As in the previous section, there will be two separate analyses: one using P_{pk}, the *Descriptive Tolerance*, and the global standard deviation statistic, s; and a second analysis using

C_{pk}, the *Predictive Tolerance,* and the within-subgroup dispersion, *Sigma(X).* In each analysis we will obtain a curve showing the *Actual Costs of Production and Use* as a function of the process average. By comparing points on these curves we can assess the benefits of different ways of operating the process. Depending upon what happens to nonconforming product we begin by using either Table Seven or Table Eight to list the *Excess Costs of Production* and the corresponding P_{pk} or C_{pk} values. Next, using the value for the *Tolerance,* we find the appropriate column of *Excess Costs of Use* values and write these in our table. Next we create a column of *ECP&U* values by adding 1.00 to the sum of the two *Excess Costs.* The result should look something like Table 9.10.

Table 9.10: Finding the *Effective Costs of Production and Use*

P_{pk}	*Excess Costs of Production*	*Excess Costs of Use @ Tolerance Value*	*ECP&U*
0.00	$ExCP_0$	$ExCU_0$	$ECP\&U_0$
0.10	$ExCP_{.1}$	$ExCU_{.1}$	$ECP\&U_{.1}$
0.20	$ExCP_{.2}$	$ExCU_{.2}$	$ECP\&U_{.2}$
0.30	$ExCP_{.3}$	$ExCU_{.3}$	$ECP\&U_{.3}$
0.40	$ExCP_{.4}$	$ExCU_{.4}$	$ECP\&U_{.4}$
0.50	$ExCP_{.5}$	$ExCU_{.5}$	$ECP\&U_{.5}$

Next we need to convert the Capability Ratios or Performance Ratios listed in Table 9.10 into process averages using the appropriate Equation 9.1, 9.2, 9.3, or 9.4 from the previous section. List these process averages as another column in the table.

Once we have the column of process averages, we can then use our process knowledge to find the appropriate nominal costs of production for each of these averages.

Table 9.11: Finding the *Actual Costs of Production and Use*

P_{pk}	*Excess Costs of Production*	*Excess Costs of Use @ Tolerance Value*	*ECP&U*	*Process Averages*	*Nominal Costs*	*Actual Costs of Production & Use*
0.00	$ExCP_0$	$ExCU_0$	$ECP\&U_0$	a_0	$NC(a_0)$	$ACP\&U_0$
0.10	$ExCP_{.1}$	$ExCU_{.1}$	$ECP\&U_{.1}$	$a_{.1}$	$NC(a_{.1})$	$ACP\&U_{.1}$
0.20	$ExCP_{.2}$	$ExCU_{.2}$	$ECP\&U_{.2}$	$a_{.2}$	$NC(a_{.2})$	$ACP\&U_{.2}$
0.30	$ExCP_{.3}$	$ExCU_{.3}$	$ECP\&U_{.3}$	$a_{.3}$	$NC(a_{.3})$	$ACP\&U_{.3}$
0.40	$ExCP_{.4}$	$ExCU_{.4}$	$ECP\&U_{.4}$	$a_{.4}$	$NC(a_{.4})$	$ACP\&U_{.4}$
0.50	$ExCP_{.5}$	$ExCU_{.5}$	$ECP\&U_{.5}$	$a_{.5}$	$NC(a_{.5})$	$ACP\&U_{.5}$

We then complete Table 9.11 by multiplying the *ECP&U* values by the nominal costs of production to obtain the *Actual Costs of Production and Use.* These *Actual Costs* are then plotted against the process averages to obtain a cost curve to use in evaluating the process.

For an example consider the Contamination Data, in parts per million, for twenty batches shown in Table 7.10 and reproduced here in Table 9.12 and Figure 9.6. We find the process to be unpredictable, with an average of 100.0 and a median moving range of 16.0. This gives a *Sigma(X)* value of 16.8. The global standard deviation statistic is s = 33.27. The maximum specification for contamination is 150 ppm, and the customer has defined a target value of 80

Table 9.12: Contamination Data and Moving Ranges in ppm

X	114	87	76	87	112	95	78	94	98	90	93	199	86	72	77	92	85	182	74	109
mR		27	11	11	25	17	17	16	4	8	3	106	113	14	5	15	7	97	108	35

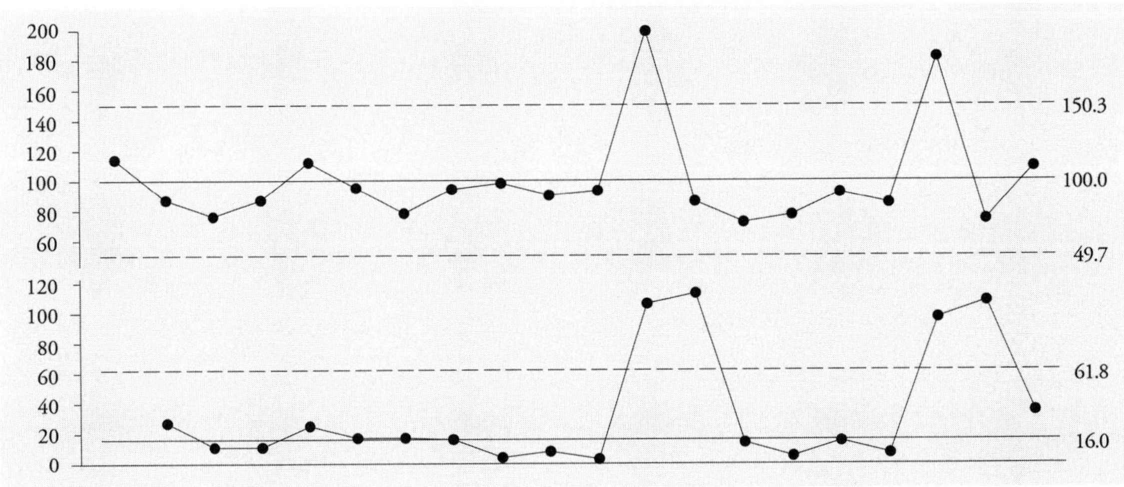

Figure 9.6: *XmR* Chart for Contamination Data

ppm. Nonconforming batches are recycled through the distillation column at 80% of nominal cost, so we will have to scale down the values from Table Eight to find our *Excess Costs.*

THE BASELINE COST CURVE

We begin Table 9.13 by scaling the *ExCP* entries in Table Eight down to 80% and listing them with the P_{pk} values. With an upper spec of 150 ppm, a target value of 80 ppm, and a global standard deviation statistic of s = 33.27, the *Descriptive Tolerance* is:

$$\text{Descriptive Tolerance} = \frac{USL - Target}{s} = \frac{150 - 80}{33.27} = 2.10$$

Using this *Tolerance* of 2.1 we scale the *ExCU* values of Table Eight down to 80% and list these values in Table 9.13. Combining these we find the *ECP&U* values shown there. Next we convert the P_{pk} values of Table 9.13 into process average values using Equation 9.1.

$$\text{Process Average} = \text{Specification Value} - [\,3\ s\ P_{pk}\,] = 150 - 99.8\ P_{pk}$$

While the Nominal Cost Function for this situation is probably similar to the one in the preceding example, where the costs increase in a nonlinear manner as we approach zero, I will use a linear Nominal Cost Function in this example for illustrative purposes. Let us assume that the nominal cost at the specification point is $NC(150ppm)$ = \$0.35/lb. Furthermore, let us assume that the nominal cost at 50 ppm is $NC(50ppm)$ = \$0.45/lb. Since two points define a straight line, we now can compute a nominal cost for any particular average ppm value. Each 10 ppm reduction corresponds to a \$0.01 increase in the nominal

Table 9.13: Finding the Baseline Curve for the *Actual Costs of Production and Use*

P_{pk}	*ExCP*	*ExCU*	*ECP&U*	*Process Average*	*Nominal Cost*	*Actual Cost of Production and Use*
−0.50	.746	.035	1.782	200	0.30	0.534
−0.40	.708	.058	1.766	190	0.31	0.547
−0.30	.653	.088	1.741	180	0.32	0.557
−0.10	.581	.122	1.703	170	0.33	0.562
−0.10	.494	.157	1.651	160	0.34	0.561
0.00	.400	.186	1.586	150	0.35	0.555
0.10	.306	.204	1.510	140	0.36	0.543
0.20	.219	.206	1.426	130	0.37	0.527
0.30	.147	.194	1.341	120	0.38	0.510
0.40	.092	.168	1.260	110	0.39	0.491
0.50	.054	.135	1.189	100	0.40	0.476
0.60	.029	.102	1.130	90	0.41	0.463
0.70	.014	.070	1.085	80	0.42	0.456
0.80	.006	.046	1.052	70	0.43	0.452
0.90	.003	.028	1.031	60	0.44	0.454
1.00	.001	.015	1.016	50	0.45	0.457
1.10	.000	.008	1.008	40	0.46	0.464
1.20	.000	.004	1.004	30	0.47	0.472
1.30	.000	.002	1.002	20	0.48	0.481
1.40	.000	.001	1.001	10	0.49	0.490
1.50	.000	.000	1.000	0	0.50	0.500

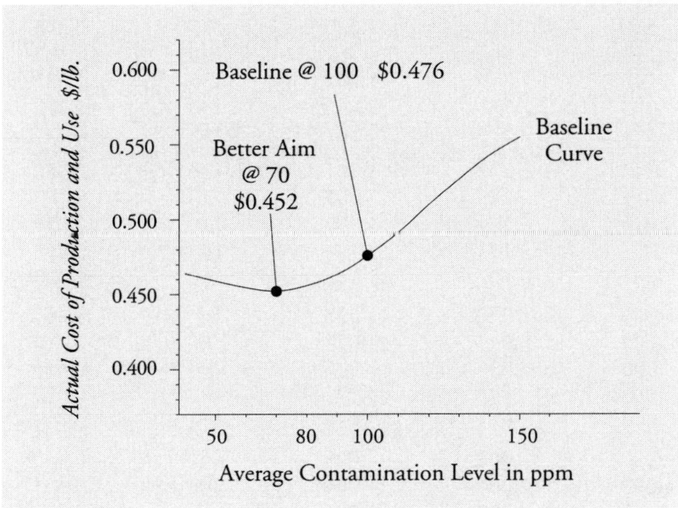

Figure 9.7: The Baseline Curve for *Actual Costs of Production and Use* for the Contamination Data

cost. Once we have the Nominal Costs, we compute the *Actual Costs of Production and Use* and plot these values to get Figure 9.7.

The average of 100 ppm from Figure 9.6 corresponds to an *Actual Cost of Production and Use* of $0.476/lb. Shifting to an average contamination level of 70 ppm would reduce this *Actual Cost of Production and Use* to $0.452/lb. With an annual production of 25 million pounds, this amounts to a savings of $ 600,000 per year. However, with unpredictable operation it will always be difficult to maintain a process at a given average.

THE PREDICTABLE COST CURVE

To see what this process has the potential to do, we now repeat the analysis above using *Sigma(X)* = 16.8:

$$\textit{Predictive Tolerance} = \frac{USL - Target}{Sigma(X)} = \frac{150 - 80}{16.8} = 4.17$$

From Table Eight, with a Tolerance of 4.2 we scale down the *ExCU* values by multiplying by 0.80, combine them with the scaled down *ExCP* values from Table 9.13 and get the *ECP&U* values. Then the process averages are found using Eq. 9.2:

$$\textit{Process Average} = \textit{Specification Value} - [\, 3\ Sigma(X)\ C_{pk}\,] = 150 - 50.4\, C_{pk}$$

These process averages are used to compute the Nominal Costs, and then the *Actual Costs* are computed and plotted in Figure 9.8.

Table 9.14: Finding the Predictable *Actual Costs of Production and Use*

C_{pk}	*ExCP*	*ExCU*	*ECP&U*	*Process Average*	*Nominal Cost*	*Actual Cost of Production and Use*
0.00	.400	.270	1.670	150	0.350	0.585
0.10	.306	.314	1.620	145	0.355	0.575
0.20	.219	.342	1.561	140	0.360	0.562
0.30	.147	.350	1.497	135	0.365	0.546
0.40	.092	.338	1.430	130	0.370	0.529
0.50	.054	.310	1.364	125	0.375	0.512
0.60	.029	.272	1.301	120	0.380	0.494
0.70	.014	.228	1.242	115	0.385	0.478
0.80	.006	.184	1.190	110	0.390	0.464
0.90	.003	.143	1.146	105	0.395	0.453
1.00	.001	.107	1.108	100	0.400	0.443
1.10	.000	.078	1.078	95	0.405	0.436
1.20	.000	.054	1.054	90	0.410	0.432
1.30	.000	.036	1.036	85	0.415	0.430
1.40	.000	.022	1.022	80	0.420	0.429
1.50	.000	.014	1.014	75	0.425	0.431
1.60	.000	.008	1.008	70	0.430	0.433
1.70	.000	.004	1.004	65	0.435	0.437
1.80	.000	.002	1.002	60	0.440	0.441
1.90	.000	.001	1.001	55	0.445	0.445
2.00	.000	.001	1.001	50	0.450	0.450

From Table 9.14 we see that operating Predictably at the current average of 100 ppm will result in an *Actual Cost of Production and Use* of $ 0.443/lb. which is about 3 cents per pound better than the baseline condition. Operating predictably with a contamination level of 80 ppm will result in an *Actual Cost of Production and Use* of $ 0.429/lb. which is 4.7 cents per

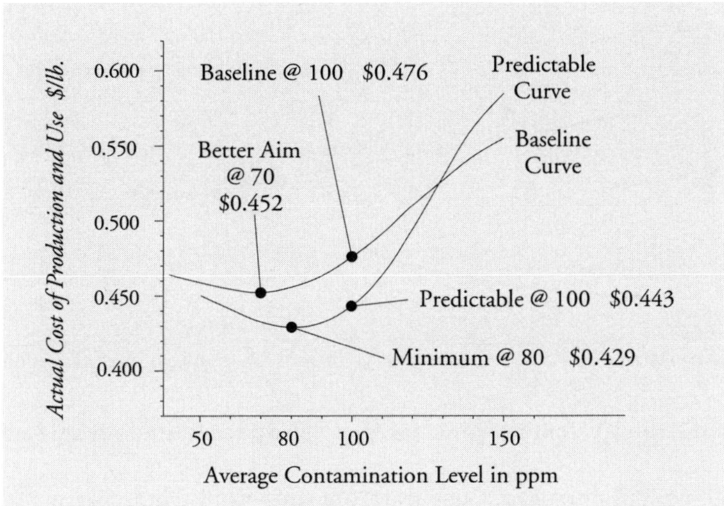

Figure 9.8: ***Actual Costs of Production and Use*** **Curves for Contamination Data**

pound less than the baseline value, for an annual savings of $ 1,175,000. Here we get to do better while operating closer to the target value of 80 ppm.

As shown in Figures 9.3 and 9.8, comparisons are made easy by marking the four points that represent:

1. the past operation (Baseline),
2. the best aim point with the historic amount of variation (Better Aim),
3. the predictable at current average (Predictable),
4. the best aim point when the process is operated predictably (Minimum).

When some actual costs are added to the graph the opportunities are made clear to all. The complexities of creating the tables are just intermediate steps in creating these simple, yet powerful graphs.

9.3 Target at Boundary

When we have a one-sided product specification with a target value at the boundary and a variable nominal cost of production, we end up with a picture similar to that in Figure 9.9.

Here the procedure is essentially the same as that outlined in the previous two sections. However, with the target at the boundary the process will always be off target, and the appropriate tables are Tables Nine and Ten. Since these two tables are formatted differently from Tables Seven and Eight, the first step of finding the *Effective Costs of Production and Use* will be slightly different than before. In this case both the *Excess Cost of Production* and the *Excess Cost of Use* will depend upon either the Centered Capability Ratio, C_{pk}, and the *Predictive Tolerance*, or the Centered Performance Ratio, P_{pk}, and the *Descriptive Tolerance.*

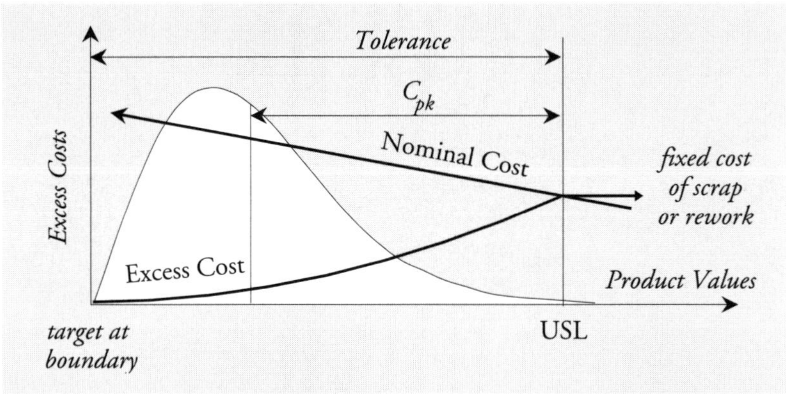

Figure 9.9: Excess Costs and Nominal Costs for One-Sided Specifications with Target at Boundary

As an example we will once again use the Contamination Data. Now, rather than having a target value of 80 ppm, we shall assume that the target value is 0 ppm, and any contamination will result in excess costs at the next process step.

Our first analysis will be for the Baseline Curve. The Contamination Data in Table 9.12 have an average of 100.0 ppm and a global standard deviation statistic of $s = 33.27$ ppm. The maximum allowable contamination is 150 ppm, and we shall use the same Nominal Cost Function defined in the previous section.

Here the *Descriptive Tolerance* becomes

$$\textit{Descriptive Tolerance} = \frac{USL - Target}{s} = \frac{150 - 0}{33.27} = 4.51$$

Since nonconforming batches are reworked at approximately \$0.35/lb. = *NC(150ppm),* we will use Table Ten to find our *Effective Costs of Production and Use.*

Since the *Descriptive Tolerance* is 4.51, we use the *ECP&U* values from the 4.5 column of Table Ten. This is the origin of the first two columns in Table 9.15.

Next we convert the P_{pk} values into process average values using Equation 9.1:

$$\textit{Process Average} = \textit{Specification Value} - [\,3\ s\ P_{pk}\,] = 150 - 99.8\,P_{pk}$$

Following this, we compute the Nominal Costs at these average values, and then multiply by the *ECP&U* to obtain the *Actual Costs of Production and Use.* These values are shown in Table 9.15. Since our example is the same as that in Table 9.13, the process averages and Nominal Costs are the same in Table 9.15 as they were in Table 9.13. However, here the *ECP&U* values are considerably larger due to the different Excess Cost Function used.

The P_{pk} values for a *Descriptive Tolerance* of 4.50 stop at $P_{pk} = 1.20$ simply because larger P_{pk} values are inconsistent with predictable operation near a boundary. These larger P_{pk} values would correspond to histograms where the mean would be considerably smaller than the standard deviation. The discussion of Figure 11.13 explains why the Best Case values of Tables Nine and Ten are what they are.

Table 9.15: Finding the Baseline *Actual Costs of Production and Use*

P_{pk}	*ECP&U*	*Process Average*	*Nominal Cost*	*Actual Cost of Production and Use*
0.00	1.847	150	0.35	0.646
0.10	1.787	140	0.36	0.643
0.20	1.717	130	0.37	0.635
0.30	1.640	120	0.38	0.623
0.40	1.560	110	0.39	0.608
0.50	1.480	100	0.40	0.592
0.60	1.401	90	0.41	0.574
0.70	1.329	80	0.42	0.558
0.80	1.264	70	0.43	0.544
0.90	1.206	60	0.44	0.531
1.00	1.157	50	0.45	0.521
1.10	1.116	40	0.46	0.513
1.20	1.081	30	0.47	0.508

To find the Predictable Curve of *Actual Costs of Production and Use* we repeat the analysis using *Sigma(X)* = 16.8 ppm and the *Predictive Tolerance* of:

$$\textit{Predictive Tolerance} = \frac{USL - Target}{Sigma(X)} = \frac{150 - 0}{16.8} = 8.93$$

Given this value, we begin Table 9.16 by listing the *ECP&U* values from the column in Table Ten having a *Tolerance* value of 9.0. However, since we do not need every value to create our curve, we reduce the size of Table 9.16 by only using every other value of C_{pk}. We compute the process averages using Equation 9.2:

$$\textit{Process Average} = \textit{Specification Value} - [\, 3\ Sigma(X)\ C_{pk}\,] = 150 - 50.4\, C_{pk}$$

We multiply the *ECP&U* values times the Nominal Costs to obtain our *Actual Costs of Production and Use,* as shown in Table 9.16.

Table 9.16: Finding the Predictable *Actual Costs of Production and Use*

C_{pk}	*ECP&U*	*Process Average*	*Nominal Cost*	*Actual Cost of Production and Use*
0.00	1.918	150	0.35	0.671
0.20	1.844	140	0.36	0.664
0.40	1.750	130	0.37	0.648
0.60	1.649	120	0.38	0.627
0.80	1.549	110	0.39	0.604
1.00	1.456	100	0.40	0.582
1.20	1.372	90	0.41	0.563
1.40	1.297	80	0.42	0.545
1.60	1.230	70	0.43	0.529
1.80	1.172	60	0.44	0.516
2.00	1.123	50	0.45	0.505
2.20	1.083	40	0.46	0.498
2.40	1.052	30	0.47	0.494
2.60	1.030	20	0.48	0.494

As before, we plot the Actual Cost Curves from Tables 9.15 and 9.16 to actually make our comparisons. The costs noted here are higher than they were in Figure 9.8 because of the different Excess Cost Functions used. However, in this case, where 0 ppm is the target value, the two Actual Cost Curves do tend to push the process down to the minimum practical aim point where the average is approximately one sigma away from the boundary.

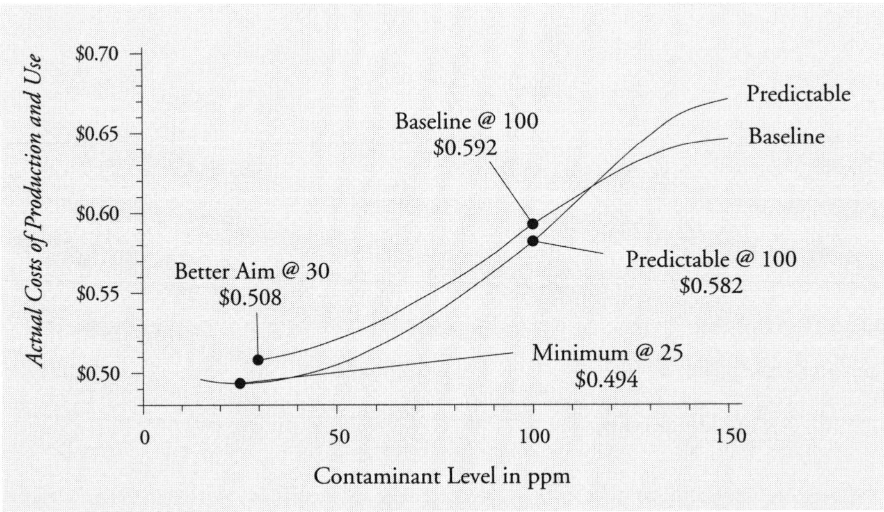

Figure 9.10: The *Actual Costs of Production and Use* for the Contamination Data when the Target is at the Boundary

9.4 Summary

Thus, when the Nominal Costs change with the Process Aim you have to compute the *Actual Costs of Production,* or the *Actual Costs of Production and Use,* over a whole range of Process Aim values. Moreover, if the process is being operated unpredictably, this has to be done twice: once to characterize the process performance and then once again to characterize the process potential. By comparing specific points on these Actual Cost curves a rational assessment of the benefits of different operating conditions can be obtained.

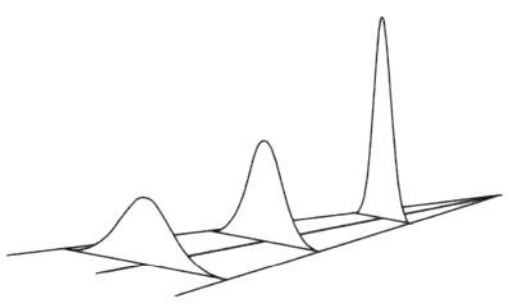

Full Potential = Minimum *ECP&U*

Chapter Ten

Excess Costs for Count Data

When it comes to capability, count data are inherently different from measurement data. While measurements will quantify some characteristic, counts will either classify an item or enumerate events. So, rather than defining where a characteristic falls within some continuum, counts describe how many items with a certain characteristic, or how many events of a given type, are found within some area of opportunity. This difference requires that we take a different approach to defining the *Effective Costs of Production and Use* for count data.

10.1 Specifications and Count Data

What happens when it is impossible to measure the items? What happens when each item may be characterized as acceptable or unacceptable, but no measurement can be made to quantify each item relative to some specification limit? We may still count the good items and the bad items, and compute the fraction of nonconforming product, but there is no underlying scale with a dividing line between acceptable product and unacceptable product. And this type of scale is part of most of the graphs in the preceding chapters. When this scale is missing, the specifications are generally written for the process rather than the product: the specifications define a maximum percentage of nonconforming product to be found in the product stream (or sometimes an average percentage of nonconforming product).

When the data are based on counts, specifications will apply to the fraction nonconforming in the product stream. This will make the specifications into *process* specifications rather than *product* specifications. While there may need to be some definition of what to count as nonconforming, this definition will not change the fact that the specification is a process specification.

So what do you do with process specifications? Consider the following data for stamped parts. Because of the complex shape of the part it was impossible to determine if the part would fit into the assembly by physical measurements; therefore, they shipped all of the parts produced, and the customer culled out the parts that did not fit. Periodically the customer would file a claim for those that were culled. The data from 22 claims are shown in Table 10.1. These values consist of the number of bad parts, the number of parts used, and the percentage of bad parts. Since the parts were shipped in containers of 40 pieces each, the numbers shipped are all multiples of 40. The specification is "no more than 2.5 percent nonfunctional."

Table 10.1: Counts of Nonfunctional Stampings

Report	No. Bad	No. Shipped	Percent Bad	Report	No. Bad	No. Shipped	Percent Bad
-1-	4	400	1.0	-12-	7	320	2.2
-2-	4	120	3.3	-13-	6	200	3.0
-3-	5	200	2.5	-14-	2	160	1.3
-4-	8	280	2.9	-15-	2	200	1.0
-5-	2	200	1.0	-16-	5	280	1.8
-6-	3	200	1.5	-17-	5	360	1.4
-7-	2	200	1.0	-18-	5	200	2.5
-8-	2	160	1.3	-19-	4	200	2.0
-9-	5	200	2.5	-20-	4	160	2.5
-10-	9	200	4.5	-21-	8	200	4.0
-11-	7	320	2.2	-22-	7	200	3.5

So how can you characterize this process? Some days are better than others? If you are going to use process specifications, then you need to characterize your process, and the premier technique for doing that is a process behavior chart.

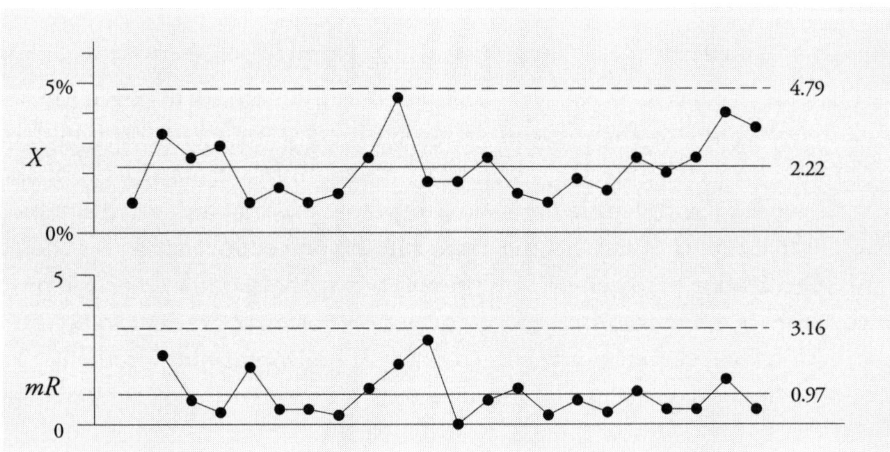

Figure 10.1: *XmR* Chart for Percentages of Nonfunctional Stampings

These percentages are placed on an *XmR* chart in Figure 10.1. There you can see that this process is predictable and is averaging about 2.2 percent nonfunctional. In any one report the fraction nonfunctional might vary up to 4.8%, but the long-term average is close to 2% nonfunctional.

So should they be concerned about report number 10 which had 4.5% nonfunctional? Does 4.5% nonfunctional signify a change in this process? Did things take a turn for the worse? The process behavior chart tells us that the 4.5% is within the bounds of routine variation and should not be interpreted, by itself, as a signal of any change in this process. Subsequent points on the chart bear out this interpretation.

But what does the specification of no more than 2.5% nonfunctional mean? If the specification is "no more than 2.5% nonfunctional on the average," then this process is capable of meeting this requirement. Since this process is predictable we can say that, unless it changes for the worse, it will average about 2.2% nonfunctional and will, therefore, meet this specification.

If the specification is "no more than 2.5% nonfunctional in any one period of time," then this process is not capable of meeting this requirement. Therefore, it is critical to clarify exactly what a process specification means. Does it define the average level of nonconformity, or does it define an absolute maximum level of nonconformity?

Finally, with process specifications, the process behavior chart for the product stream defines the actual capability of a predictable process, or the potential capability of an unpredictable process. The central line defines what to expect over the long term, and the limits show the amount of short-term, routine variation about that long-term average that you should expect in practice.

10.2 Effective Costs of Production for Counts of Items

A count is a count of items when each item out of a set of **n** items is classified as either having, or not having, some attribute. Generally the attribute being counted is the number of items that are nonconforming. The conforming items get shipped, and the nonconforming items get scrapped or reworked. Thus, the first difference we find with count data is that a count of items provides no information about how the conforming items will work at the next step. Hence we have no way to use a count of items to characterize the *Excess Costs of Use*. We are restricted to using the *Excess Costs of Production*.

When an item is classified as nonconforming it will be subject to two actions—scrap or rework. When the item is scrapped the cost of producing that item is lost. When the item is reworked the cost of the materials is salvaged while the original value added is lost. Define the proportion of the product stream that is scrapped by $\boldsymbol{p_S}$ and the proportion of the product stream that is reworked by $\boldsymbol{p_R}$. Let c_S be the excess cost of scrapping an item and c_R is the excess cost of reworking an item (both of these costs exclude the initial cost of production). (Since this differs from the definition of C_S and C_R elsewhere, the lower case c was used here for distinction.) Finally, let NC be the nominal cost of producing an item. Thus, the *Actual Cost of Production* could be written as:

$$\textit{Actual Cost of Production} = NC + c_S \boldsymbol{p_S} + c_R \boldsymbol{p_R}$$

The *Effective Cost of Production* will be the *Actual Cost of Production* divided by the nominal cost of producing the items shipped:

$$\textit{Effective Cost of Production} = \frac{\textit{Actual Cost of Production}}{\textit{Nominal Cost of Units Shipped}}$$

$$= \frac{NC + c_S\,p_S + c_R\,p_R}{NC\,[\,1.0 - p_S\,]}$$

$$= \frac{1.0 + \frac{c_S}{NC}\,p_S + \frac{c_R}{NC}\,p_R}{1.0 - p_S}$$

In this expression the cost of rework divided by the nominal cost can be thought of as the complement of the cost of the materials salvaged by the rework divided by the nominal cost:

$$\frac{c_R}{NC} = 1 - \frac{\textit{Cost of Materials Salvaged}}{\textit{Nominal Cost}}$$

When rework consists of doing a job over, and there are no materials salvaged, then we assume that $c_R \approx NC$. In the other term in the numerator, c_S is essentially the cost of disposal for the scrapped items. When this cost is negligible compared to NC, the ratio of c_S to NC may be taken to be zero. This one general equation simplifies to cover the cases where all of the nonconforming product is either scrapped or reworked, and it allows for the more complex situations where some items are scrapped and others are reworked. Given that the outcome of this equation can depend on four different inputs, tables of the *Effective Cost of Production* for Counts of Items are impractical.

10.3 Effective Costs of Production for Counts of Events

Counts that enumerate discrete events occurring within some finite area of opportunity, such as blemishes or discrepancies, are commonly used to determine the grade, or quality, of the product. When the areas of opportunity vary, the counts are converted into rates. When the areas of opportunity remain essentially the same size, the counts may be used directly.

When the count or the rate exceeds certain threshold values, the product is typically downgraded by various amounts or even scrapped. For simplicity I will denote these different grades of product by the labels Firsts, Seconds, Thirds, and Scrap. There could be more, or fewer, categories as needed.

For the product classified as Seconds, the difference between the selling price of the Firsts and that of the Seconds is a cost of production. Denote this difference as D_2. This difference is like a cost of rework since you have given up this amount of income in order to avoid scrapping the product. Likewise, for the product classified as Thirds, the difference between the selling prices of the Firsts and the Thirds is another cost of production. Denote this difference as D_3. If we denote the proportion of product that is classified as Seconds as p_2, the proportion of product classified as Thirds as p_3, and if we let p_S denote the proportion classified as Scrap, we will end up with the following equation for the *Effective Cost of Production*:

$$\text{Effective Cost of Production} \; = \; \frac{1 + \frac{D_2}{NC}\, p_2 + \frac{D_3}{NC}\, p_3}{1 - p_S}$$

For the case with more than two downgraded categories additional terms would be added in the numerator. For the case with no scrap the denominator would become 1.0. It should be noted that the differences D_2 and D_3 are differences between selling prices while NC is still the nominal cost of producing the product. This means that the ratios in the terms in the numerator can exceed 1.0.

Because the equation above depends upon the specifics of a particular situation it is not feasible to try to construct tables for these values.

Chapter Eleven

The Formulas for the Excess Costs

Having used the past six chapters to illustrate the concept and use of the *Effective Cost of Production and Use,* it is now time to consider how the values given in the tables were obtained. In most cases the probability model used was the generic model of maximum entropy, the normal distribution. While the reader may use the formulas given in this chapter to customize the computations using other probability models, in most cases the results will differ only slightly from the tabled values.

11.1 The Structure of the Effective Cost of Production and Use

The *Effective Cost of Production and Use* will be the actual cost of producing and using an item divided by the nominal cost of producing that item.

$$\textit{Effective Cost of Production and Use} = \frac{\textit{Actual Cost of Production \& Use}}{\textit{Nominal Cost of Production}}$$

The numerator of the ratio above will consist of the nominal cost of production plus the average excess costs per unit associated with producing and using such items.

$$\begin{aligned}\textit{Actual Cost of Production \& Use} \;=\; & \textit{Nominal Cost of Production} \\ & + \textit{Average Excess Cost of Production} \\ & + \textit{Average Excess Cost of Use}\end{aligned}$$

Based on the two expressions above, it appears that we can obtain a value for the *Effective Cost of Production and Use* if we can characterize the *Average Excess Costs of Production* and the *Average Excess Costs of Use.*

11.2 The Average Costs of Scrap, Rework, and Use

Since in the end, we always have to pay for our supplier's excess costs we adopt the perspective of the customer and define the excess costs to consist of three components.

1. The cost of scrapping nonconforming items. Since the producer will have to incorporate the cost of scrap into his overhead costs, and since the customer has to pay for the producer's overhead, the customer will indirectly end up paying for the scrap that is produced.

2. The cost of reworking nonconforming items. Once more, the excess costs associated with rework will have to be buried in the producer's overhead, so while the customer will pay for the reworked item, he will also have to pay for the excess costs of reworking it as well.

3. The cost of using conforming product. This is a direct cost to the customer. It consists of all of those actions taken by the customer to get his equipment and the supplier's materials to work properly together.

Thus, the job of characterizing the Average Excess Costs consists of determining each of these three components.

Since scrapping an item is a cut-your-losses action, ideally taken only when it is cheaper to scrap the item than to use it, we can think of the cost of scrap to be a fixed cost. That is, let C_S denote the cost of scrapping an item. This cost will have to equal the cost of the raw materials plus the cost of all processing done to the item, the cost of inspection, and perhaps the cost of disposal as well.

THE AVERAGE COST OF SCRAP

The Average Cost of Scrap will be the cost of scrapping an item multiplied by the proportion of the items that are scrapped. If we let the probability model shown in Figure 11.1 represent the distribution of process outcomes, and if items below the lower specification limit are to be scrapped, then the proportion scrapped will be defined by the area under the curve to the left of the lower specification limit, *LSL*.

If we let the vertical axis denote excess costs associated with the production of an item, we can see that there would be a fixed excess cost, C_S, for those items that are scrapped. As a result, the Average Excess Cost of Scrap would be:

$$\textit{Average Excess Cost of Scrap} = C_S \times \textit{Proportion Scrapped} = C_S \int_{-\infty}^{LSL} f(x)\, dx$$

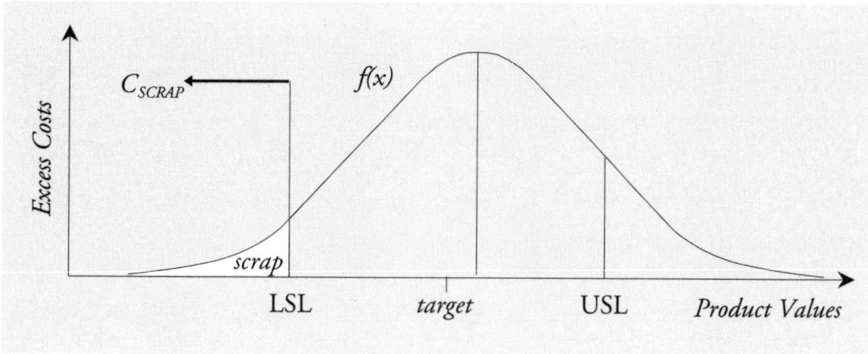

Figure 11.1: Scrap and the Cost of Scrap

THE AVERAGE COST OF REWORK

In a similar manner, the Average Cost of Rework will be the cost of reworking an item multiplied by the proportion of the items that are reworked. If items above the upper specification limit are to be reworked, then the proportion reworked will be defined by the area under the curve to the right of the upper specification limit, *USL*.

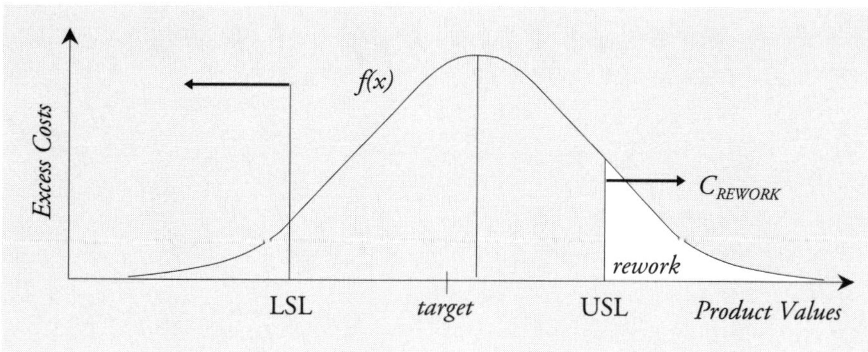

Figure 11.2: Rework and the Cost of Rework

While different items may require different amounts of rework, in our attempt to obtain the excess costs due to rework we shall, for simplicity, use the average cost of reworking an item as a fixed cost, C_R. This single value will result in a fixed excess cost for those items that are reworked. As a result, the Average Excess Cost of Rework might be written as:

$$\text{Average Excess Cost of Rework} = C_R \times \text{Proportion Reworked} = C_R \int_{USL}^{\infty} f(x)\, dx$$

Thus, we know the excess cost function in two regions—scrap and rework. To fill in the rest of the curve we need to think about the nature of excess costs. Excess costs are the result of variation. They are costs incurred because everything is not right at the target value. If we could make everything right at the target value, there would be no excess costs of production. Therefore, we define the excess cost function to be zero at the target value, as shown in Figure 11.3. In most cases the cost of scrapping an item and the average cost of reworking an item can be determined. This means that the three portions of the Excess Cost Function shown in Figure 11.3 will generally be known.

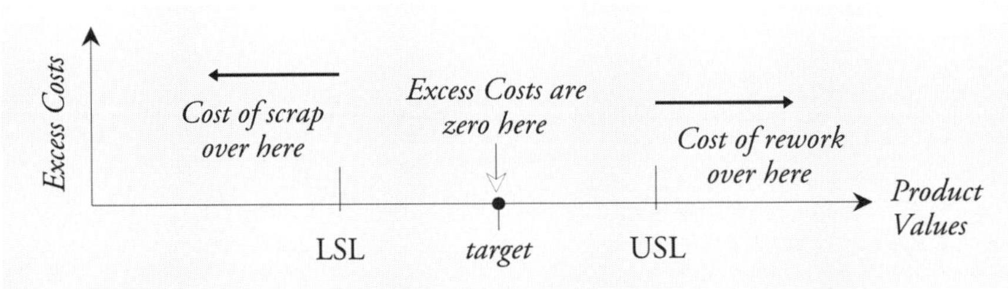

Figure 11.3: The Three Known Portions of the Excess Cost Function

But how do we fill in the gaps? Since we know that small deviations from the target can be tolerated, we prefer to use quadratic curves to connect the three known portions of the Excess Cost Function. These quadratic curves will have their minimum at the target value, and will connect with the fixed portions of the Excess Cost Function to give the piecewise continuous curve shown as $\mathcal{L}(x)$ in Figure 11.4.

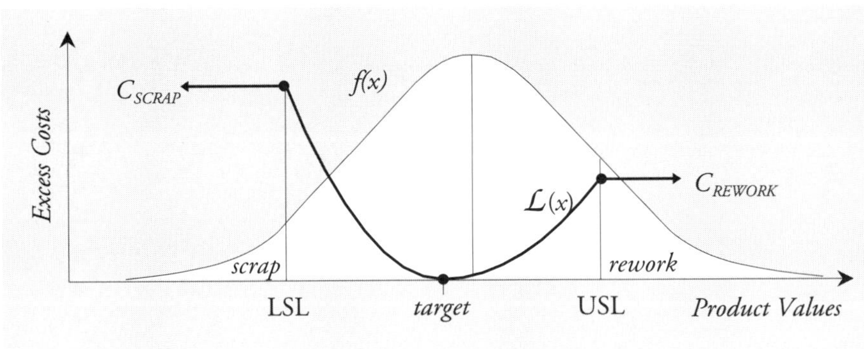

Figure 11.4: A Realistic Excess Cost Function

This Excess Cost Function suggests that when an item falls near the target there will be few excess costs associated with its use. However, as an item falls further and further away from the target the costs of using the item will increase. Moreover, these costs will increase up to that point where it will be cheaper to scrap or rework the item than it will be to try to use it. Since this is consistent with reality, the curve in Figure 11.4 is a realistic Excess Cost

Function. This Excess Cost Function will be symmetric when all nonconforming product is treated alike (either scrapped or reworked). But when nonconforming product on one side is scrapped, while that on the other side is reworked, we may have an asymmetric Excess Cost Function as shown in Figures 11.4 and 11.5.

It is important to note that the realistic Excess Cost Function in Figure 11.4 is completely determined by the three points shown as black dots. These three points have (x,y) coordinates of:

$$(\,\textit{Scrap Spec.},\ C_{SCRAP}\,) \qquad (\,\textit{target},\ 0\,) \qquad (\,\textit{Rework Spec.},\ C_{REWORK}\,)$$

THE AVERAGE COST OF USING CONFORMING PRODUCT

Using the Excess Cost Function, $\mathcal{L}(x)$, in Figure 11.5, we may find the excess costs associated with using conforming product by multiplying the realistic Excess Cost Function, $\mathcal{L}(x)$, by the model for the process outcomes, $f(x)$, and integrating between the specification limits:

$$\textit{Average Excess Cost of Use} = \int_{LSL}^{USL} \mathcal{L}(x)\, f(x)\, dx$$

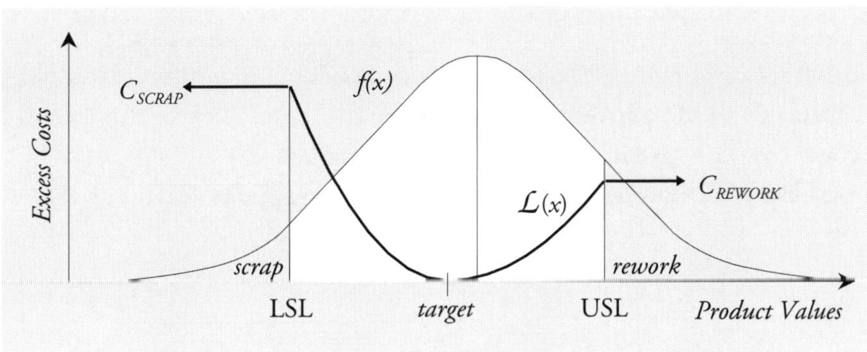

Figure 11.5: The Average Cost of Using Conforming Product

THE AVERAGE EXCESS COST

Combining the Average Excess Cost of Scrap, the Average Excess Cost of Use, and the Average Excess Cost of Rework, we find the Average Excess Cost to be:

$$\textit{Average Excess Cost} = \int_{-\infty}^{\infty} \mathcal{L}(x)\, f(x)\, dx$$

To actually evaluate this integral we will have to break it down into four parts: the region below *LSL*, the region between *LSL* and the target, the region between the target and *USL*, and the region above *USL*. By plugging in the equations for the quadratic curves we end up with a generic, four term expression for the Average Excess Cost, which can be evaluated for any particular probability model, $f(x)$, that seems appropriate:

$$\textit{Average Excess Cost} = C_S \int_{-\infty}^{LSL} f(x)\,dx \qquad \text{excess costs due to scrap}$$

$$+ \; C_S \int_{LSL}^{target} \frac{(x - target)^2}{(LSL - target)^2} f(x)\,dx \qquad \text{due to deviations below target}$$

$$+ \; C_R \int_{target}^{USL} \frac{(x - target)^2}{(USL - target)^2} f(x)\,dx \qquad \text{due to deviations above target}$$

$$+ \; C_R \int_{USL}^{\infty} f(x)\,dx \qquad \text{excess costs due to rework}$$

The first and last terms above characterize costs of production, while the middle two terms characterize costs of use. Regardless of where they originate, none of these costs are necessary. All of these costs are obstacles to economic operation.

If we replace the four integrals above with the following symbols:

$$\mathbf{IA} = \int_{-\infty}^{LSL} f(x)\,dx \qquad \text{Integral Below LSL}$$

$$\mathbf{IB} = \int_{LSL}^{target} \frac{(x - target)^2}{(LSL - target)^2} f(x)\,dx \qquad \text{Integral LSL to Target}$$

$$\mathbf{IC} = \int_{target}^{USL} \frac{(x - target)^2}{(USL - target)^2} f(x)\,dx \qquad \text{Integral Target to USL}$$

$$\mathbf{ID} = \int_{USL}^{\infty} f(x)\,dx \qquad \text{Integral Above USL}$$

we can simplify the *Average Excess Cost* expression to become:

$$\textit{Average Excess Cost} \;=\; C_S\,\mathbf{IA} + C_S\,\mathbf{IB} + C_R\,\mathbf{IC} + C_R\,\mathbf{ID}$$

To evaluate the four integrals in this expression we will need to know the relationship between the specification limits and the product distribution, $f(x)$. Since this relationship is summarized by the capability ratios, all that is needed to evaluate these integrals is a probability model and knowledge of C_p and C_{pk}.

Moreover, since the normal distribution is the distribution of maximum entropy, its use in evaluating these four integrals will provide generic, worst-case values to use in determining the *Effective Cost of Production and Use*. Table A.9 gives the normal probability values for these four integrals for 476 combinations of C_p and C_{pk}.

11.3 Finding The Effective Cost of Production and Use

In order to evaluate the *Effective Cost of Production and Use* we will need to rewrite the original equation in terms of quantities that we can compute. In Section 11.1 we wrote:

$$\textit{Effective Cost of Production and Use} \;=\; \frac{\textit{Actual Cost of Production \& Use}}{\textit{Nominal Cost of Production}}$$

Careful thought about this expression will reveal that the numerator will have to encompass all of the product produced while the denominator can only refer to the product that was actually shipped. Thus, if we let *NC* represent the nominal cost to produce each unit, and if all reworked product gets shipped, we can write the denominator as:

$$\textit{Nominal Cost of Units Shipped} \;=\; NC\,[\,1.0 - \textit{proportion scrapped}\,]$$

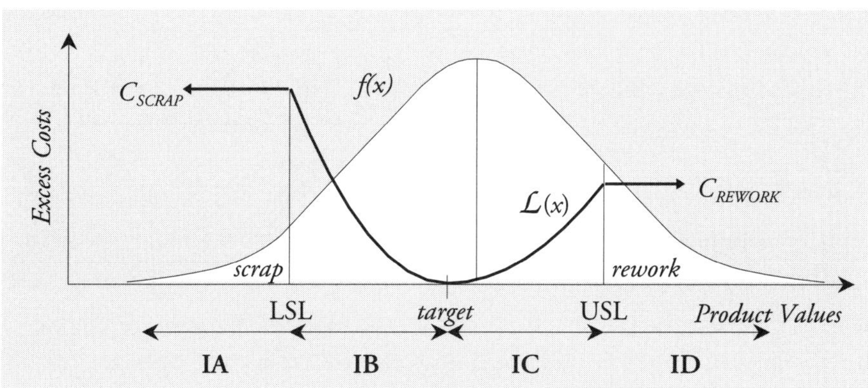

Figure 11.6: The Excess Cost Function

At the same time, since the items scrapped are already counted as part of the excess costs, we need to write the numerator of the *Effective Cost of Production and Use* in a way that will avoid double accounting for the scrap:

Thus, the numerator becomes:

$$\textit{Actual Cost of Production \& Use} = \textit{Nominal Cost of Units Shipped} + \textit{Average Excess Costs}$$

$$= NC\,[\,1.0 - \textit{proportion scrapped}\,] + \Big[\Big\{ C_S\,\mathbf{IA} + C_R\,\mathbf{ID} \Big\} + \Big\{ C_S\,\mathbf{IB} + C_R\,\mathbf{IC} \Big\}\Big]$$

By combining these new expressions for the numerator and denominator, we can write the *Effective Cost of Production and Use* as:

$$\textit{Effective Cost of Production and Use} = \frac{\textit{Actual Cost of Production and Use}}{\textit{Nominal Cost of Production}}$$

$$= \frac{NC\,[1.0 - \textit{proportion scrapped}\,] + C_S\Big[\{\,\mathbf{IA} + \frac{C_R}{C_S}\mathbf{ID}\,\} + \{\,\mathbf{IB} + \frac{C_R}{C_S}\mathbf{IC}\,\}\Big]}{NC\,[1.0 - \textit{proportion scrapped}\,]}$$

$$= \frac{1.0 - \textit{proportion scrapped}}{1.0 - \textit{proportion scrapped}} + \frac{\frac{C_S}{NC}\{\,\mathbf{IA} + \frac{C_R}{C_S}\mathbf{ID}\,\}}{1.0 - \textit{proportion scrapped}} + \frac{\frac{C_S}{NC}\{\,\mathbf{IB} + \frac{C_R}{C_S}\mathbf{IC}\,\}}{1.0 - \textit{proportion scrapped}}$$

$$= 1.0 + \textit{Excess Cost of Production} + \textit{Excess Cost of Use}$$

Thus, the *Effective Cost of Production and Use*, *ECP&U*, is defined in terms of the *Excess Costs of Production*, *ExCP*, and the *Excess Costs of Use*, *ExCU*, as:

$$ECP\&U = 1.0 + ExCP + ExCU$$

When working with an internal supplier-customer relationship, *ECP&U* will be the appropriate measure to use. This will also summarize the efficiency of your own operations and the effect of your product upon your customer.

When the focus is upon your own operations alone, The *Effective Cost of Production* may be of interest. It is found by simply adding 1.0 to the *Excess Cost of Production*:

$$\textit{Effective Cost of Production} = 1.0 + ExCP$$

(Note that in earlier books I did not separate the *Excess Costs of Production* from the *Excess Costs of Use*, so the *ECP&U* was simply called the *Effective Cost of Production*. Now that I am separating the excess costs the more complete nomenclature is necessary.)

When the focus is upon what you are doing to your customer, or what your supplier is doing to you, the *Effective Cost of Use* may be of interest. It is found by simply adding 1.0 to the *Excess Cost of Use*:

$$\textit{Effective Cost of Use} = 1.0 + \textit{ExCU}$$

This addition of 1.0 to the *Excess Costs* is necessary to convert the *Excess Costs*, which are additive, into *Effective Costs*, which are multiplicative.

TABLES FOR TWO-SIDED SPECIFICATIONS

The formulas derived above are for two-sided specifications with the target at the midpoint of the specifications. Using these formulas, it was possible to create six tables that will provide complete coverage for two-sided specifications. Table One will cover the case where all of the nonconforming product is scrapped. Table Two will cover the case where all of the nonconforming product is reworked. The remaining tables will cover the case where nonconforming material on one side is scrapped while nonconforming material on the other side is reworked. Tables Three and Four will cover the case where the average is on the scrap side of the target, while Tables Five and Six will cover the case where the average is on the rework side of the target. Each table includes both the specific formula used for the entries in that table and the general formula needed to customize the table entries to fit other situations. For examples of the use of these tables see Chapter Five.

11.4 One-Sided Process Specifications

With one-sided *process* specifications the focus is not upon the nonconforming items, all of which get shipped, but rather upon what fraction of the product stream is nonconforming. As shown in Figure 11.7, the fraction nonconforming will depend upon the Centered Capability Ratio and is given by the integral defined earlier as **IA**. Table 7.2 lists those fractions nonconforming for a one-sided process specification that correspond to 219 values of C_{pk} ranging from 1.09 to 1.09. In this case the give-away serves in lieu of the computation of an *Excess Cost* value.

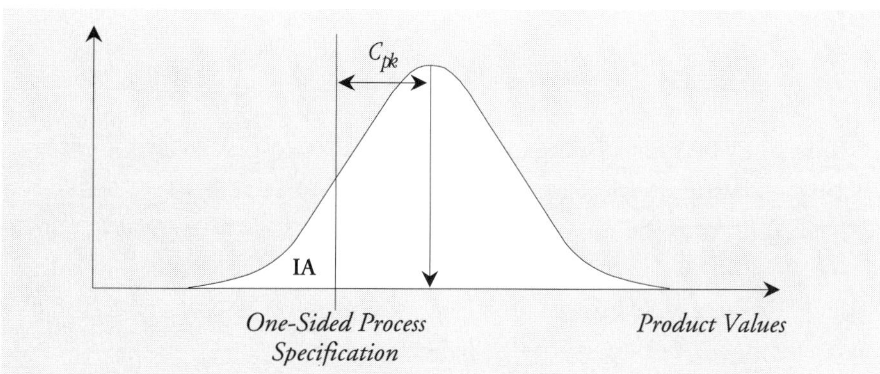

Figure 11.7: The Fraction Nonconforming Depends upon C_{pk}

11.5 The Excess Cost of Production with One-Sided Product Specs

With one-sided product specifications, where nonconforming items will be reworked or scrapped, we outlined two situations in Chapter Seven. The first was when no target value could be defined. This is pictured in Figure 11.8. Here we can only compute the *Excess Cost of Production*, and this value will depend solely upon the value of C_{pk} (or P_{pk}).

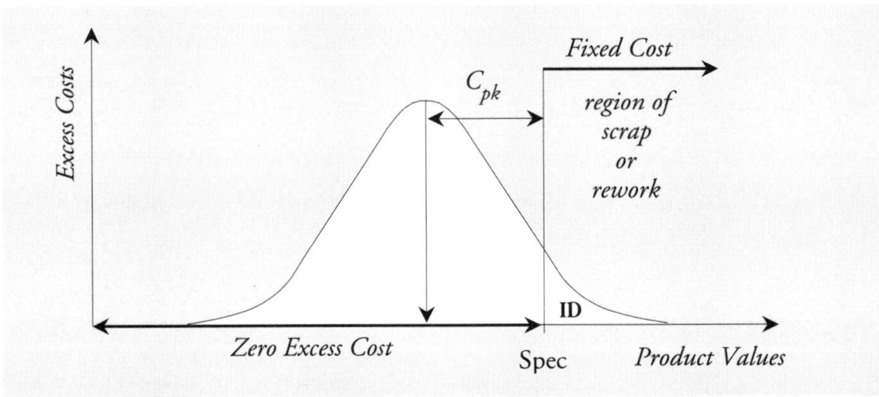

Figure 11.8: The Excess Cost Function for a One-Sided Product Specification without a Target

Using the earlier notation, when the nonconforming items are scrapped the general formula for the *Excess Cost of Production* will be:

$$ExCP \quad = \quad Excess\ Cost\ of\ Scrap \quad = \quad \frac{C_S}{NC}\,\frac{\mathbf{ID}}{1-\mathbf{ID}}$$

For the case where the nonconforming items are reworked the general formula for the *Excess Cost of Production* will be:

$$ExCP \quad = \quad Excess\ Cost\ of\ Rework \quad = \quad \frac{C_R}{NC}\,\mathbf{ID}$$

These *ExCP* values may be found in the first columns of Tables Seven and Eight respectively.

When it proves difficult to define a target value for use with a one-sided product specification (as required in the next section), an analysis may still be carried out using only the *Excess Costs of Production*. While such an analysis will be incomplete, it will still provide minimum values for the excess costs, and these minimum costs will generally allow reasonable decisions to be made. For an example see Section 7.3.

11.6 The Excess Cost of Use with One-Sided Product Specs

In order to define the *Excess Cost of Use* for one-sided product specifications there will have to be a target value. As shown in Figure 11.9, this target value should define that point where the excess cost function goes to zero. This target value should represent that point where, in a perfect world, the customer would like to see every item produced. At the same time, the specification limit should be that point where it is cheaper to scrap (or to rework) the item than it is to use it. Unfortunately, as noted in Chapter Seven, one-sided product specifications are often set at the target value rather than at the absolute cut-off point simply as a piece of insurance on the part of the customer.

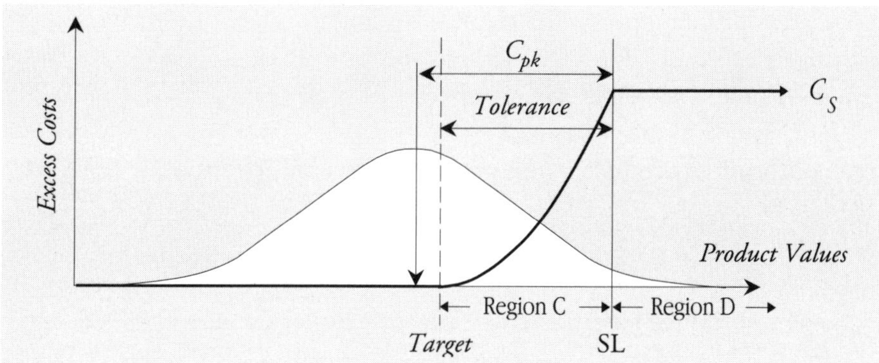

Figure 11.9: The Excess Cost of Use Will Depend upon Both C_{pk} and *Tolerance*

About the only way around this problem of the creeping one-sided specification is to get the customer to define a range of less-than-perfect values that are still acceptable. If this can be done, then one end-point of that range becomes the one-sided specification value and the target value may be taken as the other end-point of that range of acceptable values.

When defining a target value is a problem, you might use a generic value for the region labeled as Region C in Figure 11.9. By analogy with a Capability Ratio of 1.00 for two-sided specifications, we might take a generic value for Region C to be a region defined by a C_{pk} of 1.00. While this may be arbitrary, it will usually be better than completely ignoring the *Excess Cost of Use*. However, when you feel that there is insufficient information to use the *Excess Cost of Use*, or when its use presents more difficulties than you care to work with, you can still return to the previous section and focus solely upon the *Excess Cost of Production*.

In order to avoid notational difficulties, we shall denote the width of Region C, which is defined by the distance from the target to the one-sided product specification, as the *Tolerance*, and we will express this distance in standard deviation units. As seen in Figure

11.9, the *Excess Cost of Use* will depend upon both the C_{pk} value and the *Tolerance*.

As with C_{pk} and P_{pk}, there will be two versions of the *Tolerance*. The *Descriptive Tolerance* will be:

$$\textit{Descriptive Tolerance} = \frac{\textit{Distance from Specification to Target}}{s}$$

While the *Predictable Tolerance* will be:

$$\textit{Predictable Tolerance} = \frac{\textit{Distance from Specification to Target}}{\textit{Sigma(X)}}$$

As might be expected, the *Descriptive Tolerance* will be used with P_{pk} while the *Predictable Tolerance* will be used in conjunction with C_{pk}.

When nonconforming material is scrapped the general formula for the *Effective Cost of Production and Use* becomes:

$$ECP\&U = 1.0 + \frac{C_S}{NC}\frac{\mathbf{ID}}{1-\mathbf{ID}} + \frac{C_S}{NC}\frac{\mathbf{IC}}{1-\mathbf{ID}}$$

where, as before, the middle term on the right is the *Excess Cost of Production* and the last term is the *Excess Cost of Use*. Table Seven gives these values of *ExCP* and *ExCU* for the case where $C_S = NC$. There we see that while *ExCP* depends only upon C_{pk}, *ExCU* will depend upon both C_{pk} and the *Tolerance*.

When nonconforming product is reworked the general formula for the *Effective Cost of Production and Use* becomes:

$$ECP\&U = 1.0 + \frac{C_R}{NC}\,\mathbf{ID} + \frac{C_R}{NC}\,\mathbf{IC}$$

Table Eight gives these values for *ExCP* and *ExCU* for the case where $C_R = NC$.

11.7 One-Sided Product Specifications with Targets at Boundaries

As a special case of the previous section, we might consider what happens when the target value for a one-sided product specification is located at a boundary condition. As before, we shall assume that the excess cost is zero at the boundary condition, and that the cost of scrapping or reworking a nonconforming item or batch is some fixed value. In between we shall once again use a quadratic curve to characterize the excess costs associated with using the conforming product. Since the boundary restricts the range of possible values for the product characteristic, our picture changes to look like Figure 11.10.

When nonconforming product is scrapped the *Effective Cost of Production and Use* will be:

$$ECP\&U = 1.0 + \frac{C_S}{NC}\frac{\mathbf{ID}}{1-\mathbf{ID}} + \frac{C_S}{NC}\frac{\mathbf{IC}}{1-\mathbf{ID}}$$

$$= 1.0 + ExCP + ExCU$$

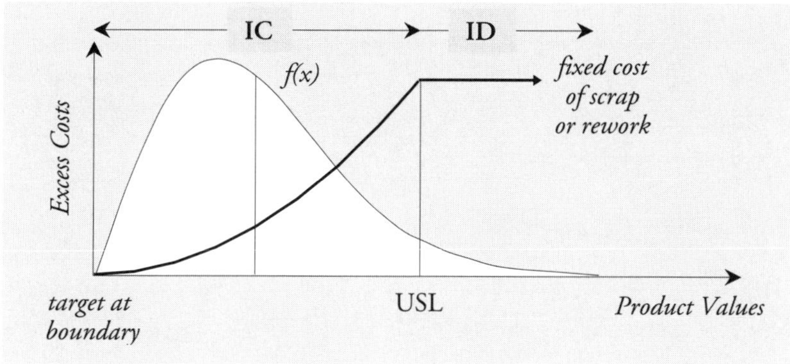

Figure 11.10: One-Sided Product Specification with Target at Boundary

When nonconforming product is reworked the *Effective Cost of Production and Use* will be:

$$ECP\&U \;=\; 1.0 \;+\; \frac{C_R}{NC}\,\mathbf{ID} \;+\; \frac{C_R}{NC}\,\mathbf{IC}$$

where, as before, the middle term defines the *Excess Costs of Production* and the last term defines the *Excess Costs of Use*. While these equations are similar to the equations found earlier, the bounded nature of the product stream will require that we evaluate these expressions using a different set of probability models than before.

Note that, as shown in Figure 11.11, the Centered Capability Ratio must be defined relative to the one-sided specification even when the average is closer to the boundary than it is to the specification value. This is of import since some boundary conditions are listed as specifications values even though they are something quite different. Boundary values are inappropriate for computing capability ratios of any sort. When using capability values generated by software you will need to verify that the ratios are properly computed.

BURR DISTRIBUTIONS

In computing the *Effective Costs of Production and Use* we can, with complete generality, work with a lower bound. The tabled values will apply equally well to those cases with an upper bound. Also, without loss of generality, we can let this lower bound be zero. The probability models that we shall use will all have a zero left-hand end point. They are all Burr distributions having a probability density function of the form:

$$f(x) \;=\; c\,k\,x^{c-1}\,[\,1 + x^c\,]^{-(k+1)} \qquad \text{for} \quad x \geq 0, \;\; c > 0, \;\; k > 0$$

For different values of c and k the family of Burr distributions cover a wide region of the shape characterization plane. Thus they provide satisfactory mathematical models for use in evaluating the integral terms in the excess cost expressions. The formulas for the mean and variance parameters for a Burr distribution involve the standard Gamma Function where $\Gamma(n) = (n-1)!$ for integer n. These formulas are:

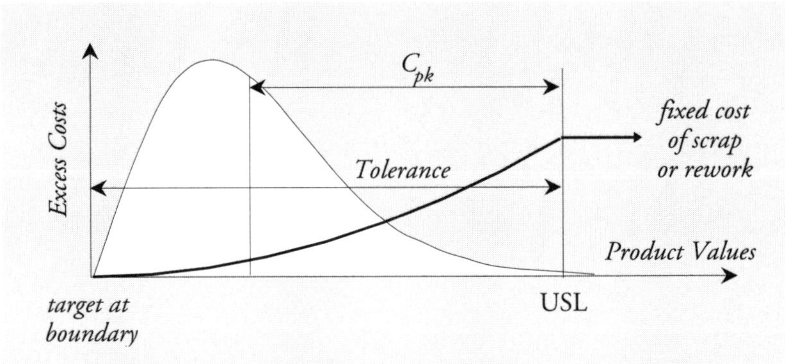

Figure 11.11: ***Tolerance*** **and Centered Capability Ratio for Target at Boundary Case**

$$\mu = \frac{\Gamma(\frac{1}{c})\,\Gamma(k-\frac{1}{c})}{c\,\Gamma(k)} \qquad\qquad \sigma^2 = \frac{2\,\Gamma(\frac{2}{c})\,\Gamma(k-\frac{2}{c})}{c\,\Gamma(k)} - \mu^2$$

All distributions that have a fixed end-point will have some relationship between the mean and standard deviation. As seen in Figure 11.11, the *Tolerance* and the C_{pk} values will effectively differ by the value of the mean of the distribution. Thus, the parameters of the Burr distribution and the *Tolerance* and C_{pk} values are all related. We will have to make use of this relationship to create a set of tables for the *Effective Cost of Production and Use* for situations having the target at a boundary value.

Inspection of Figure 11.11 will show that, for a given Burr distribution, the *Tolerance* will be equal to [the distance from the mean to the specification] plus [the distance from the boundary to the mean]. The distance from the mean to the specification, expressed in standard deviation units, is equal to 3 C_{pk}. The distance from the boundary to the mean, expressed in standard deviation units, is simply the ratio of the mean to the standard deviation. Thus, the *Tolerance* value can be written as:

$$\textit{Tolerance} = 3\,C_{pk} + \frac{\textit{mean of Burr distribution}}{\textit{standard deviation of Burr distribution}}$$

This relationship between the *Tolerance*, the C_{pk} value, and the parameters of the Burr distribution means that a single Burr distribution can be used to compute the *Effective Costs of Production and Use* for a diagonal slice through the tables for which *Tolerance* – 3 C_{pk} is constant. We are specifically interested in cases where this constant is a multiple of 0.3.

The Burr distributions used are listed in Table 11.1. Burr Distribution No. 1 was used with those cases where *Tolerance* – 3 C_{pk} = 0.9. Burr Distribution No. 2 was used where *Tolerance* – 3 C_{pk} = 1.2. etc. (See Table 11.2.) The specification values used to evaluate the integrals (in meas. units) were found by using the *Tolerance* (in s.d. units) in the equation:

$$\textit{Specification} = \textit{Tolerance} \times \textit{standard deviation of Burr distribution}$$

Table 11.1: Burr Distributions Used for Tables Nine and Ten

No.	*c*	*k*	*mean* / *std. dev.*	*skewness*	*kurtosis*
1	0.93	34.75	0.9	2.45	12.8
2	1.22	72.83	1.2	1.54	6.41
3	1.55	58.55	1.5	1.07	4.43
4	1.89	61.26	1.8	0.75	3.53
5	2.24	64.45	2.1	0.52	3.09
6	2.60	60.25	2.4	0.35	2.88
7	2.97	48.90	2.7	0.21	2.78
8	3.33	64.07	3.0	0.09	2.74

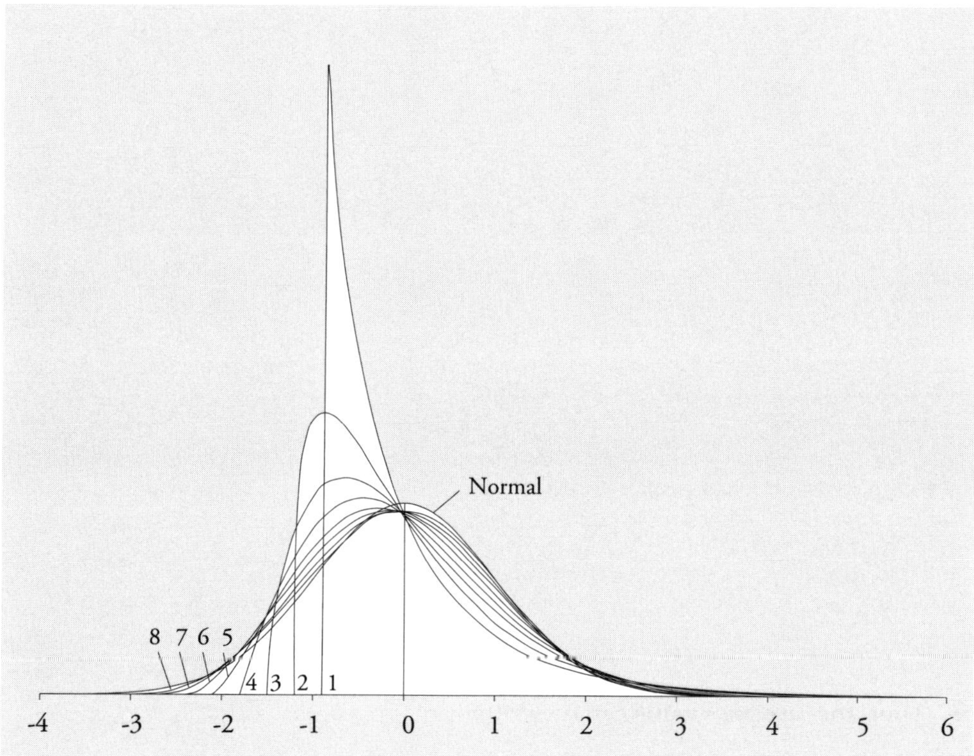

Figure 11.12: The Eight Burr Distributions of Table 11.1 Compared to the Normal Distribution

Figure 11.12 shows the eight Burr distributions of Table 11.1 in standardized form along with a standard normal distribution. By the time the *Tolerance* exceeds 3 C_{pk} + 3.0 the need for a skewed distribution has vanished since Burr Distribution No. 8 is very close to the normal distribution. Thus, after the first eight diagonals of Tables Nine and Ten, the remaining entries were computed using the standard normal distribution, as shown in Table 11.2.

Tables Nine and Ten do not contain any entries for *Tolerance* – 3 C_{pk} = 0.3 or 0.6. This is not an oversight. Any distribution that would satisfy these conditions would have a mean that was either 30% or 60% of the size of the standard deviation. While such mathematical models exist, they are not consistent with the conceptual model given in Chapter One.

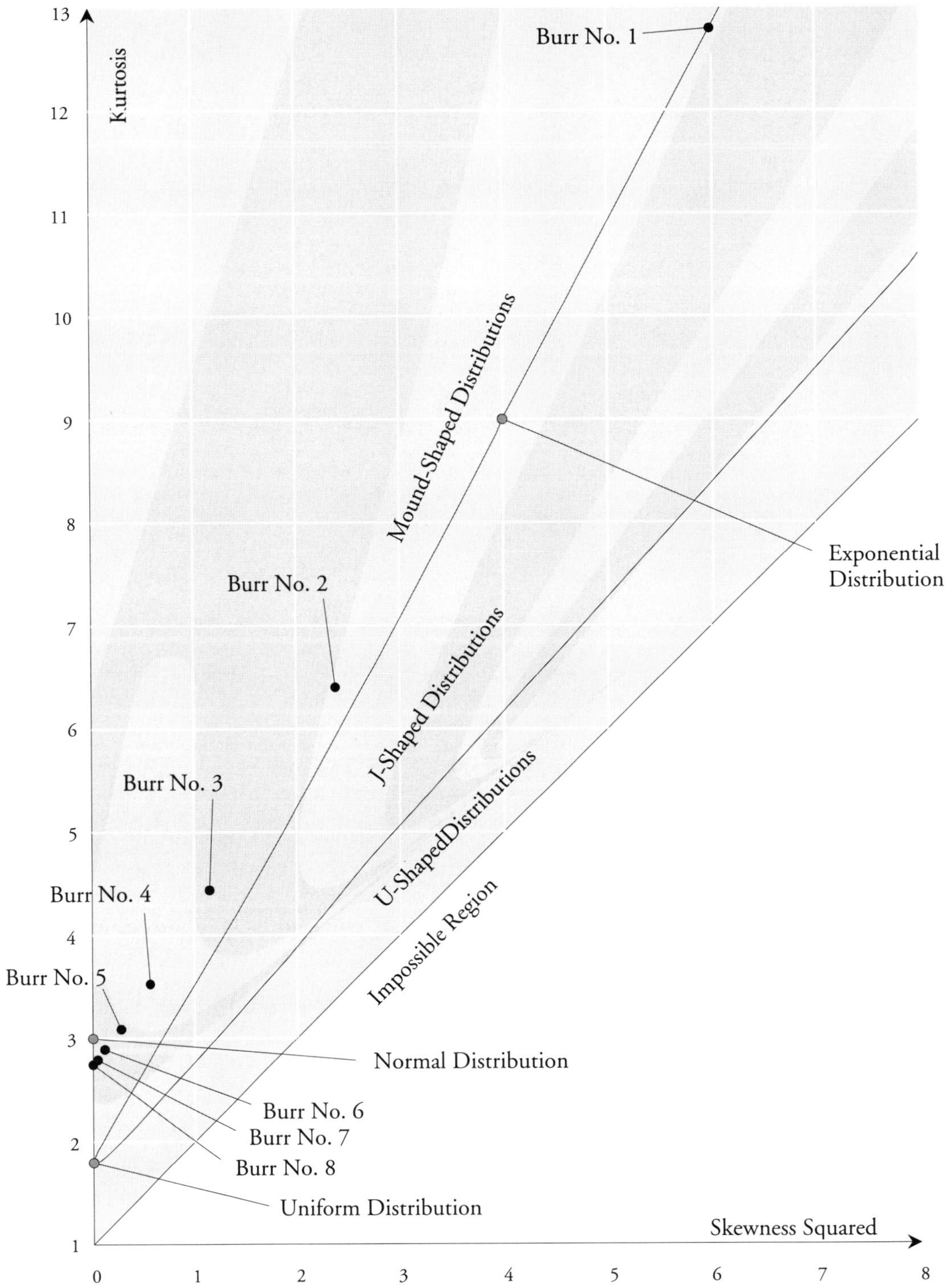

Figure 11.13: The Shape Characterization Plane

Table 11.2: Distributions Used for Tables Nine and Ten

C_{pk} or P_{pk}	*Tolerance in Standard Deviation Units* 0.3	0.6	0.9	1.2	1.5	1.8	2.1	2.4
–0.50	Burr 4	Burr 5	Burr 6	Burr 7	Burr 8	Normal	Normal	Normal
–0.40	Burr 3	Burr 4	Burr 5	Burr 6	Burr 7	Burr 8	Normal	Normal
–0.30	Burr 2	Burr 3	Burr 4	Burr 5	Burr 6	Burr 7	Burr 8	Normal
–0.20	Burr 1	Burr 2	Burr 3	Burr 4	Burr 5	Burr 6	Burr 7	Burr 8
–0.10		Burr 1	Burr 2	Burr 3	Burr 4	Burr 5	Burr 6	Burr 7
0.00			Burr 1	Burr 2	Burr 3	Burr 4	Burr 5	Burr 6
0.10				Burr 1	Burr 2	Burr 3	Burr 4	Burr 5
0.20					Burr 1	Burr 2	Burr 3	Burr 4
0.30						Burr 1	Burr 2	Burr 3
0.40							Burr 1	Burr 2
0.50								Burr 1

Specifically, whenever we use a probability model it should be a model that is a feasible model for routine variation. As was argued in Chapter One, routine variation is best thought of as the result of a large number of cause-and-effect relationships where no one cause is dominant. Thus, when a process is operated predictably, routine variation corresponds to some high-entropy condition. For phenomena that are unbounded, the distribution of maximum entropy is the normal distribution. For phenomena that are bounded on both sides, the distribution of maximum entropy is the uniform distribution. For phenomena that are bounded on one side, the distribution of maximum entropy is the exponential distribution. These three distributions are shown on the shape characterization plane in Figure 11.13, along with the eight Burr distributions of Table 11.1.

The three distributions of maximum entropy form a triangle in the shape characterization plane. Reasonable probability models for routine variation will all be found somewhere in the neighborhood of this triangle. Except for Burr Distribution No. 1, the models in Table 11.1 fall within this triangle.

Now the exponential distribution has a mean that is equal to its standard deviation. In Figure 11.12 and Figure 11.13 we see that Burr Distribution No. 1, with a mean to standard deviation ratio of 0.90, is slightly more extreme than the exponential. *This distribution is, for all practical purposes, the limit of what might be achieved in terms of predictable and on target operation when the target is at a boundary condition.* Thus, as shown in Table 11.2, Burr Distribution No. 1 was used for the boundary diagonal for Tables Nine and Ten.

Table A.10 contains the values of the integrals **IC** and **ID** used in the formulas given at the beginning of this section to create Tables Nine and Ten.

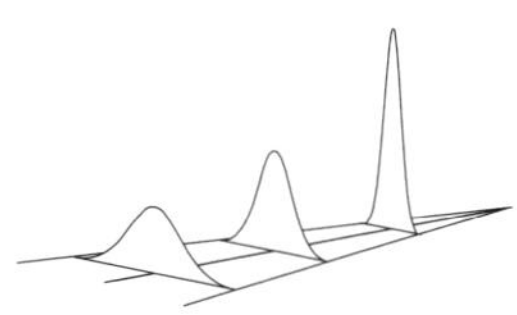

Full Potential = World Class Quality

Tables

Excess Costs of Production

Excess Costs of Use

Effective Costs of Production and Use

Tables for Two-Sided Specifications

Table One	When All Nonconforming Units Are Scrapped	170
Table Two	When All Nonconforming Units Are Reworked	178
Table Three	When the Average Is on the Scrap Side of the Target Value and the Cost of Rework Is Equal to the Cost of Scrap	186
Table Four	When the Average Is on the Scrap Side of the Target Value and the Cost of Rework Is 0.1% the Cost of Scrap	194
Table Five	When the Average Is on the Rework Side of the Target Value and the Cost of Rework Is Equal to the Cost of Scrap	202
Table Six	When the Average Is on the Rework Side of the Target Value and the Cost of Rework Is 0.1% the Cost of Scrap	210

Tables for One-Sided Specifications

Table Seven	*Excess Costs of Scrap* and *Excess Costs of Use* when Nonconforming Product Is Scrapped	218
Table Eight	*Excess Costs of Rework* and *Excess Costs of Use* when Nonconforming Product Is Reworked	222

Tables for Target at Boundary

Table Nine	When Nonconforming Product Is Scrapped	226
Table Ten	When Nonconforming Product Is Reworked	234

Table One: The *Excess Costs of Production, Excess Costs of Use,* and *Effective Cost of Production and Use* when *All* Nonconforming Units Are Scrapped

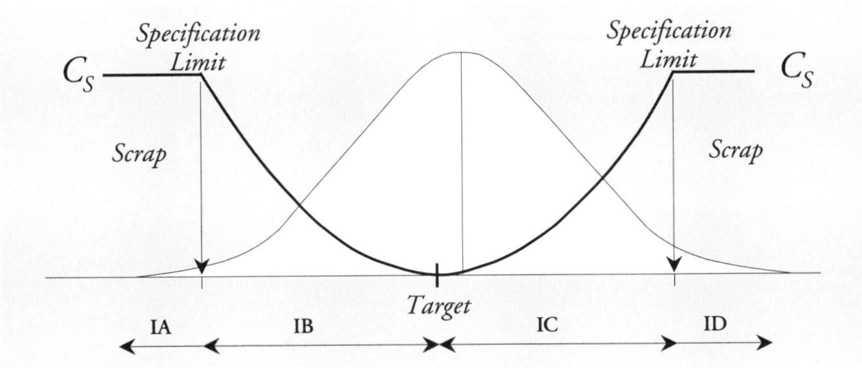

Table One: Excess Cost Function when All Nonconforming Items Are Scrapped

Corresponding to each pairing for C_p and C_{pk} there are three entries stacked in a column.

The topmost of these is the *Excess Costs of Production,* *ExCP*

The middle value is the *Excess Costs of Use,* *ExCU*

And the bottom value (in bold) is the *Effective Cost of Production and Use,* ***ECP&U***

The entries in Table One are for the case where all nonconforming product is scrapped and the Cost of Scrap, C_S, is assumed to be equal to the Nominal Cost of Production, *NC*. Table A.9 contains the normal theory values for **IA, IB, IC,** and **ID**. The proportion scrapped is given by the sum of the two integrals **IA** and **ID**, making the general and specific formulas for Table One:

$$ECP\&U = 1.0 + \frac{\frac{C_S}{NC}\{\mathbf{IA} + \mathbf{ID}\}}{1.0 - \mathbf{IA} - \mathbf{ID}} + \frac{\frac{C_S}{NC}\{\mathbf{IB} + \mathbf{IC}\}}{1.0 - \mathbf{IA} - \mathbf{ID}}$$

$$= 1.0 + \frac{\{\mathbf{IA} + \mathbf{ID}\}}{1.0 - \mathbf{IA} - \mathbf{ID}} + \frac{\{\mathbf{IB} + \mathbf{IC}\}}{1.0 - \mathbf{IA} - \mathbf{ID}}$$

$$= 1.0 + \textit{Excess Costs of Production} + \textit{Excess Costs of Use}$$

Since the Cost of Scrap can be appreciably greater than *NC*, the entries in Table One are essentially *lower bounds* for all three quantities. If greater precision is required, the entries for the *Excess Costs of Production* and the *Excess Costs of Use* may be inflated by the ratio of C_S to *NC*, as shown in the general formula above, and added to 1.00 to get a customized value for the *Effective Cost of Production and Use.*

For example, with $C_p = 0.9$ and $C_{pk} = 0.5$, and with $C_S = 1.5\ NC$ we would convert the tabled value of 1.332 for *ECP&U* into the value 1.498 (*ExCP* is 0.108 and *ExCU* is 0.390).

Table One: The *Excess Costs of Production, Excess Costs of Use,* and *Effective Cost of Production and Use* when *All* Nonconforming Units Are Scrapped

C_{pk} or P_{pk}	C_p or P_p value 0.10	0.20	0.30	0.40	0.50	0.60	0.70	0.80	0.90	1.00	1.10	1.20	1.30	1.40
	19.43	14.79	14.08	13.98	13.97	13.97	13.97	13.97	13.97	13.97	13.97	13.97	13.97	13.97
	0.342	0.377	0.438	0.506	0.567	0.618	0.660	0.694	0.722	0.746	0.766	0.783	0.798	0.811
–0.50	***20.77***	***16.17***	***15.52***	***15.49***	***15.54***	***15.59***	***15.63***	***15.66***	***15.69***	***15.71***	***15.73***	***15.75***	***15.77***	***15.78***
	11.64	8.36	7.79	7.70	7.69	7.69	7.69	7.69	7.69	7.69	7.69	7.69	7.69	7.69
	0.338	0.363	0.412	0.474	0.534	0.586	0.630	0.666	0.696	0.721	0.743	0.761	0.778	0.792
–0.40	***12.97***	***9.72***	***9.21***	***9.18***	***9.23***	***9.28***	***9.32***	***9.36***	***9.39***	***9.41***	***9.43***	***9.45***	***9.47***	***9.48***
	7.53	5.02	4.54	4.45	4.43	4.43	4.43	4.43	4.43	4.43	4.43	4.43	4.43	4.43
	0.335	0.350	0.387	0.441	0.499	0.551	0.596	0.634	0.666	0.693	0.716	0.736	0.754	0.769
–0.30	***8.86***	***6.37***	***5.92***	***5.89***	***5.93***	***5.98***	***6.03***	***6.07***	***6.10***	***6.13***	***6.15***	***6.17***	***6.19***	***6.20***
	5.28	3.20	2.76	2.66	2.65	2.65	2.65	2.65	2.65	2.65	2.65	2.65	2.65	2.65
	0.333	0.339	0.364	0.407	0.460	0.511	0.557	0.597	0.631	0.660	0.685	0.707	0.726	0.743
–0.20	***6.61***	***4.53***	***4.12***	***4.07***	***4.11***	***4.16***	***4.20***	***4.24***	***4.28***	***4.31***	***4.33***	***4.35***	***4.37***	***4.39***
	4.05	2.17	1.75	1.64	1.62	1.62	1.62	1.62	1.62	1.62	1.62	1.62	1.62	1.62
	0.331	0.330	0.342	0.374	0.419	0.468	0.514	0.556	0.591	0.623	0.650	0.673	0.694	0.713
–0.10	***5.38***	***3.50***	***3.09***	***3.02***	***3.04***	***3.09***	***3.13***	***3.18***	***3.21***	***3.24***	***3.27***	***3.29***	***3.31***	***3.33***
	3.430	1.598	1.155	1.033	1.005	1.001	1.000	1.000	1.000	1.000	1.000	1.000	1.000	1.000
	0.330	0.323	0.324	0.343	0.378	0.422	0.467	0.509	0.546	0.579	0.608	0.634	0.657	0.677
0.00	***4.759***	***2.921***	***2.479***	***2.376***	***2.383***	***2.423***	***2.467***	***2.509***	***2.546***	***2.579***	***2.608***	***2.634***	***2.657***	***2.677***
	3.240	1.305	0.815	0.667	0.627	0.620	0.618	0.618	0.618	0.618	0.618	0.618	0.618	0.618
	0.329	0.319	0.310	0.315	0.337	0.374	0.416	0.457	0.496	0.530	0.561	0.589	0.614	0.636
0.10	***4.570***	***2.624***	***2.125***	***1.981***	***1.965***	***1.993***	***2.034***	***2.075***	***2.114***	***2.149***	***2.180***	***2.207***	***2.232***	***2.254***
		1.215	0.638	0.450	0.394	0.380	0.378	0.378	0.378	0.378	0.378	0.378	0.378	0.378
		0.318	0.302	0.293	0.301	0.326	0.362	0.401	0.440	0.476	0.508	0.538	0.564	0.588
0.20		***2.533***	***1.939***	***1.743***	***1.694***	***1.707***	***1.740***	***1.779***	***1.818***	***1.854***	***1.886***	***1.916***	***1.942***	***1.966***
			0.583	0.335	0.253	0.231	0.226	0.226	0.226	0.226	0.226	0.226	0.226	0.226
			0.299	0.279	0.271	0.282	0.309	0.343	0.380	0.416	0.450	0.481	0.510	0.535
0.30 »			***1.881***	***1.614***	***1.524***	***1.513***	***1.535***	***1.569***	***1.606***	***1.642***	***1.676***	***1.707***	***1.735***	***1.761***
				0.299	0.178	0.141	0.132	0.130	0.130	0.130	0.130	0.130	0.130	0.130
				0.274	0.252	0.246	0.260	0.286	0.319	0.354	0.388	0.420	0.450	0.477
0.40 »				***1.573***	***1.430***	***1.387***	***1.391***	***1.417***	***1.449***	***1.484***	***1.518***	***1.550***	***1.580***	***1.607***
					0.154	0.093	0.076	0.072	0.072	0.072	0.072	0.072	0.072	0.072
					0.245	0.222	0.219	0.234	0.260	0.292	0.324	0.356	0.387	0.416
0.50 »					***1.399***	***1.315***	***1.295***	***1.306***	***1.332***	***1.363***	***1.396***	***1.428***	***1.459***	***1.487***
						0.077	0.046	0.039	0.037	0.037	0.037	0.037	0.037	0.037
						0.214	0.193	0.192	0.208	0.233	0.262	0.293	0.323	0.353
0.60 »						***1.292***	***1.239***	***1.231***	***1.245***	***1.270***	***1.300***	***1.330***	***1.361***	***1.390***
							0.037	0.022	0.019	0.018	0.018	0.018	0.018	0.018
							0.183	0.164	0.166	0.182	0.206	0.233	0.262	0.291
0.70 »							***1.220***	***1.186***	***1.184***	***1.200***	***1.224***	***1.251***	***1.280***	***1.309***
								0.017	0.010	0.008	0.008	0.008	0.008	0.008
								0.155	0.139	0.142	0.157	0.180	0.206	0.233
0.80 »								***1.171***	***1.148***	***1.150***	***1.166***	***1.188***	***1.214***	***1.241***
									0.007	0.004	0.004	0.003	0.003	0.003
									0.129	0.117	0.121	0.136	0.157	0.181
0.90 »									***1.136***	***1.121***	***1.125***	***1.140***	***1.160***	***1.184***
										0.003	0.002	0.001	0.001	0.001
										0.108	0.098	0.103	0.118	0.137
1.00 »										***1.111***	***1.100***	***1.105***	***1.119***	***1.138***
											0.001	0.001	0.000	0.000
											0.091	0.084	0.089	0.102
1.10 »											***1.092***	***1.084***	***1.089***	***1.103***

C_{pk}	C_p 1.20	1.30	1.40
	0.000	0.000	0.000
	0.077	0.071	0.077
1.20	***1.077***	***1.072***	***1.077***
		0.000	0.000
		0.066	0.062
1.30		***1.066***	***1.062***
			0.000
			0.057
1.40			***1.057***

Table One: The *Excess Costs of Production, Excess Costs of Use,* and *Effective Cost of Production and Use* when *All* Nonconforming Units Are Scrapped

C_{pk} or P_{pk}	C_p or P_p value 1.50	1.60	1.70	1.80	1.90	2.00	2.10	2.20	2.30	2.40	2.50	2.60	2.70	2.80
	13.97	13.97	13.97	13.97	13.97	13.97	13.97	13.97	13.97	13.97	13.97	13.97	13.97	13.97
	0.822	0.833	0.842	0.850	0.857	0.864	0.870	0.876	0.881	0.886	0.890	0.894	0.898	0.901
–0.50	***15.79***	***15.80***	***15.81***	***15.82***	***15.83***	***15.83***	***15.84***	***15.84***	***15.85***	***15.85***	***15.86***	***15.86***	***15.87***	***15.87***
	7.69	7.69	7.69	7.69	7.69	7.69	7.69	7.69	7.69	7.69	7.69	7.69	7.69	7.69
	0.804	0.815	0.825	0.834	0.842	0.850	0.856	0.862	0.868	0.873	0.878	0.883	0.887	0.891
–0.40	***9.49***	***9.51***	***9.52***	***9.52***	***9.53***	***9.54***	***9.55***	***9.55***	***9.56***	***9.56***	***9.57***	***9.57***	***9.58***	***9.58***
	4.43	4.43	4.43	4.43	4.43	4.43	4.43	4.43	4.43	4.43	4.43	4.43	4.43	4.43
	0.783	0.795	0.806	0.816	0.825	0.833	0.840	0.847	0.853	0.859	0.864	0.869	0.874	0.878
–0.30	***6.22***	***6.23***	***6.24***	***6.25***	***6.26***	***6.27***	***6.27***	***6.28***	***6.29***	***6.29***	***6.30***	***6.30***	***6.31***	***6.31***
	2.65	2.65	2.65	2.65	2.65	2.65	2.65	2.65	2.65	2.65	2.65	2.65	2.65	2.65
	0.758	0.771	0.783	0.794	0.804	0.813	0.821	0.829	0.836	0.842	0.848	0.853	0.858	0.863
–0.20	***4.40***	***4.42***	***4.43***	***4.44***	***4.45***	***4.46***	***4.47***	***4.47***	***4.48***	***4.49***	***4.49***	***4.50***	***4.50***	***4.51***
	1.62	1.62	1.62	1.62	1.62	1.62	1.62	1.62	1.62	1.62	1.62	1.62	1.62	1.62
	0.729	0.744	0.757	0.769	0.780	0.790	0.799	0.807	0.815	0.822	0.828	0.834	0.840	0.845
–0.10	***3.35***	***3.36***	***3.37***	***3.39***	***3.40***	***3.41***	***3.42***	***3.42***	***3.43***	***3.44***	***3.45***	***3.45***	***3.46***	***3.46***
	1.000	1.000	1.000	1.000	1.000	1.000	1.000	1.000	1.000	1.000	1.000	1.000	1.000	1.000
	0.695	0.711	0.726	0.739	0.751	0.762	0.772	0.782	0.790	0.798	0.805	0.812	0.819	0.825
0.00	***2.695***	***2.711***	***2.726***	***2.739***	***2.751***	***2.762***	***2.772***	***2.782***	***2.790***	***2.798***	***2.805***	***2.812***	***2.819***	***2.825***
	0.618	0.618	0.618	0.618	0.618	0.618	0.618	0.618	0.618	0.618	0.618	0.618	0.618	0.618
	0.655	0.673	0.690	0.704	0.718	0.730	0.741	0.752	0.761	0.770	0.778	0.786	0.793	0.800
0.10	***2.274***	***2.292***	***2.308***	***2.323***	***2.336***	***2.348***	***2.360***	***2.370***	***2.380***	***2.388***	***2.397***	***2.404***	***2.412***	***2.418***
	0.378	0.378	0.378	0.378	0.378	0.378	0.378	0.378	0.378	0.378	0.378	0.378	0.378	0.378
	0.610	0.630	0.648	0.664	0.679	0.693	0.705	0.717	0.728	0.738	0.747	0.756	0.764	0.771
0.20	***1.988***	***2.008***	***2.026***	***2.042***	***2.057***	***2.070***	***2.083***	***2.095***	***2.105***	***2.115***	***2.125***	***2.133***	***2.142***	***2.149***
	0.226	0.226	0.226	0.226	0.226	0.226	0.226	0.226	0.226	0.226	0.226	0.226	0.226	0.226
	0.559	0.581	0.600	0.618	0.635	0.650	0.664	0.677	0.689	0.700	0.711	0.720	0.729	0.738
0.30	***1.785***	***1.806***	***1.826***	***1.844***	***1.860***	***1.875***	***1.889***	***1.902***	***1.915***	***1.926***	***1.936***	***1.946***	***1.955***	***1.964***
	0.130	0.130	0.130	0.130	0.130	0.130	0.130	0.130	0.130	0.130	0.130	0.130	0.130	0.130
	0.503	0.526	0.547	0.567	0.585	0.602	0.618	0.632	0.645	0.658	0.670	0.681	0.691	0.700
0.40	***1.633***	***1.656***	***1.677***	***1.697***	***1.715***	***1.732***	***1.748***	***1.762***	***1.775***	***1.788***	***1.800***	***1.811***	***1.821***	***1.830***
	0.072	0.072	0.072	0.072	0.072	0.072	0.072	0.072	0.072	0.072	0.072	0.072	0.072	0.072
	0.442	0.467	0.490	0.512	0.531	0.550	0.567	0.583	0.598	0.612	0.625	0.637	0.648	0.659
0.50	***1.514***	***1.539***	***1.562***	***1.583***	***1.603***	***1.621***	***1.639***	***1.654***	***1.669***	***1.683***	***1.696***	***1.708***	***1.720***	***1.730***
	0.037	0.037	0.037	0.037	0.037	0.037	0.037	0.037	0.037	0.037	0.037	0.037	0.037	0.037
	0.380	0.406	0.431	0.453	0.475	0.495	0.513	0.530	0.547	0.562	0.576	0.590	0.602	0.614
0.60	***1.418***	***1.444***	***1.468***	***1.491***	***1.512***	***1.532***	***1.550***	***1.568***	***1.584***	***1.599***	***1.613***	***1.627***	***1.640***	***1.651***
	0.018	0.018	0.018	0.018	0.018	0.018	0.018	0.018	0.018	0.018	0.018	0.018	0.018	0.018
	0.319	0.345	0.371	0.394	0.417	0.438	0.458	0.476	0.494	0.510	0.526	0.541	0.554	0.567
0.70	***1.337***	***1.363***	***1.389***	***1.413***	***1.435***	***1.456***	***1.476***	***1.495***	***1.512***	***1.529***	***1.544***	***1.559***	***1.573***	***1.586***
	0.008	0.008	0.008	0.008	0.008	0.008	0.008	0.008	0.008	0.008	0.008	0.008	0.008	0.008
	0.260	0.286	0.312	0.336	0.360	0.382	0.403	0.422	0.441	0.459	0.475	0.491	0.506	0.520
0.80	***1.268***	***1.295***	***1.320***	***1.345***	***1.368***	***1.390***	***1.411***	***1.431***	***1.449***	***1.467***	***1.483***	***1.499***	***1.514***	***1.528***
	0.003	0.003	0.003	0.003	0.003	0.003	0.003	0.003	0.003	0.003	0.003	0.003	0.003	0.003
	0.206	0.232	0.257	0.281	0.305	0.328	0.349	0.370	0.389	0.408	0.425	0.442	0.458	0.473
0.90	***1.210***	***1.235***	***1.260***	***1.285***	***1.308***	***1.331***	***1.353***	***1.373***	***1.393***	***1.411***	***1.429***	***1.445***	***1.461***	***1.476***
	0.001	0.001	0.001	0.001	0.001	0.001	0.001	0.001	0.001	0.001	0.001	0.001	0.001	0.001
	0.159	0.183	0.207	0.231	0.254	0.277	0.299	0.319	0.339	0.359	0.377	0.394	0.411	0.427
1.00	***1.161***	***1.184***	***1.208***	***1.232***	***1.255***	***1.278***	***1.300***	***1.321***	***1.341***	***1.360***	***1.378***	***1.396***	***1.412***	***1.428***
	0.000	0.000	0.000	0.000	0.000	0.000	0.000	0.000	0.000	0.000	0.000	0.000	0.000	0.000
	0.120	0.141	0.163	0.185	0.208	0.230	0.252	0.273	0.293	0.312	0.331	0.349	0.366	0.382
1.10	***1.120***	***1.141***	***1.163***	***1.186***	***1.208***	***1.230***	***1.252***	***1.273***	***1.293***	***1.313***	***1.332***	***1.349***	***1.367***	***1.383***

Table One: The *Excess Costs of Production, Excess Costs of Use,* and *Effective Cost of Production and Use* when *All* Nonconforming Units Are Scrapped

C_{pk} or P_{pk}	C_p or P_p value 2.90	3.00	3.20	3.40	3.60	3.80	4.00	4.50	5.00	5.50	6.0	8.0	12.0	20.0
	13.97	13.97	13.97	13.97	13.97	13.97	13.97	13.97	13.97	13.97	13.97	13.97	13.97	13.97
	0.905	0.908	0.913	0.919	0.923	0.927	0.931	0.939	0.945	0.950	0.954	0.967	0.980	0.993
-0.50	***15.87***	***15.88***	***15.88***	***15.89***	***15.89***	***15.90***	***15.90***	***15.91***	***15.91***	***15.92***	***15.92***	***15.94***	***15.95***	***15.96***
	7.69	7.69	7.69	7.69	7.69	7.69	7.69	7.69	7.69	7.69	7.69	7.69	7.69	7.69
	0.894	0.898	0.904	0.909	0.914	0.919	0.923	0.931	0.938	0.944	0.949	0.963	0.977	0.990
-0.40	***9.58***	***9.59***	***9.59***	***9.60***	***9.60***	***9.61***	***9.61***	***9.62***	***9.63***	***9.63***	***9.64***	***9.65***	***9.67***	***9.68***
	4.43	4.43	4.43	4.43	4.43	4.43	4.43	4.43	4.43	4.43	4.43	4.43	4.43	4.43
	0.882	0.886	0.893	0.899	0.904	0.909	0.914	0.923	0.931	0.937	0.943	0.958	0.973	0.987
-0.30	***6.32***	***6.32***	***6.33***	***6.33***	***6.34***	***6.34***	***6.35***	***6.36***	***6.36***	***6.37***	***6.38***	***6.39***	***6.41***	***6.42***
	2.65	2.65	2.65	2.65	2.65	2.65	2.65	2.65	2.65	2.65	2.65	2.65	2.65	2.65
	0.868	0.872	0.879	0.886	0.892	0.898	0.903	0.913	0.922	0.929	0.935	0.952	0.969	0.984
-0.20	***4.51***	***4.52***	***4.53***	***4.53***	***4.54***	***4.54***	***4.55***	***4.56***	***4.57***	***4.58***	***4.58***	***4.60***	***4.62***	***4.63***
	1.62	1.62	1.62	1.62	1.62	1.62	1.62	1.62	1.62	1.62	1.62	1.62	1.62	1.62
	0.850	0.855	0.864	0.871	0.878	0.884	0.890	0.902	0.911	0.919	0.926	0.945	0.964	0.981
-0.10	***3.47***	***3.47***	***3.48***	***3.49***	***3.50***	***3.50***	***3.51***	***3.52***	***3.53***	***3.54***	***3.54***	***3.56***	***3.58***	***3.60***
	1.000	1.000	1.000	1.000	1.000	1.000	1.000	1.000	1.000	1.000	1.000	1.000	1.000	1.000
	0.830	0.835	0.845	0.854	0.861	0.868	0.875	0.888	0.899	0.908	0.915	0.936	0.958	0.977
0.00	***2.830***	***2.835***	***2.845***	***2.854***	***2.861***	***2.868***	***2.875***	***2.888***	***2.899***	***2.908***	***2.915***	***2.936***	***2.958***	***2.977***
	0.618	0.618	0.618	0.618	0.618	0.618	0.618	0.618	0.618	0.618	0.618	0.618	0.618	0.618
	0.806	0.812	0.823	0.833	0.841	0.849	0.856	0.872	0.884	0.894	0.903	0.927	0.951	0.972
0.10	***2.425***	***2.431***	***2.441***	***2.451***	***2.460***	***2.468***	***2.475***	***2.490***	***2.502***	***2.512***	***2.521***	***2.545***	***2.570***	***2.590***
	0.378	0.378	0.378	0.378	0.378	0.378	0.378	0.378	0.378	0.378	0.378	0.378	0.378	0.378
	0.778	0.785	0.797	0.808	0.818	0.827	0.835	0.852	0.866	0.878	0.888	0.915	0.943	0.967
0.20	***2.156***	***2.163***	***2.175***	***2.186***	***2.196***	***2.205***	***2.213***	***2.230***	***2.244***	***2.256***	***2.266***	***2.293***	***2.321***	***2.345***
	0.226	0.226	0.226	0.226	0.226	0.226	0.226	0.226	0.226	0.226	0.226	0.226	0.226	0.226
	0.746	0.754	0.768	0.780	0.791	0.801	0.811	0.830	0.846	0.859	0.871	0.902	0.934	0.961
0.30	***1.972***	***1.979***	***1.993***	***2.006***	***2.017***	***2.027***	***2.036***	***2.056***	***2.072***	***2.085***	***2.096***	***2.128***	***2.160***	***2.187***
	0.130	0.130	0.130	0.130	0.130	0.130	0.130	0.130	0.130	0.130	0.130	0.130	0.130	0.130
	0.710	0.718	0.734	0.748	0.760	0.772	0.782	0.805	0.823	0.838	0.851	0.887	0.924	0.954
0.40	***1.840***	***1.848***	***1.864***	***1.878***	***1.890***	***1.902***	***1.912***	***1.935***	***1.953***	***1.968***	***1.981***	***2.017***	***2.054***	***2.084***
	0.072	0.072	0.072	0.072	0.072	0.072	0.072	0.072	0.072	0.072	0.072	0.072	0.072	0.072
	0.669	0.679	0.696	0.712	0.726	0.739	0.751	0.776	0.797	0.814	0.829	0.870	0.912	0.947
0.50	***1.741***	***1.750***	***1.768***	***1.784***	***1.798***	***1.811***	***1.823***	***1.848***	***1.869***	***1.886***	***1.900***	***1.941***	***1.984***	***2.018***
	0.037	0.037	0.037	0.037	0.037	0.037	0.037	0.037	0.037	0.037	0.037	0.037	0.037	0.037
	0.625	0.636	0.656	0.673	0.689	0.704	0.717	0.745	0.769	0.788	0.805	0.851	0.899	0.939
0.60	***1.663***	***1.673***	***1.693***	***1.710***	***1.726***	***1.741***	***1.754***	***1.783***	***1.806***	***1.825***	***1.842***	***1.888***	***1.936***	***1.976***
	0.018	0.018	0.018	0.018	0.018	0.018	0.018	0.018	0.018	0.018	0.018	0.018	0.018	0.018
	0.580	0.591	0.613	0.632	0.650	0.666	0.681	0.712	0.739	0.760	0.779	0.831	0.885	0.930
0.70	***1.598***	***1.610***	***1.631***	***1.651***	***1.668***	***1.684***	***1.699***	***1.731***	***1.757***	***1.778***	***1.797***	***1.849***	***1.903***	***1.948***
	0.008	0.008	0.008	0.008	0.008	0.008	0.008	0.008	0.008	0.008	0.008	0.008	0.008	0.008
	0.533	0.546	0.569	0.590	0.610	0.627	0.644	0.679	0.707	0.731	0.752	0.810	0.871	0.921
0.80	***1.541***	***1.554***	***1.578***	***1.599***	***1.618***	***1.636***	***1.652***	***1.687***	***1.716***	***1.740***	***1.760***	***1.818***	***1.879***	***1.929***
	0.003	0.003	0.003	0.003	0.003	0.003	0.003	0.003	0.003	0.003	0.003	0.003	0.003	0.003
	0.487	0.500	0.526	0.548	0.569	0.588	0.606	0.644	0.676	0.702	0.725	0.789	0.856	0.912
0.90	***1.490***	***1.504***	***1.529***	***1.552***	***1.573***	***1.592***	***1.610***	***1.648***	***1.679***	***1.706***	***1.728***	***1.792***	***1.859***	***1.915***
	0.001	0.001	0.001	0.001	0.001	0.001	0.001	0.001	0.001	0.001	0.001	0.001	0.001	0.001
	0.442	0.456	0.483	0.507	0.529	0.550	0.569	0.610	0.644	0.673	0.697	0.767	0.841	0.903
1.00	***1.443***	***1.457***	***1.484***	***1.508***	***1.531***	***1.551***	***1.570***	***1.611***	***1.645***	***1.674***	***1.698***	***1.768***	***1.842***	***1.904***
	0.000	0.000	0.000	0.000	0.000	0.000	0.000	0.000	0.000	0.000	0.000	0.000	0.000	0.000
	0.398	0.413	0.441	0.467	0.491	0.512	0.532	0.576	0.613	0.643	0.670	0.746	0.826	0.893
1.10	***1.399***	***1.414***	***1.442***	***1.467***	***1.491***	***1.513***	***1.533***	***1.577***	***1.613***	***1.644***	***1.670***	***1.746***	***1.826***	***1.894***

Table One: The *Excess Costs of Production, Excess Costs of Use,* and *Effective Cost of Production and Use* when *All* Nonconforming Units Are Scrapped

C_{pk} or P_{pk}	C_p or P_p value 1.50	1.60	1.70	1.80	1.90	2.00	2.10	2.20	2.30	2.40	2.50	2.60	2.70	2.80
	0.000	0.000	0.000	0.000	0.000	0.000	0.000	0.000	0.000	0.000	0.000	0.000	0.000	0.000
	0.089	0.106	0.125	0.145	0.166	0.188	0.209	0.229	0.250	0.269	0.288	0.306	0.324	0.341
1.20	***1.089***	***1.106***	***1.125***	***1.145***	***1.167***	***1.188***	***1.209***	***1.230***	***1.250***	***1.269***	***1.288***	***1.306***	***1.324***	***1.341***
	0.000	0.000	0.000	0.000	0.000	0.000	0.000	0.000	0.000	0.000	0.000	0.000	0.000	0.000
	0.067	0.079	0.094	0.111	0.130	0.150	0.170	0.190	0.210	0.229	0.248	0.266	0.284	0.301
1.30	***1.067***	***1.079***	***1.094***	***1.111***	***1.131***	***1.150***	***1.170***	***1.190***	***1.210***	***1.229***	***1.248***	***1.266***	***1.284***	***1.301***
	0.000	0.000	0.000	0.000	0.000	0.000	0.000	0.000	0.000	0.000	0.000	0.000	0.000	0.000
	0.054	0.059	0.070	0.084	0.100	0.118	0.136	0.155	0.174	0.193	0.211	0.229	0.247	0.264
1.40	***1.054***	***1.059***	***1.070***	***1.084***	***1.100***	***1.118***	***1.136***	***1.155***	***1.174***	***1.193***	***1.211***	***1.229***	***1.247***	***1.264***
	0.000	0.000	0.000	0.000	0.000	0.000	0.000	0.000	0.000	0.000	0.000	0.000	0.000	0.000
	0.049	0.047	0.052	0.062	0.075	0.090	0.107	0.124	0.142	0.160	0.178	0.195	0.213	0.230
1.50	***1.049***	***1.047***	***1.052***	***1.062***	***1.075***	***1.090***	***1.107***	***1.124***	***1.142***	***1.160***	***1.178***	***1.195***	***1.213***	***1.230***
		0.000	0.000	0.000	0.000	0.000	0.000	0.000	0.000	0.000	0.000	0.000	0.000	0.000
		0.043	0.042	0.047	0.056	0.068	0.082	0.097	0.114	0.130	0.147	0.164	0.181	0.198
1.60		***1.043***	***1.042***	***1.047***	***1.056***	***1.068***	***1.082***	***1.097***	***1.114***	***1.130***	***1.147***	***1.164***	***1.181***	***1.198***
			0.000	0.000	0.000	0.000	0.000	0.000	0.000	0.000	0.000	0.000	0.000	0.000
			0.038	0.037	0.042	0.050	0.061	0.075	0.089	0.104	0.120	0.136	0.152	0.169
1.70 »			***1.038***	***1.037***	***1.042***	***1.050***	***1.061***	***1.075***	***1.089***	***1.104***	***1.120***	***1.136***	***1.152***	***1.169***
				0.000	0.000	0.000	0.000	0.000	0.000	0.000	0.000	0.000	0.000	0.000
				0.034	0.034	0.038	0.046	0.056	0.068	0.082	0.096	0.111	0.126	0.142
1.80 »				***1.034***	***1.034***	***1.038***	***1.046***	***1.056***	***1.068***	***1.082***	***1.096***	***1.111***	***1.126***	***1.142***
					0.000	0.000	0.000	0.000	0.000	0.000	0.000	0.000	0.000	0.000
					0.031	0.030	0.034	0.042	0.051	0.063	0.075	0.089	0.103	0.117
1.90 »					***1.031***	***1.030***	***1.034***	***1.042***	***1.051***	***1.063***	***1.075***	***1.089***	***1.103***	***1.117***
						0.000	0.000	0.000	0.000	0.000	0.000	0.000	0.000	0.000
						0.028	0.027	0.031	0.038	0.047	0.058	0.070	0.082	0.096
2.00 »						***1.028***	***1.027***	***1.031***	***1.038***	***1.047***	***1.058***	***1.070***	***1.082***	***1.096***
							0.000	0.000	0.000	0.000	0.000	0.000	0.000	0.000
							0.025	0.025	0.029	0.035	0.043	0.053	0.065	0.077
2.10 »							***1.025***	***1.025***	***1.029***	***1.035***	***1.043***	***1.053***	***1.065***	***1.077***
								0.000	0.000	0.000	0.000	0.000	0.000	0.000
								0.023	0.023	0.026	0.032	0.040	0.050	0.060
2.20 »								***1.023***	***1.023***	***1.026***	***1.032***	***1.040***	***1.050***	***1.060***
									0.000	0.000	0.000	0.000	0.000	0.000
									0.021	0.021	0.024	0.030	0.037	0.046
2.30 »									***1.021***	***1.021***	***1.024***	***1.030***	***1.037***	***1.046***
										0.000	0.000	0.000	0.000	0.000
										0.019	0.019	0.022	0.028	0.035
2.40 »										***1.019***	***1.019***	***1.022***	***1.028***	***1.035***
											0.000	0.000	0.000	0.000
											0.018	0.018	0.021	0.026
2.50 »											***1.018***	***1.018***	***1.021***	***1.026***
												0.000	0.000	0.000
												0.016	0.017	0.019
2.60 »												***1.016***	***1.017***	***1.019***
													0.000	0.000
													0.015	0.015
2.70 »													***1.015***	***1.015***
														0.000
														0.014
2.80 »														***1.014***

Table One: The *Excess Costs of Production, Excess Costs of Use,* and *Effective Cost of Production and Use* when *All* Nonconforming Units Are Scrapped

C_{pk} or P_{pk}	C_p or P_p *value*													
	2.90	*3.00*	*3.20*	*3.40*	*3.60*	*3.80*	*4.00*	*4.50*	*5.00*	*5.50*	*6.0*	*8.0*	*12.0*	*20.0*
	0.000	0.000	0.000	0.000	0.000	0.000	0.000	0.000	0.000	0.000	0.000	0.000	0.000	0.000
	0.357	0.372	0.401	0.428	0.453	0.476	0.497	0.543	0.582	0.615	0.643	0.724	0.811	0.884
1.20	***1.357***	***1.372***	***1.402***	***1.428***	***1.453***	***1.476***	***1.497***	***1.543***	***1.582***	***1.615***	***1.643***	***1.724***	***1.811***	***1.884***
	0.000	0.000	0.000	0.000	0.000	0.000	0.000	0.000	0.000	0.000	0.000	0.000	0.000	0.000
	0.318	0.333	0.363	0.391	0.417	0.440	0.463	0.511	0.552	0.587	0.617	0.703	0.796	0.874
1.30	***1.318***	***1.333***	***1.363***	***1.391***	***1.417***	***1.441***	***1.463***	***1.511***	***1.552***	***1.587***	***1.617***	***1.703***	***1.796***	***1.875***
	0.000	0.000	0.000	0.000	0.000	0.000	0.000	0.000	0.000	0.000	0.000	0.000	0.000	0.000
	0.281	0.297	0.327	0.356	0.382	0.407	0.429	0.480	0.523	0.559	0.591	0.682	0.781	0.865
1.40	***1.281***	***1.297***	***1.327***	***1.356***	***1.382***	***1.407***	***1.429***	***1.480***	***1.523***	***1.559***	***1.591***	***1.682***	***1.781***	***1.865***
	0.000	0.000	0.000	0.000	0.000	0.000	0.000	0.000	0.000	0.000	0.000	0.000	0.000	0.000
	0.246	0.262	0.293	0.322	0.349	0.374	0.398	0.450	0.494	0.533	0.566	0.662	0.766	0.856
1.50	***1.246***	***1.262***	***1.293***	***1.322***	***1.349***	***1.374***	***1.398***	***1.450***	***1.494***	***1.533***	***1.566***	***1.662***	***1.766***	***1.856***
	0.000	0.000	0.000	0.000	0.000	0.000	0.000	0.000	0.000	0.000	0.000	0.000	0.000	0.000
	0.214	0.230	0.261	0.290	0.317	0.343	0.367	0.421	0.467	0.506	0.541	0.642	0.752	0.847
1.60	***1.214***	***1.230***	***1.261***	***1.290***	***1.317***	***1.343***	***1.367***	***1.421***	***1.467***	***1.506***	***1.541***	***1.642***	***1.752***	***1.847***
	0.000	0.000	0.000	0.000	0.000	0.000	0.000	0.000	0.000	0.000	0.000	0.000	0.000	0.000
	0.184	0.200	0.231	0.260	0.287	0.313	0.338	0.393	0.440	0.481	0.517	0.622	0.738	0.838
1.70	***1.184***	***1.200***	***1.231***	***1.260***	***1.287***	***1.313***	***1.338***	***1.393***	***1.440***	***1.481***	***1.517***	***1.622***	***1.738***	***1.838***
	0.000	0.000	0.000	0.000	0.000	0.000	0.000	0.000	0.000	0.000	0.000	0.000	0.000	0.000
	0.157	0.172	0.202	0.231	0.259	0.285	0.309	0.365	0.414	0.456	0.493	0.602	0.723	0.828
1.80	***1.157***	***1.172***	***1.202***	***1.231***	***1.259***	***1.285***	***1.309***	***1.365***	***1.414***	***1.456***	***1.493***	***1.602***	***1.723***	***1.828***
	0.000	0.000	0.000	0.000	0.000	0.000	0.000	0.000	0.000	0.000	0.000	0.000	0.000	0.000
	0.132	0.147	0.176	0.204	0.232	0.258	0.283	0.339	0.389	0.432	0.470	0.583	0.709	0.819
1.90	***1.132***	***1.147***	***1.176***	***1.204***	***1.232***	***1.258***	***1.283***	***1.339***	***1.389***	***1.432***	***1.470***	***1.583***	***1.709***	***1.819***
	0.000	0.000	0.000	0.000	0.000	0.000	0.000	0.000	0.000	0.000	0.000	0.000	0.000	0.000
	0.110	0.123	0.151	0.179	0.206	0.232	0.257	0.314	0.364	0.409	0.448	0.564	0.695	0.810
2.00	***1.110***	***1.123***	***1.151***	***1.179***	***1.206***	***1.232***	***1.257***	***1.314***	***1.364***	***1.409***	***1.448***	***1.564***	***1.695***	***1.810***
	0.000	0.000	0.000	0.000	0.000	0.000	0.000	0.000	0.000	0.000	0.000	0.000	0.000	0.000
	0.089	0.102	0.129	0.156	0.182	0.208	0.233	0.290	0.341	0.386	0.426	0.546	0.681	0.801
2.10	***1.089***	***1.102***	***1.129***	***1.156***	***1.182***	***1.208***	***1.233***	***1.290***	***1.341***	***1.386***	***1.426***	***1.546***	***1.681***	***1.801***
	0.000	0.000	0.000	0.000	0.000	0.000	0.000	0.000	0.000	0.000	0.000	0.000	0.000	0.000
	0.071	0.083	0.109	0.134	0.160	0.185	0.209	0.267	0.318	0.364	0.404	0.527	0.668	0.792
2.20	***1.071***	***1.083***	***1.109***	***1.134***	***1.160***	***1.185***	***1.209***	***1.267***	***1.318***	***1.364***	***1.404***	***1.527***	***1.668***	***1.792***
	0.000	0.000	0.000	0.000	0.000	0.000	0.000	0.000	0.000	0.000	0.000	0.000	0.000	0.000
	0.056	0.067	0.090	0.114	0.139	0.164	0.188	0.244	0.296	0.342	0.383	0.509	0.654	0.784
2.30	***1.056***	***1.067***	***1.090***	***1.114***	***1.139***	***1.164***	***1.188***	***1.244***	***1.296***	***1.342***	***1.383***	***1.509***	***1.654***	***1.784***
	0.000	0.000	0.000	0.000	0.000	0.000	0.000	0.000	0.000	0.000	0.000	0.000	0.000	0.000
	0.043	0.052	0.073	0.096	0.120	0.143	0.167	0.223	0.275	0.321	0.363	0.492	0.641	0.775
2.40	***1.043***	***1.052***	***1.073***	***1.096***	***1.120***	***1.143***	***1.167***	***1.223***	***1.275***	***1.321***	***1.363***	***1.492***	***1.641***	***1.775***
	0.000	0.000	0.000	0.000	0.000	0.000	0.000	0.000	0.000	0.000	0.000	0.000	0.000	0.000
	0.032	0.040	0.059	0.080	0.102	0.125	0.148	0.203	0.254	0.301	0.343	0.474	0.628	0.766
2.50	***1.032***	***1.040***	***1.059***	***1.080***	***1.102***	***1.125***	***1.148***	***1.203***	***1.254***	***1.301***	***1.343***	***1.474***	***1.628***	***1.766***
	0.000	0.000	0.000	0.000	0.000	0.000	0.000	0.000	0.000	0.000	0.000	0.000	0.000	0.000
	0.024	0.030	0.046	0.065	0.086	0.107	0.129	0.184	0.235	0.282	0.324	0.457	0.614	0.757
2.60	***1.024***	***1.030***	***1.046***	***1.065***	***1.086***	***1.107***	***1.129***	***1.184***	***1.235***	***1.282***	***1.324***	***1.457***	***1.614***	***1.757***
	0.000	0.000	0.000	0.000	0.000	0.000	0.000	0.000	0.000	0.000	0.000	0.000	0.000	0.000
	0.018	0.022	0.035	0.052	0.071	0.091	0.113	0.165	0.216	0.263	0.306	0.441	0.601	0.749
2.70	***1.018***	***1.022***	***1.035***	***1.052***	***1.071***	***1.091***	***1.113***	***1.165***	***1.216***	***1.263***	***1.306***	***1.441***	***1.601***	***1.749***
	0.000	0.000	0.000	0.000	0.000	0.000	0.000	0.000	0.000	0.000	0.000	0.000	0.000	0.000
	0.014	0.017	0.026	0.041	0.058	0.077	0.097	0.148	0.198	0.245	0.288	0.424	0.589	0.740
2.80	***1.014***	***1.017***	***1.026***	***1.041***	***1.058***	***1.077***	***1.097***	***1.148***	***1.198***	***1.245***	***1.288***	***1.424***	***1.589***	***1.740***

Table One: The *Excess Costs of Production, Excess Costs of Use,* and *Effective Cost of Production and Use* when *All* Nonconforming Units Are Scrapped

C_{pk} or P_{pk}	*C_p or P_p value* 2.90	3.00	3.20	3.40	3.60	3.80	4.00	4.50	5.00	5.50	6.0	8.0	12.0	20.0
	0.000	0.000	0.000	0.000	0.000	0.000	0.000	0.000	0.000	0.000	0.000	0.000	0.000	0.000
	0.013	0.013	0.020	0.031	0.046	0.064	0.083	0.132	0.181	0.227	0.270	0.408	0.576	0.731
2.90	***1.013***	***1.013***	***1.020***	***1.031***	***1.046***	***1.064***	***1.083***	***1.132***	***1.181***	***1.227***	***1.270***	***1.408***	***1.576***	***1.731***
		0.000	0.000	0.000	0.000	0.000	0.000	0.000	0.000	0.000	0.000	0.000	0.000	0.000
		0.012	0.015	0.023	0.036	0.052	0.069	0.117	0.164	0.210	0.253	0.392	0.563	0.723
3.00		***1.012***	***1.015***	***1.023***	***1.036***	***1.052***	***1.069***	***1.117***	***1.164***	***1.210***	***1.253***	***1.392***	***1.563***	***1.723***
			0.000	0.000	0.000	0.000	0.000	0.000	0.000	0.000	0.000	0.000	0.000	0.000
			0.011	0.013	0.021	0.033	0.047	0.089	0.134	0.179	0.221	0.362	0.539	0.706
3.20 »			***1.011***	***1.013***	***1.021***	***1.033***	***1.047***	***1.089***	***1.134***	***1.179***	***1.221***	***1.362***	***1.539***	***1.706***
				0.000	0.000	0.000	0.000	0.000	0.000	0.000	0.000	0.000	0.000	0.000
				0.010	0.012	0.019	0.029	0.065	0.107	0.149	0.191	0.332	0.514	0.689
3.40 »				***1.010***	***1.012***	***1.019***	***1.029***	***1.065***	***1.107***	***1.149***	***1.191***	***1.332***	***1.514***	***1.689***
					0.000	0.000	0.000	0.000	0.000	0.000	0.000	0.000	0.000	0.000
					0.009	0.010	0.017	0.045	0.083	0.123	0.163	0.304	0.491	0.673
3.60 »					***1.009***	***1.010***	***1.017***	***1.045***	***1.083***	***1.123***	***1.163***	***1.304***	***1.491***	***1.673***
						0.000	0.000	0.000	0.000	0.000	0.000	0.000	0.000	0.000
						0.008	0.009	0.030	0.062	0.099	0.138	0.277	0.468	0.656
3.80 »						***1.008***	***1.009***	***1.030***	***1.062***	***1.099***	***1.138***	***1.277***	***1.468***	***1.656***
							0.000	0.000	0.000	0.000	0.000	0.000	0.000	0.000
							0.007	0.018	0.044	0.078	0.114	0.252	0.445	0.640
4.00 »							***1.007***	***1.018***	***1.044***	***1.078***	***1.114***	***1.252***	***1.445***	***1.640***
								0.000	0.000	0.000	0.000	0.000	0.000	0.000
								0.005	0.014	0.037	0.066	0.193	0.391	0.601
4.50 »								***1.005***	***1.014***	***1.037***	***1.066***	***1.193***	***1.391***	***1.601***
									0.000	0.000	0.000	0.000	0.000	0.000
									0.004	0.012	0.031	0.142	0.341	0.563
5.00 »									***1.004***	***1.012***	***1.031***	***1.142***	***1.341***	***1.563***
										0.000	0.000	0.000	0.000	0.000
										0.004	0.010	0.099	0.294	0.526
5.50 »										***1.004***	***1.010***	***1.099***	***1.294***	***1.526***
											0.000	0.000	0.000	0.000
											0.003	0.064	0.251	0.490
6.00 »											***1.003***	***1.064***	***1.251***	***1.490***
												0.000	0.000	0.000
												0.002	0.112	0.360
8.00 »												***1.002***	***1.112***	***1.360***
													0.000	0.000
													0.001	0.160
12.0 »													***1.001***	***1.160***
														0.000
														0.000
20.0 »														***1.000***

Table Two: The *Excess Costs of Production, Excess Costs of Use,* and *Effective Cost of Production and Use* when *All* Nonconforming Units Are Reworked

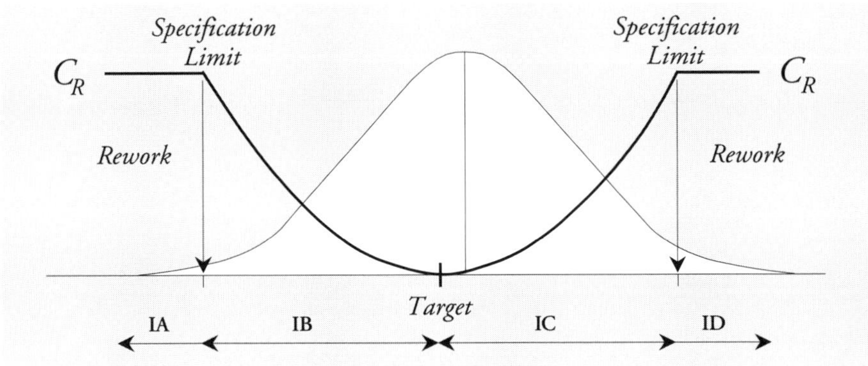

Table Two: Excess Cost Function when All Nonconforming Items Are Reworked

Corresponding to each pairing for C_p and C_{pk} there are three entries stacked in a column.

The topmost of these is the *Excess Costs of Production,* *ExCP*

The middle value is the *Excess Costs of Use,* *ExCU*

And the bottom value (in bold) is the *Effective Cost of Production and Use,* ***ECP&U***

The entries in Table Two are for the case where all nonconforming product is reworked and the Cost of Rework, C_R, is assumed to be equal to the Nominal Cost of Production, *NC*. Table A.9 contains the normal theory values for **IA**, **IB**, **IC**, and **ID**. The general formula for Table Two is:

$$ECP\&U = 1.0 + \frac{C_R}{NC}\{\mathbf{IA} + \mathbf{ID}\} + \frac{C_R}{NC}\{\mathbf{IB} + \mathbf{IC}\}$$

$$= 1.0 + \textit{Excess Costs of Production} + \textit{Excess Costs of Use}$$

Since the Cost of Rework will generally be less than *NC*, the entries in Table Two are essentially *upper bounds* for all three quantities. If greater precision is required, the entries for the *Excess Costs of Production* and the *Excess Costs of Use* may be deflated by the ratio of C_R to *NC*, as shown in the formula above, and added to 1.00 to get a customized value for the *Effective Cost of Production and Use.*

For example, with $C_p = 0.9$ and $C_{pk} = 0.5$, and with $C_R = 0.50\ NC$ we would convert the tabled value of 1.310 for *ECP&U* into the value 1.155. The *Excess Cost of Production* would be 0.034, while the *Excess Cost of Use* would be 0.121,

Table Two: The *Excess Costs of Production, Excess Costs of Use,* and *Effective Cost of Production and Use* when *All* Nonconforming Units Are Reworked

C_{pk} or P_{pk}	C_p or P_p values 0.10	0.20	0.30	0.40	0.50	0.60	0.70	0.80	0.90	1.00	1.10	1.20	1.30	1.40
	0.951	0.937	0.934	0.933	0.933	0.933	0.933	0.933	0.933	0.933	0.933	0.933	0.933	0.933
	0.017	0.024	0.029	0.034	0.038	0.041	0.044	0.046	0.048	0.050	0.051	0.052	0.053	0.054
–0.50	***1.968***	***1.961***	***1.963***	***1.967***	***1.971***	***1.975***	***1.977***	***1.980***	***1.981***	***1.983***	***1.984***	***1.986***	***1.986***	***1.987***
	0.921	0.893	0.886	0.885	0.885	0.885	0.885	0.885	0.885	0.885	0.885	0.885	0.885	0.885
	0.027	0.039	0.047	0.054	0.061	0.067	0.072	0.077	0.080	0.083	0.085	0.088	0.089	0.091
–0.40	***1.948***	***1.932***	***1.933***	***1.940***	***1.946***	***1.952***	***1.957***	***1.962***	***1.965***	***1.968***	***1.970***	***1.973***	***1.974***	***1.976***
	0.883	0.834	0.819	0.816	0.816	0.816	0.816	0.816	0.816	0.816	0.816	0.816	0.816	0.816
	0.039	0.058	0.070	0.081	0.092	0.101	0.110	0.117	0.123	0.128	0.132	0.136	0.139	0.142
–0.30	***1.922***	***1.892***	***1.889***	***1.897***	***1.908***	***1.917***	***1.926***	***1.933***	***1.938***	***1.943***	***1.948***	***1.951***	***1.955***	***1.958***
	0.841	0.762	0.734	0.727	0.726	0.726	0.726	0.726	0.726	0.726	0.726	0.726	0.726	0.726
	0.053	0.081	0.097	0.111	0.126	0.140	0.153	0.164	0.173	0.181	0.188	0.194	0.199	0.204
–0.20	***1.894***	***1.842***	***1.831***	***1.838***	***1.852***	***1.866***	***1.879***	***1.890***	***1.899***	***1.907***	***1.914***	***1.920***	***1.925***	***1.930***
	0.802	0.685	0.636	0.621	0.618	0.618	0.618	0.618	0.618	0.618	0.618	0.618	0.618	0.618
	0.066	0.104	0.125	0.142	0.160	0.179	0.197	0.212	0.226	0.238	0.248	0.257	0.265	0.272
–0.10	***1.867***	***1.789***	***1.760***	***1.763***	***1.778***	***1.797***	***1.814***	***1.830***	***1.844***	***1.856***	***1.866***	***1.875***	***1.883***	***1.890***
	0.774	0.615	0.536	0.508	0.501	0.500	0.500	0.500	0.500	0.500	0.500	0.500	0.500	0.500
	0.074	0.124	0.150	0.168	0.188	0.211	0.233	0.254	0.273	0.290	0.304	0.317	0.328	0.338
0.00	***1.849***	***1.739***	***1.686***	***1.677***	***1.690***	***1.711***	***1.733***	***1.754***	***1.773***	***1.790***	***1.804***	***1.817***	***1.828***	***1.838***
	0.764	0.566	0.449	0.400	0.386	0.383	0.382	0.382	0.382	0.382	0.382	0.382	0.382	0.382
	0.078	0.138	0.171	0.189	0.207	0.231	0.257	0.282	0.306	0.328	0.347	0.364	0.379	0.393
0.10	***1.842***	***1.705***	***1.620***	***1.589***	***1.593***	***1.613***	***1.639***	***1.665***	***1.688***	***1.710***	***1.729***	***1.746***	***1.761***	***1.775***
		0.549	0.389	0.310	0.282	0.276	0.274	0.274	0.274	0.274	0.274	0.274	0.274	0.274
		0.143	0.184	0.202	0.216	0.236	0.263	0.291	0.319	0.345	0.369	0.390	0.410	0.427
0.20		***1.692***	***1.574***	***1.512***	***1.498***	***1.512***	***1.537***	***1.566***	***1.593***	***1.619***	***1.643***	***1.665***	***1.684***	***1.701***
			0.368	0.251	0.202	0.188	0.185	0.184	0.184	0.184	0.184	0.184	0.184	0.184
			0.189	0.209	0.216	0.229	0.252	0.280	0.310	0.340	0.367	0.393	0.416	0.437
	0.30 »		***1.557***	***1.460***	***1.418***	***1.417***	***1.436***	***1.464***	***1.494***	***1.524***	***1.551***	***1.577***	***1.600***	***1.621***
				0.230	0.151	0.123	0.116	0.115	0.115	0.115	0.115	0.115	0.115	0.115
				0.211	0.214	0.216	0.229	0.253	0.283	0.313	0.343	0.372	0.398	0.422
		0.40 »		***1.441***	***1.365***	***1.339***	***1.346***	***1.369***	***1.398***	***1.428***	***1.458***	***1.487***	***1.513***	***1.538***
					0.134	0.085	0.070	0.067	0.067	0.067	0.067	0.067	0.067	0.067
					0.212	0.201	0.204	0.218	0.243	0.272	0.303	0.333	0.361	0.388
			0.50 »		***1.346***	***1.288***	***1.274***	***1.286***	***1.310***	***1.339***	***1.370***	***1.399***	***1.428***	***1.455***
						0.072	0.044	0.037	0.036	0.036	0.036	0.036	0.036	0.036
						0.199	0.184	0.185	0.200	0.225	0.253	0.283	0.312	0.340
				0.60 »		***1.271***	***1.228***	***1.222***	***1.236***	***1.260***	***1.289***	***1.318***	***1.348***	***1.376***
							0.036	0.021	0.018	0.018	0.018	0.018	0.018	0.018
							0.177	0.161	0.163	0.178	0.202	0.229	0.257	0.286
					0.70 »		***1.213***	***1.182***	***1.181***	***1.196***	***1.220***	***1.247***	***1.275***	***1.303***
								0.016	0.010	0.008	0.008	0.008	0.008	0.008
								0.152	0.137	0.141	0.156	0.178	0.204	0.231
						0.80 »		***1.169***	***1.147***	***1.149***	***1.164***	***1.187***	***1.212***	***1.239***
									0.007	0.004	0.004	0.003	0.003	0.003
									0.129	0.116	0.121	0.136	0.156	0.180
							0.90 »		***1.135***	***1.120***	***1.124***	***1.139***	***1.160***	***1.184***
										0.003	0.002	0.001	0.001	0.001
										0.108	0.098	0.103	0.117	0.137
								1.00 »		***1.111***	***1.100***	***1.105***	***1.119***	***1.138***
											0.001	0.001	0.000	0.000
											0.091	0.083	0.089	0.102
									1.10 »		***1.092***	***1.084***	***1.089***	***1.103***

C_{pk}	C_p 1.20	1.30	1.40
	0.000	0.000	0.000
	0.077	0.071	0.077
1.20	***1.077***	***1.072***	***1.077***
		0.000	0.000
		0.066	0.062
1.30		***1.066***	***1.062***
			0.000
			0.057
1.40			***1.057***

Table Two: The *Excess Costs of Production, Excess Costs of Use,* and *Effective Cost of Production and Use* when *All* Nonconforming Units Are Reworked

C_{pk} or P_{pk}	C_p or P_p value 1.50	1.60	1.70	1.80	1.90	2.00	2.10	2.20	2.30	2.40	2.50	2.60	2.70	2.80
	0.933	0.933	0.933	0.933	0.933	0.933	0.933	0.933	0.933	0.933	0.933	0.933	0.933	0.933
	0.055	0.056	0.056	0.057	0.057	0.058	0.058	0.059	0.059	0.059	0.059	0.060	0.060	0.060
–0.50	***1.988***	***1.989***	***1.989***	***1.990***	***1.990***	***1.991***	***1.991***	***1.992***	***1.992***	***1.992***	***1.993***	***1.993***	***1.993***	***1.993***
	0.885	0.885	0.885	0.885	0.885	0.885	0.885	0.885	0.885	0.885	0.885	0.885	0.885	0.885
	0.093	0.094	0.095	0.096	0.097	0.098	0.099	0.099	0.100	0.100	0.101	0.102	0.102	0.102
–0.40	***1.977***	***1.979***	***1.980***	***1.981***	***1.982***	***1.983***	***1.983***	***1.984***	***1.985***	***1.985***	***1.986***	***1.986***	***1.987***	***1.987***
	0.816	0.816	0.816	0.816	0.816	0.816	0.816	0.816	0.816	0.816	0.816	0.816	0.816	0.816
	0.144	0.146	0.148	0.150	0.152	0.153	0.155	0.156	0.157	0.158	0.159	0.160	0.161	0.162
–0.30	***1.960***	***1.962***	***1.964***	***1.966***	***1.968***	***1.969***	***1.971***	***1.972***	***1.973***	***1.974***	***1.975***	***1.976***	***1.977***	***1.978***
	0.726	0.726	0.726	0.726	0.726	0.726	0.726	0.726	0.726	0.726	0.726	0.726	0.726	0.726
	0.208	0.212	0.215	0.218	0.221	0.223	0.225	0.227	0.229	0.231	0.233	0.234	0.235	0.237
–0.20	***1.934***	***1.937***	***1.941***	***1.944***	***1.946***	***1.949***	***1.951***	***1.953***	***1.955***	***1.957***	***1.958***	***1.960***	***1.961***	***1.962***
	0.618	0.618	0.618	0.618	0.618	0.618	0.618	0.618	0.618	0.618	0.618	0.618	0.618	0.618
	0.279	0.284	0.289	0.294	0.298	0.302	0.305	0.308	0.311	0.314	0.317	0.319	0.321	0.323
–0.10	***1.896***	***1.902***	***1.907***	***1.912***	***1.916***	***1.920***	***1.923***	***1.926***	***1.929***	***1.932***	***1.934***	***1.937***	***1.939***	***1.941***
	0.500	0.500	0.500	0.500	0.500	0.500	0.500	0.500	0.500	0.500	0.500	0.500	0.500	0.500
	0.347	0.356	0.363	0.370	0.376	0.381	0.386	0.391	0.395	0.399	0.403	0.406	0.409	0.412
0.00	***1.847***	***1.856***	***1.863***	***1.870***	***1.876***	***1.881***	***1.886***	***1.891***	***1.895***	***1.899***	***1.903***	***1.906***	***1.909***	***1.912***
	0.382	0.382	0.382	0.382	0.382	0.382	0.382	0.382	0.382	0.382	0.382	0.382	0.382	0.382
	0.405	0.416	0.426	0.435	0.443	0.451	0.458	0.464	0.470	0.476	0.481	0.486	0.490	0.494
0.10	***1.787***	***1.798***	***1.808***	***1.817***	***1.826***	***1.833***	***1.840***	***1.847***	***1.852***	***1.858***	***1.863***	***1.868***	***1.872***	***1.876***
	0.274	0.274	0.274	0.274	0.274	0.274	0.274	0.274	0.274	0.274	0.274	0.274	0.274	0.274
	0.443	0.457	0.470	0.482	0.493	0.503	0.512	0.520	0.528	0.535	0.542	0.548	0.554	0.560
0.20	***1.717***	***1.731***	***1.744***	***1.756***	***1.767***	***1.777***	***1.786***	***1.794***	***1.802***	***1.810***	***1.816***	***1.823***	***1.828***	***1.834***
	0.184	0.184	0.184	0.184	0.184	0.184	0.184	0.184	0.184	0.184	0.184	0.184	0.184	0.184
	0.456	0.474	0.490	0.504	0.518	0.530	0.542	0.552	0.562	0.571	0.580	0.588	0.595	0.602
0.30	***1.640***	***1.658***	***1.674***	***1.688***	***1.702***	***1.714***	***1.726***	***1.736***	***1.746***	***1.755***	***1.764***	***1.772***	***1.779***	***1.786***
	0.115	0.115	0.115	0.115	0.115	0.115	0.115	0.115	0.115	0.115	0.115	0.115	0.115	0.115
	0.445	0.465	0.484	0.502	0.518	0.533	0.547	0.559	0.571	0.582	0.593	0.602	0.611	0.620
0.40	***1.560***	***1.581***	***1.599***	***1.617***	***1.633***	***1.648***	***1.662***	***1.674***	***1.686***	***1.697***	***1.708***	***1.717***	***1.726***	***1.735***
	0.067	0.067	0.067	0.067	0.067	0.067	0.067	0.067	0.067	0.067	0.067	0.067	0.067	0.067
	0.413	0.436	0.458	0.477	0.496	0.513	0.529	0.544	0.558	0.571	0.583	0.594	0.605	0.615
0.50	***1.480***	***1.503***	***1.524***	***1.544***	***1.563***	***1.580***	***1.596***	***1.611***	***1.625***	***1.638***	***1.650***	***1.661***	***1.672***	***1.682***
	0.036	0.036	0.036	0.036	0.036	0.036	0.036	0.036	0.036	0.036	0.036	0.036	0.036	0.036
	0.367	0.392	0.415	0.437	0.458	0.477	0.495	0.511	0.527	0.542	0.555	0.568	0.581	0.592
0.60	***1.403***	***1.428***	***1.451***	***1.473***	***1.494***	***1.513***	***1.531***	***1.547***	***1.563***	***1.578***	***1.591***	***1.604***	***1.617***	***1.628***
	0.018	0.018	0.018	0.018	0.018	0.018	0.018	0.018	0.018	0.018	0.018	0.018	0.018	0.018
	0.313	0.339	0.364	0.387	0.409	0.430	0.450	0.468	0.485	0.501	0.517	0.531	0.544	0.557
0.70	***1.331***	***1.357***	***1.382***	***1.405***	***1.427***	***1.448***	***1.467***	***1.486***	***1.503***	***1.519***	***1.534***	***1.549***	***1.562***	***1.575***
	0.008	0.008	0.008	0.008	0.008	0.008	0.008	0.008	0.008	0.008	0.008	0.008	0.008	0.008
	0.258	0.284	0.309	0.334	0.357	0.379	0.399	0.419	0.437	0.455	0.471	0.487	0.502	0.516
0.80	***1.266***	***1.292***	***1.318***	***1.342***	***1.365***	***1.387***	***1.408***	***1.427***	***1.446***	***1.463***	***1.479***	***1.495***	***1.510***	***1.524***
	0.003	0.003	0.003	0.003	0.003	0.003	0.003	0.003	0.003	0.003	0.003	0.003	0.003	0.003
	0.205	0.231	0.256	0.280	0.304	0.326	0.348	0.368	0.388	0.406	0.424	0.440	0.456	0.471
0.90	***1.209***	***1.234***	***1.259***	***1.284***	***1.307***	***1.330***	***1.351***	***1.372***	***1.391***	***1.410***	***1.427***	***1.444***	***1.459***	***1.474***
	0.001	0.001	0.001	0.001	0.001	0.001	0.001	0.001	0.001	0.001	0.001	0.001	0.001	0.001
	0.159	0.183	0.206	0.230	0.254	0.276	0.298	0.319	0.339	0.358	0.376	0.394	0.410	0.426
1.00	***1.160***	***1.184***	***1.208***	***1.232***	***1.255***	***1.278***	***1.299***	***1.320***	***1.340***	***1.359***	***1.378***	***1.395***	***1.412***	***1.427***
	0.000	0.000	0.000	0.000	0.000	0.000	0.000	0.000	0.000	0.000	0.000	0.000	0.000	0.000
	0.120	0.141	0.162	0.185	0.208	0.230	0.251	0.272	0.293	0.312	0.331	0.349	0.366	0.382
1.10	***1.120***	***1.141***	***1.163***	***1.185***	***1.208***	***1.230***	***1.252***	***1.273***	***1.293***	***1.313***	***1.331***	***1.349***	***1.366***	***1.383***

Table Two: The *Excess Costs of Production, Excess Costs of Use,* and *Effective Cost of Production and Use* when *All* Nonconforming Units Are Reworked

C_{pk} or P_{pk}	C_p or P_p value 2.90	3.00	3.20	3.40	3.60	3.80	4.00	4.50	5.00	5.50	6.0	8.0	12.0	20.0
	0.933	0.933	0.933	0.933	0.933	0.933	0.933	0.933	0.933	0.933	0.933	0.933	0.933	0.933
	0.060	0.061	0.061	0.061	0.062	0.062	0.062	0.063	0.063	0.063	0.064	0.065	0.065	0.066
-0.50	***1.994***	***1.994***	***1.994***	***1.995***	***1.995***	***1.995***	***1.995***	***1.996***	***1.996***	***1.997***	***1.997***	***1.998***	***1.999***	***2.000***
	0.885	0.885	0.885	0.885	0.885	0.885	0.885	0.885	0.885	0.885	0.885	0.885	0.885	0.885
	0.103	0.103	0.104	0.105	0.105	0.106	0.106	0.107	0.108	0.109	0.109	0.111	0.112	0.114
-0.40	***1.988***	***1.988***	***1.989***	***1.990***	***1.990***	***1.991***	***1.991***	***1.992***	***1.993***	***1.994***	***1.994***	***1.996***	***1.997***	***1.999***
	0.816	0.816	0.816	0.816	0.816	0.816	0.816	0.816	0.816	0.816	0.816	0.816	0.816	0.816
	0.162	0.163	0.164	0.165	0.166	0.167	0.168	0.170	0.171	0.173	0.173	0.176	0.179	0.182
-0.30	***1.978***	***1.979***	***1.980***	***1.981***	***1.982***	***1.983***	***1.984***	***1.986***	***1.987***	***1.988***	***1.989***	***1.992***	***1.995***	***1.998***
	0.726	0.726	0.726	0.726	0.726	0.726	0.726	0.726	0.726	0.726	0.726	0.726	0.726	0.726
	0.238	0.239	0.241	0.243	0.245	0.246	0.248	0.250	0.253	0.255	0.256	0.261	0.266	0.270
-0.20	***1.964***	***1.965***	***1.967***	***1.969***	***1.970***	***1.972***	***1.973***	***1.976***	***1.979***	***1.981***	***1.982***	***1.987***	***1.992***	***1.996***
	0.618	0.618	0.618	0.618	0.618	0.618	0.618	0.618	0.618	0.618	0.618	0.618	0.618	0.618
	0.325	0.327	0.330	0.333	0.336	0.338	0.340	0.345	0.348	0.351	0.354	0.361	0.368	0.375
-0.10	***1.943***	***1.945***	***1.948***	***1.951***	***1.953***	***1.956***	***1.958***	***1.962***	***1.966***	***1.969***	***1.972***	***1.979***	***1.986***	***1.993***
	0.500	0.500	0.500	0.500	0.500	0.500	0.500	0.500	0.500	0.500	0.500	0.500	0.500	0.500
	0.415	0.418	0.423	0.427	0.431	0.434	0.437	0.444	0.449	0.454	0.458	0.468	0.479	0.488
0.00	***1.915***	***1.918***	***1.923***	***1.927***	***1.931***	***1.934***	***1.937***	***1.944***	***1.949***	***1.954***	***1.958***	***1.968***	***1.979***	***1.988***
	0.382	0.382	0.382	0.382	0.382	0.382	0.382	0.382	0.382	0.382	0.382	0.382	0.382	0.382
	0.498	0.502	0.509	0.515	0.520	0.525	0.529	0.539	0.546	0.553	0.558	0.573	0.588	0.601
0.10	***1.880***	***1.884***	***1.891***	***1.897***	***1.902***	***1.907***	***1.911***	***1.921***	***1.928***	***1.935***	***1.940***	***1.955***	***1.970***	***1.983***
	0.274	0.274	0.274	0.274	0.274	0.274	0.274	0.274	0.274	0.274	0.274	0.274	0.274	0.274
	0.565	0.570	0.579	0.587	0.594	0.600	0.606	0.619	0.629	0.637	0.644	0.664	0.685	0.702
0.20	***1.839***	***1.844***	***1.853***	***1.861***	***1.868***	***1.875***	***1.880***	***1.893***	***1.903***	***1.912***	***1.919***	***1.939***	***1.959***	***1.976***
	0.184	0.184	0.184	0.184	0.184	0.184	0.184	0.184	0.184	0.184	0.184	0.184	0.184	0.184
	0.609	0.615	0.626	0.636	0.646	0.654	0.661	0.677	0.690	0.701	0.710	0.736	0.762	0.784
0.30	***1.793***	***1.799***	***1.810***	***1.821***	***1.830***	***1.838***	***1.845***	***1.861***	***1.874***	***1.885***	***1.894***	***1.920***	***1.946***	***1.968***
	0.115	0.115	0.115	0.115	0.115	0.115	0.115	0.115	0.115	0.115	0.115	0.115	0.115	0.115
	0.628	0.635	0.649	0.662	0.673	0.683	0.692	0.712	0.728	0.742	0.753	0.785	0.817	0.844
0.40	***1.743***	***1.751***	***1.764***	***1.777***	***1.788***	***1.798***	***1.807***	***1.827***	***1.843***	***1.857***	***1.868***	***1.900***	***1.932***	***1.960***
	0.067	0.067	0.067	0.067	0.067	0.067	0.067	0.067	0.067	0.067	0.067	0.067	0.067	0.067
	0.624	0.633	0.650	0.664	0.678	0.690	0.701	0.724	0.744	0.760	0.773	0.812	0.851	0.884
0.50	***1.691***	***1.700***	***1.716***	***1.731***	***1.745***	***1.757***	***1.768***	***1.791***	***1.811***	***1.827***	***1.840***	***1.878***	***1.918***	***1.950***
	0.036	0.036	0.036	0.036	0.036	0.036	0.036	0.036	0.036	0.036	0.036	0.036	0.036	0.036
	0.603	0.613	0.632	0.649	0.664	0.678	0.691	0.719	0.741	0.760	0.776	0.820	0.867	0.905
0.60	***1.639***	***1.649***	***1.668***	***1.685***	***1.700***	***1.714***	***1.727***	***1.754***	***1.777***	***1.796***	***1.812***	***1.856***	***1.903***	***1.941***
	0.018	0.018	0.018	0.018	0.018	0.018	0.018	0.018	0.018	0.018	0.018	0.018	0.018	0.018
	0.569	0.581	0.602	0.621	0.638	0.654	0.669	0.700	0.725	0.747	0.765	0.816	0.869	0.914
0.70	***1.587***	***1.599***	***1.620***	***1.639***	***1.656***	***1.672***	***1.686***	***1.718***	***1.743***	***1.765***	***1.783***	***1.834***	***1.887***	***1.931***
	0.008	0.008	0.008	0.008	0.008	0.008	0.008	0.008	0.008	0.008	0.008	0.008	0.008	0.008
	0.529	0.541	0.565	0.586	0.605	0.622	0.638	0.673	0.701	0.725	0.746	0.803	0.864	0.914
0.80	***1.537***	***1.550***	***1.573***	***1.594***	***1.613***	***1.630***	***1.646***	***1.681***	***1.710***	***1.734***	***1.754***	***1.812***	***1.872***	***1.922***
	0.003	0.003	0.003	0.003	0.003	0.003	0.003	0.003	0.003	0.003	0.003	0.003	0.003	0.003
	0.485	0.499	0.524	0.547	0.567	0.586	0.604	0.642	0.673	0.700	0.722	0.786	0.853	0.909
0.90	***1.489***	***1.502***	***1.527***	***1.550***	***1.571***	***1.590***	***1.607***	***1.645***	***1.677***	***1.703***	***1.725***	***1.789***	***1.856***	***1.912***
	0.001	0.001	0.001	0.001	0.001	0.001	0.001	0.001	0.001	0.001	0.001	0.001	0.001	0.001
	0.441	0.455	0.482	0.506	0.529	0.549	0.568	0.609	0.643	0.672	0.696	0.766	0.840	0.901
1.00	***1.442***	***1.457***	***1.483***	***1.508***	***1.530***	***1.551***	***1.569***	***1.610***	***1.644***	***1.673***	***1.697***	***1.767***	***1.841***	***1.903***
	0.000	0.000	0.000	0.000	0.000	0.000	0.000	0.000	0.000	0.000	0.000	0.000	0.000	0.000
	0.398	0.413	0.441	0.467	0.490	0.512	0.532	0.576	0.612	0.643	0.670	0.745	0.825	0.893
1.10	***1.398***	***1.413***	***1.441***	***1.467***	***1.491***	***1.513***	***1.533***	***1.576***	***1.613***	***1.644***	***1.670***	***1.746***	***1.826***	***1.893***

Table Two: The *Excess Costs of Production, Excess Costs of Use,* and *Effective Cost of Production and Use* when *All* Nonconforming Units Are Reworked

C_{pk} or P_{pk}	C_p or P_p value 1.50	1.60	1.70	1.80	1.90	2.00	2.10	2.20	2.30	2.40	2.50	2.60	2.70	2.80
	0.000	0.000	0.000	0.000	0.000	0.000	0.000	0.000	0.000	0.000	0.000	0.000	0.000	0.000
	0.089	0.106	0.125	0.145	0.166	0.188	0.209	0.229	0.250	0.269	0.288	0.306	0.324	0.341
1.20	***1.089***	***1.106***	***1.125***	***1.145***	***1.166***	***1.188***	***1.209***	***1.230***	***1.250***	***1.269***	***1.288***	***1.306***	***1.324***	***1.341***
	0.000	0.000	0.000	0.000	0.000	0.000	0.000	0.000	0.000	0.000	0.000	0.000	0.000	0.000
	0.067	0.079	0.094	0.111	0.130	0.150	0.170	0.190	0.210	0.229	0.248	0.266	0.284	0.301
1.30	***1.067***	***1.079***	***1.094***	***1.111***	***1.130***	***1.150***	***1.170***	***1.190***	***1.210***	***1.229***	***1.248***	***1.266***	***1.284***	***1.301***
	0.000	0.000	0.000	0.000	0.000	0.000	0.000	0.000	0.000	0.000	0.000	0.000	0.000	0.000
	0.054	0.059	0.070	0.084	0.100	0.118	0.136	0.155	0.174	0.193	0.211	0.229	0.247	0.264
1.40	***1.054***	***1.059***	***1.070***	***1.084***	***1.100***	***1.118***	***1.136***	***1.155***	***1.174***	***1.193***	***1.211***	***1.229***	***1.247***	***1.264***
	0.000	0.000	0.000	0.000	0.000	0.000	0.000	0.000	0.000	0.000	0.000	0.000	0.000	0.000
	0.049	0.047	0.052	0.062	0.075	0.090	0.107	0.124	0.142	0.160	0.178	0.195	0.213	0.230
1.50	***1.049***	***1.047***	***1.052***	***1.062***	***1.075***	***1.090***	***1.107***	***1.124***	***1.142***	***1.160***	***1.178***	***1.195***	***1.213***	***1.230***
		0.000	0.000	0.000	0.000	0.000	0.000	0.000	0.000	0.000	0.000	0.000	0.000	0.000
		0.043	0.042	0.047	0.056	0.068	0.082	0.097	0.114	0.130	0.147	0.164	0.181	0.198
1.60		***1.043***	***1.042***	***1.047***	***1.056***	***1.068***	***1.082***	***1.097***	***1.114***	***1.130***	***1.147***	***1.164***	***1.181***	***1.198***
			0.000	0.000	0.000	0.000	0.000	0.000	0.000	0.000	0.000	0.000	0.000	0.000
			0.038	0.037	0.042	0.050	0.061	0.075	0.089	0.104	0.120	0.136	0.152	0.169
1.70 »			***1.038***	***1.037***	***1.042***	***1.050***	***1.061***	***1.075***	***1.089***	***1.104***	***1.120***	***1.136***	***1.152***	***1.169***
				0.000	0.000	0.000	0.000	0.000	0.000	0.000	0.000	0.000	0.000	0.000
				0.034	0.034	0.038	0.046	0.056	0.068	0.082	0.096	0.111	0.126	0.142
1.80 »				***1.034***	***1.034***	***1.038***	***1.046***	***1.056***	***1.068***	***1.082***	***1.096***	***1.111***	***1.126***	***1.142***
					0.000	0.000	0.000	0.000	0.000	0.000	0.000	0.000	0.000	0.000
					0.031	0.030	0.034	0.042	0.051	0.063	0.075	0.089	0.103	0.117
1.90 »					***1.031***	***1.030***	***1.034***	***1.042***	***1.051***	***1.063***	***1.075***	***1.089***	***1.103***	***1.117***
						0.000	0.000	0.000	0.000	0.000	0.000	0.000	0.000	0.000
						0.028	0.027	0.031	0.038	0.047	0.058	0.070	0.082	0.096
2.00 »						***1.028***	***1.027***	***1.031***	***1.038***	***1.047***	***1.058***	***1.070***	***1.082***	***1.096***
							0.000	0.000	0.000	0.000	0.000	0.000	0.000	0.000
							0.025	0.025	0.029	0.035	0.043	0.053	0.065	0.077
2.10 »							***1.025***	***1.025***	***1.029***	***1.035***	***1.043***	***1.053***	***1.065***	***1.077***
								0.000	0.000	0.000	0.000	0.000	0.000	0.000
								0.023	0.023	0.026	0.032	0.040	0.050	0.060
2.20 »								***1.023***	***1.023***	***1.026***	***1.032***	***1.040***	***1.050***	***1.060***
									0.000	0.000	0.000	0.000	0.000	0.000
									0.021	0.021	0.024	0.030	0.037	0.046
2.30 »									***1.021***	***1.021***	***1.024***	***1.030***	***1.037***	***1.046***
										0.000	0.000	0.000	0.000	0.000
										0.019	0.019	0.022	0.028	0.035
2.40 »										***1.019***	***1.019***	***1.022***	***1.028***	***1.035***
											0.000	0.000	0.000	0.000
											0.018	0.018	0.021	0.026
2.50 »											***1.018***	***1.018***	***1.021***	***1.026***
												0.000	0.000	0.000
												0.016	0.017	0.019
2.60 »												***1.016***	***1.017***	***1.019***
													0.000	0.000
													0.015	0.015
2.70 »													***1.015***	***1.015***
														0.000
														0.014
2.80 »														***1.014***

Table Two: The *Excess Costs of Production, Excess Costs of Use,* and *Effective Cost of Production and Use* when *All* Nonconforming Units Are Reworked

C_{pk} or P_{pk}	C_p or P_p value 2.90	3.00	3.20	3.40	3.60	3.80	4.00	4.50	5.00	5.50	6.0	8.0	12.0	20.0
	0.000	0.000	0.000	0.000	0.000	0.000	0.000	0.000	0.000	0.000	0.000	0.000	0.000	0.000
	0.357	0.372	0.401	0.428	0.453	0.476	0.497	0.543	0.582	0.615	0.643	0.724	0.811	0.884
1.20	***1.357***	***1.372***	***1.402***	***1.428***	***1.453***	***1.476***	***1.497***	***1.543***	***1.582***	***1.615***	***1.643***	***1.724***	***1.811***	***1.884***
	0.000	0.000	0.000	0.000	0.000	0.000	0.000	0.000	0.000	0.000	0.000	0.000	0.000	0.000
	0.318	0.333	0.363	0.391	0.417	0.440	0.463	0.511	0.552	0.587	0.617	0.703	0.796	0.874
1.30	***1.318***	***1.333***	***1.363***	***1.391***	***1.417***	***1.441***	***1.463***	***1.511***	***1.552***	***1.587***	***1.617***	***1.703***	***1.796***	***1.875***
	0.000	0.000	0.000	0.000	0.000	0.000	0.000	0.000	0.000	0.000	0.000	0.000	0.000	0.000
	0.281	0.297	0.327	0.356	0.382	0.407	0.429	0.480	0.523	0.559	0.591	0.682	0.781	0.865
1.40	***1.281***	***1.297***	***1.327***	***1.356***	***1.382***	***1.407***	***1.429***	***1.480***	***1.523***	***1.559***	***1.591***	***1.682***	***1.781***	***1.865***
	0.000	0.000	0.000	0.000	0.000	0.000	0.000	0.000	0.000	0.000	0.000	0.000	0.000	0.000
	0.246	0.262	0.293	0.322	0.349	0.374	0.398	0.450	0.494	0.533	0.566	0.662	0.766	0.856
1.50	***1.246***	***1.262***	***1.293***	***1.322***	***1.349***	***1.374***	***1.398***	***1.450***	***1.494***	***1.533***	***1.566***	***1.662***	***1.766***	***1.856***
	0.000	0.000	0.000	0.000	0.000	0.000	0.000	0.000	0.000	0.000	0.000	0.000	0.000	0.000
	0.214	0.230	0.261	0.290	0.317	0.343	0.367	0.421	0.467	0.506	0.541	0.642	0.752	0.847
1.60	***1.214***	***1.230***	***1.261***	***1.290***	***1.317***	***1.343***	***1.367***	***1.421***	***1.467***	***1.506***	***1.541***	***1.642***	***1.752***	***1.847***
	0.000	0.000	0.000	0.000	0.000	0.000	0.000	0.000	0.000	0.000	0.000	0.000	0.000	0.000
	0.184	0.200	0.231	0.260	0.287	0.313	0.338	0.393	0.440	0.481	0.517	0.622	0.738	0.838
1.70	***1.184***	***1.200***	***1.231***	***1.260***	***1.287***	***1.313***	***1.338***	***1.393***	***1.440***	***1.481***	***1.517***	***1.622***	***1.738***	***1.838***
	0.000	0.000	0.000	0.000	0.000	0.000	0.000	0.000	0.000	0.000	0.000	0.000	0.000	0.000
	0.157	0.172	0.202	0.231	0.259	0.285	0.309	0.365	0.414	0.456	0.493	0.602	0.723	0.828
1.80	***1.157***	***1.172***	***1.202***	***1.231***	***1.259***	***1.285***	***1.309***	***1.365***	***1.414***	***1.456***	***1.493***	***1.602***	***1.723***	***1.828***
	0.000	0.000	0.000	0.000	0.000	0.000	0.000	0.000	0.000	0.000	0.000	0.000	0.000	0.000
	0.132	0.147	0.176	0.204	0.232	0.258	0.283	0.339	0.389	0.432	0.470	0.583	0.709	0.819
1.90	***1.132***	***1.147***	***1.176***	***1.204***	***1.232***	***1.258***	***1.283***	***1.339***	***1.389***	***1.432***	***1.470***	***1.583***	***1.709***	***1.819***
	0.000	0.000	0.000	0.000	0.000	0.000	0.000	0.000	0.000	0.000	0.000	0.000	0.000	0.000
	0.110	0.123	0.151	0.179	0.206	0.232	0.257	0.314	0.364	0.409	0.448	0.564	0.695	0.810
2.00	***1.110***	***1.123***	***1.151***	***1.179***	***1.206***	***1.232***	***1.257***	***1.314***	***1.364***	***1.409***	***1.448***	***1.564***	***1.695***	***1.810***
	0.000	0.000	0.000	0.000	0.000	0.000	0.000	0.000	0.000	0.000	0.000	0.000	0.000	0.000
	0.089	0.102	0.129	0.156	0.182	0.208	0.233	0.290	0.341	0.386	0.426	0.546	0.681	0.801
2.10	***1.089***	***1.102***	***1.129***	***1.156***	***1.182***	***1.208***	***1.233***	***1.290***	***1.341***	***1.386***	***1.426***	***1.546***	***1.681***	***1.801***
	0.000	0.000	0.000	0.000	0.000	0.000	0.000	0.000	0.000	0.000	0.000	0.000	0.000	0.000
	0.071	0.083	0.109	0.134	0.160	0.185	0.209	0.267	0.318	0.364	0.404	0.527	0.668	0.792
2.20	***1.071***	***1.083***	***1.109***	***1.134***	***1.160***	***1.185***	***1.209***	***1.267***	***1.318***	***1.364***	***1.404***	***1.527***	***1.668***	***1.792***
	0.000	0.000	0.000	0.000	0.000	0.000	0.000	0.000	0.000	0.000	0.000	0.000	0.000	0.000
	0.056	0.067	0.090	0.114	0.139	0.164	0.188	0.244	0.296	0.342	0.383	0.509	0.654	0.784
2.30	***1.056***	***1.067***	***1.090***	***1.114***	***1.139***	***1.164***	***1.188***	***1.244***	***1.296***	***1.342***	***1.383***	***1.509***	***1.654***	***1.784***
	0.000	0.000	0.000	0.000	0.000	0.000	0.000	0.000	0.000	0.000	0.000	0.000	0.000	0.000
	0.043	0.052	0.073	0.096	0.120	0.143	0.167	0.223	0.275	0.321	0.363	0.492	0.641	0.775
2.40	***1.043***	***1.052***	***1.073***	***1.096***	***1.120***	***1.143***	***1.167***	***1.223***	***1.275***	***1.321***	***1.363***	***1.492***	***1.641***	***1.775***
	0.000	0.000	0.000	0.000	0.000	0.000	0.000	0.000	0.000	0.000	0.000	0.000	0.000	0.000
	0.032	0.040	0.059	0.080	0.102	0.125	0.148	0.203	0.254	0.301	0.343	0.474	0.628	0.766
2.50	***1.032***	***1.040***	***1.059***	***1.080***	***1.102***	***1.125***	***1.148***	***1.203***	***1.254***	***1.301***	***1.343***	***1.474***	***1.628***	***1.766***
	0.000	0.000	0.000	0.000	0.000	0.000	0.000	0.000	0.000	0.000	0.000	0.000	0.000	0.000
	0.024	0.030	0.046	0.065	0.086	0.107	0.129	0.184	0.235	0.282	0.324	0.457	0.614	0.757
2.60	***1.024***	***1.030***	***1.046***	***1.065***	***1.086***	***1.107***	***1.129***	***1.184***	***1.235***	***1.282***	***1.324***	***1.457***	***1.614***	***1.757***
	0.000	0.000	0.000	0.000	0.000	0.000	0.000	0.000	0.000	0.000	0.000	0.000	0.000	0.000
	0.018	0.022	0.035	0.052	0.071	0.091	0.113	0.165	0.216	0.263	0.306	0.441	0.601	0.749
2.70	***1.018***	***1.022***	***1.035***	***1.052***	***1.071***	***1.091***	***1.113***	***1.165***	***1.216***	***1.263***	***1.306***	***1.441***	***1.601***	***1.749***
	0.000	0.000	0.000	0.000	0.000	0.000	0.000	0.000	0.000	0.000	0.000	0.000	0.000	0.000
	0.014	0.017	0.026	0.041	0.058	0.077	0.097	0.148	0.198	0.245	0.288	0.424	0.589	0.740
2.80	***1.014***	***1.017***	***1.026***	***1.041***	***1.058***	***1.077***	***1.097***	***1.148***	***1.198***	***1.245***	***1.288***	***1.424***	***1.589***	***1.740***

Table Two: The *Excess Costs of Production, Excess Costs of Use,* and *Effective Cost of Production and Use* when *All* Nonconforming Units Are Reworked

C_{pk} or P_{pk}	C_p or P_p value 2.90	3.00	3.20	3.40	3.60	3.80	4.00	4.50	5.00	5.50	6.0	8.0	12.0	20.0
	0.000	0.000	0.000	0.000	0.000	0.000	0.000	0.000	0.000	0.000	0.000	0.000	0.000	0.000
	0.013	0.013	0.020	0.031	0.046	0.064	0.083	0.132	0.181	0.227	0.270	0.408	0.576	0.731
2.90	***1.013***	***1.013***	***1.020***	***1.031***	***1.046***	***1.064***	***1.083***	***1.132***	***1.181***	***1.227***	***1.270***	***1.408***	***1.576***	***1.731***
		0.000	0.000	0.000	0.000	0.000	0.000	0.000	0.000	0.000	0.000	0.000	0.000	0.000
		0.012	0.015	0.023	0.036	0.052	0.069	0.117	0.164	0.210	0.253	0.392	0.563	0.723
3.00		***1.012***	***1.015***	***1.023***	***1.036***	***1.052***	***1.069***	***1.117***	***1.164***	***1.210***	***1.253***	***1.392***	***1.563***	***1.723***
			0.000	0.000	0.000	0.000	0.000	0.000	0.000	0.000	0.000	0.000	0.000	0.000
			0.011	0.013	0.021	0.033	0.047	0.089	0.134	0.179	0.221	0.362	0.539	0.706
3.20 »			***1.011***	***1.013***	***1.021***	***1.033***	***1.047***	***1.089***	***1.134***	***1.179***	***1.221***	***1.362***	***1.539***	***1.706***
				0.000	0.000	0.000	0.000	0.000	0.000	0.000	0.000	0.000	0.000	0.000
				0.010	0.012	0.019	0.029	0.065	0.107	0.149	0.191	0.332	0.514	0.689
3.40 »				***1.010***	***1.012***	***1.019***	***1.029***	***1.065***	***1.107***	***1.149***	***1.191***	***1.332***	***1.514***	***1.689***
					0.000	0.000	0.000	0.000	0.000	0.000	0.000	0.000	0.000	0.000
					0.009	0.010	0.017	0.045	0.083	0.123	0.163	0.304	0.491	0.673
3.60 »					***1.009***	***1.010***	***1.017***	***1.045***	***1.083***	***1.123***	***1.163***	***1.304***	***1.491***	***1.673***
						0.000	0.000	0.000	0.000	0.000	0.000	0.000	0.000	0.000
						0.008	0.009	0.030	0.062	0.099	0.138	0.277	0.468	0.656
3.80 »						***1.008***	***1.009***	***1.030***	***1.062***	***1.099***	***1.138***	***1.277***	***1.468***	***1.656***
							0.000	0.000	0.000	0.000	0.000	0.000	0.000	0.000
							0.007	0.018	0.044	0.078	0.114	0.252	0.445	0.640
4.00 »							***1.007***	***1.018***	***1.044***	***1.078***	***1.114***	***1.252***	***1.445***	***1.640***
								0.000	0.000	0.000	0.000	0.000	0.000	0.000
								0.005	0.014	0.037	0.066	0.193	0.391	0.601
4.50 »								***1.005***	***1.014***	***1.037***	***1.066***	***1.193***	***1.391***	***1.601***
									0.000	0.000	0.000	0.000	0.000	0.000
									0.004	0.012	0.031	0.142	0.341	0.563
5.00 »									***1.004***	***1.012***	***1.031***	***1.142***	***1.341***	***1.563***
										0.000	0.000	0.000	0.000	0.000
										0.004	0.010	0.099	0.294	0.526
5.50 »										***1.004***	***1.010***	***1.099***	***1.294***	***1.526***
											0.000	0.000	0.000	0.000
											0.003	0.064	0.251	0.490
6.00 »											***1.003***	***1.064***	***1.251***	***1.490***
												0.000	0.000	0.000
												0.002	0.112	0.360
8.00 »												***1.002***	***1.112***	***1.360***
													0.000	0.000
													0.001	0.160
12.0 »													***1.001***	***1.160***
														0.000
														0.000
20.0 »														***1.000***

Table Three: The *Excess Costs of Production, Excess Costs of Use,* and *Effective Cost of Production and Use* when Average Is on the Scrap Side of Target and the Cost of Rework Is Equal to the Cost of Scrap

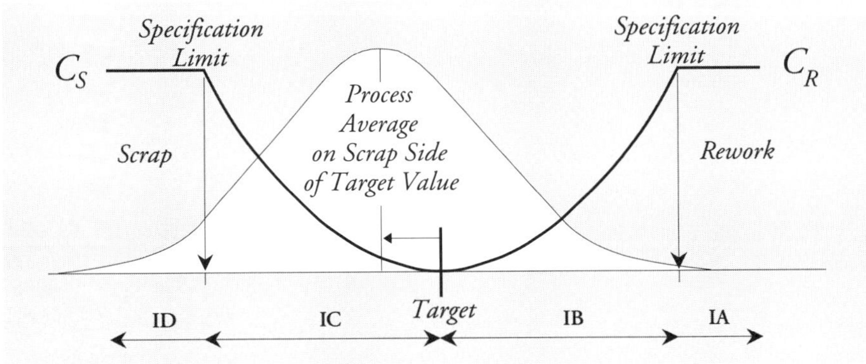

Table Three: Excess Cost Function when Average Is on Scrap Side of Target and Cost of Rework Is Equal to Cost of Scrap

Corresponding to each pairing for C_p and C_{pk} there are three entries stacked in a column.

The topmost of these is the *Excess Costs of Production,* *ExCP*

The middle value is the *Excess Costs of Use,* *ExCU*

And the bottom value (in bold) is the *Effective Cost of Production and Use,* ***ECP&U***

The entries in Table Three assume that the Cost of Rework, C_R, the Cost of Scrap, C_S, and the Nominal Cost of Production, *NC*, are all equal. Table A.9 contains the normal theory values for **IA, IB, IC,** and **ID**. Those computing values should note the reversal of **IA, IB, IC** and **ID** in the Figure above compared to the order shown in Figure 11.6. The general and specific formulas used for Table Three are:

$$ECP\&U = 1.0 + \frac{\frac{C_S}{NC}\{\mathbf{ID} + \frac{C_R}{C_S}\mathbf{IA}\}}{1.0 - \mathbf{ID}} + \frac{\frac{C_S}{NC}\{\mathbf{IC} + \frac{C_R}{C_S}\mathbf{IB}\}}{1.0 - \mathbf{ID}}$$

$$= 1.0 + \frac{\{\mathbf{ID} + \mathbf{IA}\}}{1.0 - \mathbf{ID}} + \frac{\{\mathbf{IC} + \mathbf{IB}\}}{1.0 - \mathbf{ID}}$$

$$= 1.0 + \textit{Excess Costs of Production} + \textit{Excess Costs of Use}$$

When the Cost of Rework is less than the Cost of Scrap appropriate values may be obtained by interpolating between the entry in Table Three and the corresponding entry in Table Four.

When the Cost of Scrap exceeds the Nominal Cost, the *Excess Costs* in Table Three, or any interpolated values, may be inflated as shown in the general formula above.

Table Three: *ExCP, ExCU, and ECP&U* when Average Is on Scrap Side of Target and the Cost of Rework Is Equal to the Cost of Scrap

C_{pk} or P_{pk}	C_p or P_p values 0.10	0.20	0.30	0.40	0.50	0.60	0.70	0.80	0.90	1.00	1.10	1.20	1.30	1.40
–0.50	14.24 0.250 ***15.49***	14.02 0.358 ***15.38***	13.98 0.435 ***15.41***	13.97 0.505 ***15.47***	13.97 0.567 ***15.54***	13.97 0.618 ***15.59***	13.97 0.660 ***15.63***	13.97 0.694 ***15.66***	13.97 0.722 ***15.69***	13.97 0.746 ***15.71***	13.97 0.766 ***15.73***	13.97 0.783 ***15.75***	13.97 0.798 ***15.77***	13.97 0.811 ***15.78***
–0.40	8.003 0.233 ***9.235***	7.762 0.337 ***9.099***	7.702 0.408 ***9.110***	7.692 0.474 ***9.165***	7.690 0.534 ***9.225***	7.690 0.586 ***9.277***	7.690 0.630 ***9.320***	7.690 0.666 ***9.356***	7.690 0.696 ***9.386***	7.690 0.721 ***9.412***	7.690 0.743 ***9.433***	7.690 0.761 ***9.452***	7.690 0.778 ***9.468***	7.690 0.792 ***9.482***
–0.30	4.796 0.213 ***6.009***	4.530 0.316 ***5.846***	4.452 0.380 ***5.832***	4.436 0.440 ***5.876***	4.433 0.498 ***5.932***	4.433 0.551 ***5.984***	4.433 0.596 ***6.029***	4.433 0.634 ***6.067***	4.433 0.666 ***6.099***	4.433 0.693 ***6.126***	4.433 0.716 ***6.149***	4.433 0.736 ***6.169***	4.433 0.754 ***6.187***	4.433 0.769 ***6.202***
–0.20	3.066 0.193 ***4.259***	2.777 0.295 ***4.072***	2.676 0.353 ***4.029***	2.651 0.405 ***4.057***	2.647 0.460 ***4.107***	2.646 0.511 ***4.158***	2.646 0.557 ***4.204***	2.646 0.597 ***4.243***	2.646 0.631 ***4.277***	2.646 0.660 ***4.307***	2.646 0.685 ***4.332***	2.646 0.707 ***4.353***	2.646 0.726 ***4.373***	2.646 0.743 ***4.389***
–0.10	2.099 0.171 ***3.270***	1.792 0.272 ***3.064***	1.664 0.326 ***2.990***	1.626 0.371 ***2.997***	1.618 0.419 ***3.037***	1.617 0.468 ***3.086***	1.617 0.514 ***3.132***	1.617 0.556 ***3.173***	1.617 0.591 ***3.209***	1.617 0.623 ***3.240***	1.617 0.650 ***3.267***	1.617 0.673 ***3.291***	1.617 0.694 ***3.311***	1.617 0.713 ***3.330***
0.00	1.549 0.149 ***2.697***	1.230 0.249 ***2.479***	1.072 0.301 ***2.373***	1.016 0.337 ***2.353***	1.003 0.377 ***2.379***	1.000 0.422 ***2.422***	1.000 0.467 ***2.467***	1.000 0.509 ***2.509***	1.000 0.546 ***2.546***	1.000 0.579 ***2.579***	1.000 0.608 ***2.608***	1.000 0.634 ***2.634***	1.000 0.657 ***2.657***	1.000 0.677 ***2.677***
0.10	1.237 0.126 ***2.362***	0.916 0.224 ***2.140***	0.726 0.277 ***2.003***	0.647 0.306 ***1.953***	0.624 0.335 ***1.959***	0.619 0.373 ***1.993***	0.618 0.416 ***2.034***	0.618 0.457 ***2.075***	0.618 0.496 ***2.114***	0.618 0.530 ***2.149***	0.618 0.561 ***2.180***	0.618 0.589 ***2.207***	0.618 0.614 ***2.232***	0.618 0.636 ***2.254***
0.20		0.756 0.198 ***1.953***	0.536 0.254 ***1.790***	0.427 0.278 ***1.706***	0.389 0.297 ***1.687***	0.380 0.325 ***1.705***	0.378 0.362 ***1.740***	0.378 0.401 ***1.779***	0.378 0.440 ***1.818***	0.378 0.476 ***1.854***	0.378 0.508 ***1.886***	0.378 0.538 ***1.916***	0.378 0.564 ***1.942***	0.378 0.588 ***1.966***
	0.30 »		0.451 0.231 ***1.683***	0.307 0.256 ***1.564***	0.247 0.265 ***1.513***	0.230 0.281 ***1.511***	0.226 0.309 ***1.535***	0.226 0.343 ***1.569***	0.226 0.380 ***1.606***	0.226 0.416 ***1.642***	0.226 0.450 ***1.676***	0.226 0.481 ***1.707***	0.226 0.510 ***1.735***	0.226 0.535 ***1.761***
		0.40 »		0.260 0.238 ***1.499***	0.171 0.242 ***1.412***	0.139 0.244 ***1.383***	0.132 0.259 ***1.391***	0.130 0.286 ***1.416***	0.130 0.319 ***1.449***	0.130 0.354 ***1.484***	0.130 0.388 ***1.518***	0.130 0.420 ***1.550***	0.130 0.450 ***1.580***	0.130 0.477 ***1.607***
			0.50 »		0.143 0.228 ***1.371***	0.091 0.218 ***1.309***	0.075 0.218 ***1.294***	0.072 0.234 ***1.306***	0.072 0.260 ***1.332***	0.072 0.292 ***1.363***	0.072 0.321 ***1.396***	0.072 0.356 ***1.428***	0.072 0.387 ***1.459***	0.072 0.416 ***1.487***
				0.60 »		0.075 0.206 ***1.281***	0.046 0.191 ***1.237***	0.039 0.192 ***1.230***	0.037 0.208 ***1.245***	0.037 0.233 ***1.270***	0.037 0.262 ***1.300***	0.037 0.293 ***1.330***	0.037 0.323 ***1.361***	0.037 0.353 ***1.390***
					0.70 »		0.036 0.180 ***1.216***	0.022 0.164 ***1.185***	0.019 0.166 ***1.184***	0.018 0.182 ***1.200***	0.018 0.206 ***1.224***	0.018 0.233 ***1.251***	0.018 0.262 ***1.280***	0.018 0.291 ***1.309***
						0.80 »		0.017 0.153 ***1.170***	0.010 0.139 ***1.148***	0.008 0.142 ***1.150***	0.008 0.157 ***1.166***	0.008 0.180 ***1.188***	0.008 0.206 ***1.214***	0.008 0.233 ***1.241***
							0.90 »		0.007 0.129 ***1.136***	0.004 0.117 ***1.121***	0.004 0.121 ***1.125***	0.003 0.136 ***1.140***	0.003 0.157 ***1.160***	0.003 0.181 ***1.184***
								1.00 »		0.003 0.108 ***1.111***	0.002 0.098 ***1.100***	0.001 0.103 ***1.105***	0.001 0.118 ***1.119***	0.001 0.137 ***1.138***
									1.10 »		0.001 0.091 ***1.092***	0.001 0.084 ***1.084***	0.000 0.089 ***1.089***	0.000 0.102 ***1.103***

C_{pk}	C_p 1.20	1.30	1.40
1.20	0.000 0.077 ***1.077***	0.000 0.071 ***1.072***	0.000 0.077 ***1.077***
1.30		0.000 0.066 ***1.066***	0.000 0.062 ***1.062***
1.40			0.000 0.057 ***1.057***

Table Three: *ExCP, ExCU, and ECP&U* when Average Is on Scrap Side of Target and the Cost of Rework Is Equal to the Cost of Scrap

C_{pk} or P_{pk}	*C_p or P_p value* 1.50	1.60	1.70	1.80	1.90	2.00	2.10	2.20	2.30	2.40	2.50	2.60	2.70	2.80
	13.97	13.97	13.97	13.97	13.97	13.97	13.97	13.97	13.97	13.97	13.97	13.97	13.97	13.97
	0.822	0.833	0.842	0.850	0.857	0.864	0.870	0.876	0.881	0.886	0.890	0.894	0.898	0.901
−0.50	***15.79***	***15.80***	***15.81***	***15.82***	***15.83***	***15.83***	***15.84***	***15.84***	***15.85***	***15.85***	***15.86***	***15.86***	***15.87***	***15.87***
	7.690	7.690	7.690	7.690	7.690	7.690	7.690	7.690	7.690	7.690	7.690	7.690	7.690	7.690
	0.804	0.815	0.825	0.834	0.842	0.850	0.856	0.862	0.868	0.873	0.878	0.883	0.887	0.891
−0.40	***9.495***	***9.506***	***9.516***	***9.525***	***9.533***	***9.540***	***9.547***	***9.553***	***9.559***	***9.564***	***9.569***	***9.573***	***9.577***	***9.581***
	4.433	4.433	4.433	4.433	4.433	4.433	4.433	4.433	4.433	4.433	4.433	4.433	4.433	4.433
	0.783	0.795	0.806	0.816	0.825	0.833	0.840	0.847	0.853	0.859	0.864	0.869	0.874	0.878
−0.30	***6.216***	***6.228***	***6.239***	***6.249***	***6.258***	***6.266***	***6.273***	***6.280***	***6.286***	***6.292***	***6.297***	***6.302***	***6.307***	***6.311***
	2.646	2.646	2.646	2.646	2.646	2.646	2.646	2.646	2.646	2.646	2.646	2.646	2.646	2.646
	0.758	0.771	0.783	0.794	0.804	0.813	0.821	0.829	0.836	0.842	0.848	0.853	0.858	0.863
−0.20	***4.404***	***4.418***	***4.430***	***4.441***	***4.450***	***4.459***	***4.467***	***4.475***	***4.482***	***4.488***	***4.494***	***4.500***	***4.505***	***4.509***
	1.617	1.617	1.617	1.617	1.617	1.617	1.617	1.617	1.617	1.617	1.617	1.617	1.617	1.617
	0.729	0.744	0.757	0.769	0.780	0.790	0.799	0.807	0.815	0.822	0.828	0.834	0.840	0.845
−0.10	***3.346***	***3.361***	***3.374***	***3.386***	***3.397***	***3.407***	***3.416***	***3.424***	***3.432***	***3.439***	***3.446***	***3.452***	***3.457***	***3.463***
	1.000	1.000	1.000	1.000	1.000	1.000	1.000	1.000	1.000	1.000	1.000	1.000	1.000	1.000
	0.695	0.711	0.726	0.739	0.751	0.762	0.772	0.782	0.790	0.798	0.805	0.812	0.819	0.825
0.00	***2.695***	***2.711***	***2.726***	***2.739***	***2.751***	***2.762***	***2.772***	***2.782***	***2.790***	***2.798***	***2.805***	***2.812***	***2.819***	***2.825***
	0.618	0.618	0.618	0.618	0.618	0.618	0.618	0.618	0.618	0.618	0.618	0.618	0.618	0.618
	0.655	0.673	0.690	0.704	0.718	0.730	0.741	0.752	0.761	0.770	0.778	0.786	0.793	0.800
0.10	***2.274***	***2.292***	***2.308***	***2.323***	***2.336***	***2.348***	***2.360***	***2.370***	***2.380***	***2.388***	***2.397***	***2.404***	***2.412***	***2.418***
	0.378	0.378	0.378	0.378	0.378	0.378	0.378	0.378	0.378	0.378	0.378	0.378	0.378	0.378
	0.610	0.630	0.648	0.664	0.679	0.693	0.705	0.717	0.728	0.738	0.747	0.756	0.764	0.771
0.20	***1.988***	***2.008***	***2.026***	***2.042***	***2.057***	***2.070***	***2.083***	***2.095***	***2.105***	***2.115***	***2.125***	***2.133***	***2.142***	***2.149***
	0.226	0.226	0.226	0.226	0.226	0.226	0.226	0.226	0.226	0.226	0.226	0.226	0.226	0.226
	0.559	0.581	0.600	0.618	0.635	0.650	0.664	0.677	0.689	0.700	0.711	0.720	0.729	0.738
0.30	***1.785***	***1.806***	***1.826***	***1.844***	***1.860***	***1.875***	***1.889***	***1.902***	***1.915***	***1.926***	***1.936***	***1.946***	***1.955***	***1.964***
	0.130	0.130	0.130	0.130	0.130	0.130	0.130	0.130	0.130	0.130	0.130	0.130	0.130	0.130
	0.503	0.526	0.547	0.567	0.585	0.602	0.618	0.632	0.645	0.658	0.670	0.681	0.691	0.700
0.40	***1.633***	***1.656***	***1.677***	***1.697***	***1.715***	***1.732***	***1.748***	***1.762***	***1.775***	***1.788***	***1.800***	***1.811***	***1.821***	***1.830***
	0.072	0.072	0.072	0.072	0.072	0.072	0.072	0.072	0.072	0.072	0.072	0.072	0.072	0.072
	0.442	0.467	0.490	0.512	0.531	0.550	0.567	0.583	0.598	0.612	0.625	0.637	0.648	0.659
0.50	***1.514***	***1.539***	***1.562***	***1.583***	***1.603***	***1.621***	***1.639***	***1.654***	***1.669***	***1.683***	***1.696***	***1.708***	***1.720***	***1.730***
	0.037	0.037	0.037	0.037	0.037	0.037	0.037	0.037	0.037	0.037	0.037	0.037	0.037	0.037
	0.380	0.406	0.431	0.453	0.475	0.495	0.513	0.530	0.547	0.562	0.576	0.590	0.602	0.614
0.60	***1.418***	***1.444***	***1.468***	***1.491***	***1.512***	***1.532***	***1.550***	***1.568***	***1.584***	***1.599***	***1.613***	***1.627***	***1.640***	***1.651***
	0.018	0.018	0.018	0.018	0.018	0.018	0.018	0.018	0.018	0.018	0.018	0.018	0.018	0.018
	0.319	0.345	0.371	0.394	0.417	0.438	0.458	0.476	0.494	0.510	0.526	0.541	0.554	0.567
0.70	***1.337***	***1.363***	***1.389***	***1.413***	***1.435***	***1.456***	***1.476***	***1.495***	***1.512***	***1.529***	***1.544***	***1.559***	***1.573***	***1.586***
	0.008	0.008	0.008	0.008	0.008	0.008	0.008	0.008	0.008	0.008	0.008	0.008	0.008	0.008
	0.260	0.286	0.312	0.336	0.360	0.382	0.403	0.422	0.441	0.459	0.475	0.491	0.506	0.520
0.80	***1.268***	***1.295***	***1.320***	***1.345***	***1.368***	***1.390***	***1.411***	***1.431***	***1.449***	***1.467***	***1.483***	***1.499***	***1.514***	***1.528***
	0.003	0.003	0.003	0.003	0.003	0.003	0.003	0.003	0.003	0.003	0.003	0.003	0.003	0.003
	0.206	0.232	0.257	0.281	0.305	0.328	0.349	0.370	0.389	0.408	0.425	0.442	0.458	0.473
0.90	***1.210***	***1.235***	***1.260***	***1.285***	***1.308***	***1.331***	***1.353***	***1.373***	***1.393***	***1.411***	***1.429***	***1.445***	***1.461***	***1.476***
	0.001	0.001	0.001	0.001	0.001	0.001	0.001	0.001	0.001	0.001	0.001	0.001	0.001	0.001
	0.159	0.183	0.207	0.231	0.254	0.277	0.299	0.319	0.339	0.359	0.377	0.394	0.411	0.427
1.00	***1.161***	***1.184***	***1.208***	***1.232***	***1.255***	***1.278***	***1.300***	***1.321***	***1.341***	***1.360***	***1.378***	***1.396***	***1.412***	***1.428***
	0.000	0.000	0.000	0.000	0.000	0.000	0.000	0.000	0.000	0.000	0.000	0.000	0.000	0.000
	0.120	0.141	0.163	0.185	0.208	0.230	0.252	0.273	0.293	0.312	0.331	0.349	0.366	0.382
1.10	***1.120***	***1.141***	***1.163***	***1.186***	***1.208***	***1.230***	***1.252***	***1.273***	***1.293***	***1.313***	***1.332***	***1.349***	***1.367***	***1.383***

Table Three: *ExCP, ExCU, and ECP&U* when Average Is on Scrap Side of Target and the Cost of Rework Is Equal to the Cost of Scrap

C_{pk} or P_{pk}	C_p or P_p value 2.90	3.00	3.20	3.40	3.60	3.80	4.00	4.50	5.00	5.50	6.0	8.0	12.0	20.0
	13.97	13.97	13.97	13.97	13.97	13.97	13.97	13.97	13.97	13.97	13.97	13.97	13.97	13.97
	0.905	0.908	0.913	0.919	0.923	0.927	0.931	0.939	0.945	0.950	0.954	0.967	0.980	0.993
-0.50	***15.87***	***15.88***	***15.88***	***15.89***	***15.89***	***15.90***	***15.90***	***15.91***	***15.91***	***15.92***	***15.92***	***15.94***	***15.95***	***15.96***
	7.690	7.690	7.690	7.690	7.690	7.690	7.690	7.690	7.690	7.690	7.690	7.690	7.690	7.690
	0.894	0.898	0.904	0.909	0.914	0.919	0.923	0.931	0.938	0.944	0.949	0.963	0.977	0.990
-0.40	***9.585***	***9.588***	***9.594***	***9.600***	***9.605***	***9.609***	***9.613***	***9.622***	***9.629***	***9.635***	***9.639***	***9.653***	***9.667***	***9.681***
	4.433	4.433	4.433	4.433	4.433	4.433	4.433	4.433	4.433	4.433	4.433	4.433	4.433	4.433
	0.882	0.886	0.893	0.899	0.904	0.909	0.914	0.923	0.931	0.937	0.943	0.958	0.973	0.987
-0.30	***6.315***	***6.319***	***6.326***	***6.332***	***6.337***	***6.342***	***6.347***	***6.356***	***6.364***	***6.370***	***6.376***	***6.391***	***6.406***	***6.420***
	2.646	2.646	2.646	2.646	2.646	2.646	2.646	2.646	2.646	2.646	2.646	2.646	2.646	2.646
	0.868	0.872	0.879	0.886	0.892	0.898	0.903	0.913	0.922	0.929	0.935	0.952	0.969	0.984
-0.20	***4.514***	***4.518***	***4.526***	***4.533***	***4.539***	***4.544***	***4.549***	***4.560***	***4.568***	***4.575***	***4.581***	***4.598***	***4.615***	***4.631***
	1.617	1.617	1.617	1.617	1.617	1.617	1.617	1.617	1.617	1.617	1.617	1.617	1.617	1.617
	0.850	0.855	0.864	0.871	0.878	0.884	0.890	0.902	0.911	0.919	0.926	0.945	0.964	0.981
-0.10	***3.468***	***3.472***	***3.481***	***3.489***	***3.495***	***3.502***	***3.507***	***3.519***	***3.529***	***3.537***	***3.543***	***3.562***	***3.581***	***3.598***
	1.000	1.000	1.000	1.000	1.000	1.000	1.000	1.000	1.000	1.000	1.000	1.000	1.000	1.000
	0.830	0.835	0.845	0.854	0.861	0.868	0.875	0.888	0.899	0.908	0.915	0.936	0.958	0.977
0.00	***2.830***	***2.835***	***2.845***	***2.854***	***2.861***	***2.868***	***2.875***	***2.888***	***2.899***	***2.908***	***2.915***	***2.936***	***2.958***	***2.977***
	0.618	0.618	0.618	0.618	0.618	0.618	0.618	0.618	0.618	0.618	0.618	0.618	0.618	0.618
	0.806	0.812	0.823	0.833	0.841	0.849	0.856	0.872	0.884	0.894	0.903	0.927	0.951	0.972
0.10	***2.425***	***2.431***	***2.441***	***2.451***	***2.460***	***2.468***	***2.475***	***2.490***	***2.502***	***2.512***	***2.521***	***2.545***	***2.570***	***2.590***
	0.378	0.378	0.378	0.378	0.378	0.378	0.378	0.378	0.378	0.378	0.378	0.378	0.378	0.378
	0.778	0.785	0.797	0.808	0.818	0.827	0.835	0.852	0.866	0.878	0.888	0.915	0.943	0.967
0.20	***2.156***	***2.163***	***2.175***	***2.186***	***2.196***	***2.205***	***2.213***	***2.230***	***2.244***	***2.256***	***2.266***	***2.293***	***2.321***	***2.345***
	0.226	0.226	0.226	0.226	0.226	0.226	0.226	0.226	0.226	0.226	0.226	0.226	0.226	0.226
	0.746	0.754	0.768	0.780	0.791	0.801	0.811	0.830	0.846	0.859	0.871	0.902	0.934	0.961
0.30	***1.972***	***1.979***	***1.993***	***2.006***	***2.017***	***2.027***	***2.036***	***2.056***	***2.072***	***2.085***	***2.096***	***2.128***	***2.160***	***2.187***
	0.130	0.130	0.130	0.130	0.130	0.130	0.130	0.130	0.130	0.130	0.130	0.130	0.130	0.130
	0.710	0.718	0.734	0.748	0.760	0.772	0.782	0.805	0.823	0.838	0.851	0.887	0.924	0.954
0.40	***1.840***	***1.848***	***1.864***	***1.878***	***1.890***	***1.902***	***1.912***	***1.935***	***1.953***	***1.968***	***1.981***	***2.017***	***2.054***	***2.084***
	0.072	0.072	0.072	0.072	0.072	0.072	0.072	0.072	0.072	0.072	0.072	0.072	0.072	0.072
	0.669	0.679	0.696	0.712	0.726	0.739	0.751	0.776	0.797	0.814	0.829	0.870	0.912	0.947
0.50	***1.741***	***1.750***	***1.768***	***1.784***	***1.798***	***1.811***	***1.823***	***1.848***	***1.869***	***1.886***	***1.900***	***1.941***	***1.984***	***2.018***
	0.037	0.037	0.037	0.037	0.037	0.037	0.037	0.037	0.037	0.037	0.037	0.037	0.037	0.037
	0.625	0.636	0.656	0.673	0.689	0.704	0.717	0.745	0.769	0.788	0.805	0.851	0.899	0.939
0.60	***1.663***	***1.673***	***1.693***	***1.710***	***1.726***	***1.741***	***1.754***	***1.783***	***1.806***	***1.825***	***1.842***	***1.888***	***1.936***	***1.976***
	0.018	0.018	0.018	0.018	0.018	0.018	0.018	0.018	0.018	0.018	0.018	0.018	0.018	0.018
	0.580	0.591	0.613	0.632	0.650	0.666	0.681	0.712	0.739	0.760	0.779	0.831	0.885	0.930
0.70	***1.598***	***1.610***	***1.631***	***1.651***	***1.668***	***1.684***	***1.699***	***1.731***	***1.757***	***1.778***	***1.797***	***1.849***	***1.903***	***1.948***
	0.008	0.008	0.008	0.008	0.008	0.008	0.008	0.008	0.008	0.008	0.008	0.008	0.008	0.008
	0.533	0.546	0.569	0.590	0.610	0.627	0.644	0.679	0.707	0.731	0.752	0.810	0.871	0.921
0.80	***1.541***	***1.554***	***1.578***	***1.599***	***1.618***	***1.636***	***1.652***	***1.687***	***1.716***	***1.740***	***1.760***	***1.818***	***1.879***	***1.929***
	0.003	0.003	0.003	0.003	0.003	0.003	0.003	0.003	0.003	0.003	0.003	0.003	0.003	0.003
	0.487	0.500	0.526	0.548	0.569	0.588	0.606	0.644	0.676	0.702	0.725	0.789	0.856	0.912
0.90	***1.490***	***1.504***	***1.529***	***1.552***	***1.573***	***1.592***	***1.610***	***1.648***	***1.679***	***1.706***	***1.728***	***1.792***	***1.859***	***1.915***
	0.001	0.001	0.001	0.001	0.001	0.001	0.001	0.001	0.001	0.001	0.001	0.001	0.001	0.001
	0.442	0.456	0.483	0.507	0.529	0.550	0.569	0.610	0.644	0.673	0.697	0.767	0.841	0.903
1.00	***1.443***	***1.457***	***1.484***	***1.508***	***1.531***	***1.551***	***1.570***	***1.611***	***1.645***	***1.674***	***1.698***	***1.768***	***1.842***	***1.904***
	0.000	0.000	0.000	0.000	0.000	0.000	0.000	0.000	0.000	0.000	0.000	0.000	0.000	0.000
	0.398	0.413	0.441	0.467	0.491	0.512	0.532	0.576	0.613	0.643	0.670	0.746	0.826	0.893
1.10	***1.399***	***1.414***	***1.442***	***1.467***	***1.491***	***1.513***	***1.533***	***1.577***	***1.613***	***1.644***	***1.670***	***1.746***	***1.826***	***1.894***

Table Three: *ExCP, ExCU, and ECP&U* when Average Is on Scrap Side of Target and the Cost of Rework Is Equal to the Cost of Scrap

C_{pk} or P_{pk}	C_p or P_p value 1.50	1.60	1.70	1.80	1.90	2.00	2.10	2.20	2.30	2.40	2.50	2.60	2.70	2.80
	0.000	0.000	0.000	0.000	0.000	0.000	0.000	0.000	0.000	0.000	0.000	0.000	0.000	0.000
	0.089	0.106	0.125	0.145	0.166	0.188	0.209	0.229	0.250	0.269	0.288	0.306	0.324	0.341
1.20	**1.089**	**1.106**	**1.125**	**1.145**	**1.167**	**1.188**	**1.209**	**1.230**	**1.250**	**1.269**	**1.288**	**1.306**	**1.324**	**1.341**
	0.000	0.000	0.000	0.000	0.000	0.000	0.000	0.000	0.000	0.000	0.000	0.000	0.000	0.000
	0.067	0.079	0.094	0.111	0.130	0.150	0.170	0.190	0.210	0.229	0.248	0.266	0.284	0.301
1.30	**1.067**	**1.079**	**1.094**	**1.111**	**1.131**	**1.150**	**1.170**	**1.190**	**1.210**	**1.229**	**1.248**	**1.266**	**1.284**	**1.301**
	0.000	0.000	0.000	0.000	0.000	0.000	0.000	0.000	0.000	0.000	0.000	0.000	0.000	0.000
	0.054	0.059	0.070	0.084	0.100	0.118	0.136	0.155	0.174	0.193	0.211	0.229	0.247	0.264
1.40	**1.054**	**1.059**	**1.070**	**1.084**	**1.100**	**1.118**	**1.136**	**1.155**	**1.174**	**1.193**	**1.211**	**1.229**	**1.247**	**1.264**
	0.000	0.000	0.000	0.000	0.000	0.000	0.000	0.000	0.000	0.000	0.000	0.000	0.000	0.000
	0.049	0.047	0.052	0.062	0.075	0.090	0.107	0.124	0.142	0.160	0.178	0.195	0.213	0.230
1.50	**1.049**	**1.047**	**1.052**	**1.062**	**1.075**	**1.090**	**1.107**	**1.124**	**1.142**	**1.160**	**1.178**	**1.195**	**1.213**	**1.230**
		0.000	0.000	0.000	0.000	0.000	0.000	0.000	0.000	0.000	0.000	0.000	0.000	0.000
		0.043	0.042	0.047	0.056	0.068	0.082	0.097	0.114	0.130	0.147	0.164	0.181	0.198
1.60		**1.043**	**1.042**	**1.047**	**1.056**	**1.068**	**1.082**	**1.097**	**1.114**	**1.130**	**1.147**	**1.164**	**1.181**	**1.198**
			0.000	0.000	0.000	0.000	0.000	0.000	0.000	0.000	0.000	0.000	0.000	0.000
			0.038	0.037	0.042	0.050	0.061	0.075	0.089	0.104	0.120	0.136	0.152	0.169
1.70 »			**1.038**	**1.037**	**1.042**	**1.050**	**1.061**	**1.075**	**1.089**	**1.104**	**1.120**	**1.136**	**1.152**	**1.169**
				0.000	0.000	0.000	0.000	0.000	0.000	0.000	0.000	0.000	0.000	0.000
				0.034	0.034	0.038	0.046	0.056	0.068	0.082	0.096	0.111	0.126	0.142
1.80 »				**1.034**	**1.034**	**1.038**	**1.046**	**1.056**	**1.068**	**1.082**	**1.096**	**1.111**	**1.126**	**1.142**
					0.000	0.000	0.000	0.000	0.000	0.000	0.000	0.000	0.000	0.000
					0.031	0.030	0.034	0.042	0.051	0.063	0.075	0.089	0.103	0.117
1.90 »					**1.031**	**1.030**	**1.034**	**1.042**	**1.051**	**1.063**	**1.075**	**1.089**	**1.103**	**1.117**
						0.000	0.000	0.000	0.000	0.000	0.000	0.000	0.000	0.000
						0.028	0.027	0.031	0.038	0.047	0.058	0.070	0.082	0.096
2.00 »						**1.028**	**1.027**	**1.031**	**1.038**	**1.047**	**1.058**	**1.070**	**1.082**	**1.096**
							0.000	0.000	0.000	0.000	0.000	0.000	0.000	0.000
							0.025	0.025	0.029	0.035	0.043	0.053	0.065	0.077
2.10 »							**1.025**	**1.025**	**1.029**	**1.035**	**1.043**	**1.053**	**1.065**	**1.077**
								0.000	0.000	0.000	0.000	0.000	0.000	0.000
								0.023	0.023	0.026	0.032	0.040	0.050	0.060
2.20 »								**1.023**	**1.023**	**1.026**	**1.032**	**1.040**	**1.050**	**1.060**
									0.000	0.000	0.000	0.000	0.000	0.000
									0.021	0.021	0.024	0.030	0.037	0.046
2.30 »									**1.021**	**1.021**	**1.024**	**1.030**	**1.037**	**1.046**
										0.000	0.000	0.000	0.000	0.000
										0.019	0.019	0.022	0.028	0.035
2.40 »										**1.019**	**1.019**	**1.022**	**1.028**	**1.035**
											0.000	0.000	0.000	0.000
											0.018	0.018	0.021	0.026
2.50 »											**1.018**	**1.018**	**1.021**	**1.026**
												0.000	0.000	0.000
												0.016	0.017	0.019
2.60 »												**1.016**	**1.017**	**1.019**
													0.000	0.000
													0.015	0.015
2.70 »													**1.015**	**1.015**
														0.000
														0.014
2.80 »														**1.014**

Table Three: *ExCP, ExCU, and ECP&U* when Average Is on Scrap Side of Target and the Cost of Rework Is Equal to the Cost of Scrap

C_{pk} or P_{pk}	C_p or P_p value 2.90	3.00	3.20	3.40	3.60	3.80	4.00	4.50	5.00	5.50	6.0	8.0	12.0	20.0
	0.000	0.000	0.000	0.000	0.000	0.000	0.000	0.000	0.000	0.000	0.000	0.000	0.000	0.000
	0.357	0.372	0.401	0.428	0.453	0.476	0.497	0.543	0.582	0.615	0.643	0.724	0.811	0.884
1.20	***1.357***	***1.372***	***1.402***	***1.428***	***1.453***	***1.476***	***1.497***	***1.543***	***1.582***	***1.615***	***1.643***	***1.724***	***1.811***	***1.884***
	0.000	0.000	0.000	0.000	0.000	0.000	0.000	0.000	0.000	0.000	0.000	0.000	0.000	0.000
	0.318	0.333	0.363	0.391	0.417	0.440	0.463	0.511	0.552	0.587	0.617	0.703	0.796	0.874
1.30	***1.318***	***1.333***	***1.363***	***1.391***	***1.417***	***1.441***	***1.463***	***1.511***	***1.552***	***1.587***	***1.617***	***1.703***	***1.796***	***1.875***
	0.000	0.000	0.000	0.000	0.000	0.000	0.000	0.000	0.000	0.000	0.000	0.000	0.000	0.000
	0.281	0.297	0.327	0.356	0.382	0.407	0.429	0.480	0.523	0.559	0.591	0.682	0.781	0.865
1.40	***1.281***	***1.297***	***1.327***	***1.356***	***1.382***	***1.407***	***1.429***	***1.480***	***1.523***	***1.559***	***1.591***	***1.682***	***1.781***	***1.865***
	0.000	0.000	0.000	0.000	0.000	0.000	0.000	0.000	0.000	0.000	0.000	0.000	0.000	0.000
	0.246	0.262	0.293	0.322	0.349	0.374	0.398	0.450	0.494	0.533	0.566	0.662	0.766	0.856
1.50	***1.246***	***1.262***	***1.293***	***1.322***	***1.349***	***1.374***	***1.398***	***1.450***	***1.494***	***1.533***	***1.566***	***1.662***	***1.766***	***1.856***
	0.000	0.000	0.000	0.000	0.000	0.000	0.000	0.000	0.000	0.000	0.000	0.000	0.000	0.000
	0.214	0.230	0.261	0.290	0.317	0.343	0.367	0.421	0.467	0.506	0.541	0.642	0.752	0.847
1.60	***1.214***	***1.230***	***1.261***	***1.290***	***1.317***	***1.343***	***1.367***	***1.421***	***1.467***	***1.506***	***1.541***	***1.642***	***1.752***	***1.847***
	0.000	0.000	0.000	0.000	0.000	0.000	0.000	0.000	0.000	0.000	0.000	0.000	0.000	0.000
	0.184	0.200	0.231	0.260	0.287	0.313	0.338	0.393	0.440	0.481	0.517	0.622	0.738	0.838
1.70	***1.184***	***1.200***	***1.231***	***1.260***	***1.287***	***1.313***	***1.338***	***1.393***	***1.440***	***1.481***	***1.517***	***1.622***	***1.738***	***1.838***
	0.000	0.000	0.000	0.000	0.000	0.000	0.000	0.000	0.000	0.000	0.000	0.000	0.000	0.000
	0.157	0.172	0.202	0.231	0.259	0.285	0.309	0.365	0.414	0.456	0.493	0.602	0.723	0.828
1.80	***1.157***	***1.172***	***1.202***	***1.231***	***1.259***	***1.285***	***1.309***	***1.365***	***1.414***	***1.456***	***1.493***	***1.602***	***1.723***	***1.828***
	0.000	0.000	0.000	0.000	0.000	0.000	0.000	0.000	0.000	0.000	0.000	0.000	0.000	0.000
	0.132	0.147	0.176	0.204	0.232	0.258	0.283	0.339	0.389	0.432	0.470	0.583	0.709	0.819
1.90	***1.132***	***1.147***	***1.176***	***1.204***	***1.232***	***1.258***	***1.283***	***1.339***	***1.389***	***1.432***	***1.470***	***1.583***	***1.709***	***1.819***
	0.000	0.000	0.000	0.000	0.000	0.000	0.000	0.000	0.000	0.000	0.000	0.000	0.000	0.000
	0.110	0.123	0.151	0.179	0.206	0.232	0.257	0.314	0.364	0.409	0.448	0.564	0.695	0.810
2.00	***1.110***	***1.123***	***1.151***	***1.179***	***1.206***	***1.232***	***1.257***	***1.314***	***1.364***	***1.409***	***1.448***	***1.564***	***1.695***	***1.810***
	0.000	0.000	0.000	0.000	0.000	0.000	0.000	0.000	0.000	0.000	0.000	0.000	0.000	0.000
	0.089	0.102	0.129	0.156	0.182	0.208	0.233	0.290	0.341	0.386	0.426	0.546	0.681	0.801
2.10	***1.089***	***1.102***	***1.129***	***1.156***	***1.182***	***1.208***	***1.233***	***1.290***	***1.341***	***1.386***	***1.426***	***1.546***	***1.681***	***1.801***
	0.000	0.000	0.000	0.000	0.000	0.000	0.000	0.000	0.000	0.000	0.000	0.000	0.000	0.000
	0.071	0.083	0.109	0.134	0.160	0.185	0.209	0.267	0.318	0.364	0.404	0.527	0.668	0.792
2.20	***1.071***	***1.083***	***1.109***	***1.134***	***1.160***	***1.185***	***1.209***	***1.267***	***1.318***	***1.364***	***1.404***	***1.527***	***1.668***	***1.792***
	0.000	0.000	0.000	0.000	0.000	0.000	0.000	0.000	0.000	0.000	0.000	0.000	0.000	0.000
	0.056	0.067	0.090	0.114	0.139	0.164	0.188	0.244	0.296	0.342	0.383	0.509	0.654	0.784
2.30	***1.056***	***1.067***	***1.090***	***1.114***	***1.139***	***1.164***	***1.188***	***1.244***	***1.296***	***1.342***	***1.383***	***1.509***	***1.654***	***1.784***
	0.000	0.000	0.000	0.000	0.000	0.000	0.000	0.000	0.000	0.000	0.000	0.000	0.000	0.000
	0.043	0.052	0.073	0.096	0.120	0.143	0.167	0.223	0.275	0.321	0.363	0.492	0.641	0.775
2.40	***1.043***	***1.052***	***1.073***	***1.096***	***1.120***	***1.143***	***1.167***	***1.223***	***1.275***	***1.321***	***1.363***	***1.492***	***1.641***	***1.775***
	0.000	0.000	0.000	0.000	0.000	0.000	0.000	0.000	0.000	0.000	0.000	0.000	0.000	0.000
	0.032	0.040	0.059	0.080	0.102	0.125	0.148	0.203	0.254	0.301	0.343	0.474	0.628	0.766
2.50	***1.032***	***1.040***	***1.059***	***1.080***	***1.102***	***1.125***	***1.148***	***1.203***	***1.254***	***1.301***	***1.343***	***1.474***	***1.628***	***1.766***
	0.000	0.000	0.000	0.000	0.000	0.000	0.000	0.000	0.000	0.000	0.000	0.000	0.000	0.000
	0.024	0.030	0.046	0.065	0.086	0.107	0.129	0.184	0.235	0.282	0.324	0.457	0.614	0.757
2.60	***1.024***	***1.030***	***1.046***	***1.065***	***1.086***	***1.107***	***1.129***	***1.184***	***1.235***	***1.282***	***1.324***	***1.457***	***1.614***	***1.757***
	0.000	0.000	0.000	0.000	0.000	0.000	0.000	0.000	0.000	0.000	0.000	0.000	0.000	0.000
	0.018	0.022	0.035	0.052	0.071	0.091	0.113	0.165	0.216	0.263	0.306	0.441	0.601	0.749
2.70	***1.018***	***1.022***	***1.035***	***1.052***	***1.071***	***1.091***	***1.113***	***1.165***	***1.216***	***1.263***	***1.306***	***1.441***	***1.601***	***1.749***
	0.000	0.000	0.000	0.000	0.000	0.000	0.000	0.000	0.000	0.000	0.000	0.000	0.000	0.000
	0.014	0.017	0.026	0.041	0.058	0.077	0.097	0.148	0.198	0.245	0.288	0.424	0.589	0.740
2.80	***1.014***	***1.017***	***1.026***	***1.041***	***1.058***	***1.077***	***1.097***	***1.148***	***1.198***	***1.245***	***1.288***	***1.424***	***1.589***	***1.740***

Table Three: *ExCP, ExCU, and ECP&U* when Average Is on Scrap Side of Target and the Cost of Rework Is Equal to the Cost of Scrap

C_{pk} or P_{pk}	*C_p or P_p value* 2.90	3.00	3.20	3.40	3.60	3.80	4.00	4.50	5.00	5.50	6.0	8.0	12.0	20.0
	0.000	0.000	0.000	0.000	0.000	0.000	0.000	0.000	0.000	0.000	0.000	0.000	0.000	0.000
	0.013	0.013	0.020	0.031	0.046	0.064	0.083	0.132	0.181	0.227	0.270	0.408	0.576	0.731
2.90	***1.013***	***1.013***	***1.020***	***1.031***	***1.046***	***1.064***	***1.083***	***1.132***	***1.181***	***1.227***	***1.270***	***1.408***	***1.576***	***1.731***
		0.000	0.000	0.000	0.000	0.000	0.000	0.000	0.000	0.000	0.000	0.000	0.000	0.000
		0.012	0.015	0.023	0.036	0.052	0.069	0.117	0.164	0.210	0.253	0.392	0.563	0.723
3.00		***1.012***	***1.015***	***1.023***	***1.036***	***1.052***	***1.069***	***1.117***	***1.164***	***1.210***	***1.253***	***1.392***	***1.563***	***1.723***
			0.000	0.000	0.000	0.000	0.000	0.000	0.000	0.000	0.000	0.000	0.000	0.000
			0.011	0.013	0.021	0.033	0.047	0.089	0.134	0.179	0.221	0.362	0.539	0.706
3.20 »			***1.011***	***1.013***	***1.021***	***1.033***	***1.047***	***1.089***	***1.134***	***1.179***	***1.221***	***1.362***	***1.539***	***1.706***
				0.000	0.000	0.000	0.000	0.000	0.000	0.000	0.000	0.000	0.000	0.000
				0.010	0.012	0.019	0.029	0.065	0.107	0.149	0.191	0.332	0.514	0.689
3.40 »				***1.010***	***1.012***	***1.019***	***1.029***	***1.065***	***1.107***	***1.149***	***1.191***	***1.332***	***1.514***	***1.689***
					0.000	0.000	0.000	0.000	0.000	0.000	0.000	0.000	0.000	0.000
					0.009	0.010	0.017	0.045	0.083	0.123	0.163	0.304	0.491	0.673
3.60 »					***1.009***	***1.010***	***1.017***	***1.045***	***1.083***	***1.123***	***1.163***	***1.304***	***1.491***	***1.673***
						0.000	0.000	0.000	0.000	0.000	0.000	0.000	0.000	0.000
						0.008	0.009	0.030	0.062	0.099	0.138	0.277	0.468	0.656
3.80 »						***1.008***	***1.009***	***1.030***	***1.062***	***1.099***	***1.138***	***1.277***	***1.468***	***1.656***
							0.000	0.000	0.000	0.000	0.000	0.000	0.000	0.000
							0.007	0.018	0.044	0.078	0.114	0.252	0.445	0.640
4.00 »							***1.007***	***1.018***	***1.044***	***1.078***	***1.114***	***1.252***	***1.445***	***1.640***
								0.000	0.000	0.000	0.000	0.000	0.000	0.000
								0.005	0.014	0.037	0.066	0.193	0.391	0.601
4.50 »								***1.005***	***1.014***	***1.037***	***1.066***	***1.193***	***1.391***	***1.601***
									0.000	0.000	0.000	0.000	0.000	0.000
									0.004	0.012	0.031	0.142	0.341	0.563
5.00 »									***1.004***	***1.012***	***1.031***	***1.142***	***1.341***	***1.563***
										0.000	0.000	0.000	0.000	0.000
										0.004	0.010	0.099	0.294	0.526
5.50 »										***1.004***	***1.010***	***1.099***	***1.294***	***1.526***
											0.000	0.000	0.000	0.000
											0.003	0.064	0.251	0.490
6.00 »											***1.003***	***1.064***	***1.251***	***1.490***
												0.000	0.000	0.000
												0.002	0.112	0.360
8.00 »												***1.002***	***1.112***	***1.360***
													0.000	0.000
													0.001	0.160
12.0 »													***1.001***	***1.160***
														0.000
														0.000
20.0 »														***1.000***

Table Four: The *Excess Costs of Production, Excess Costs of Use,* and *Effective Cost of Production and Use* when Average Is on the Scrap Side of Target and the Cost of Rework Is 0.1% of the Cost of Scrap

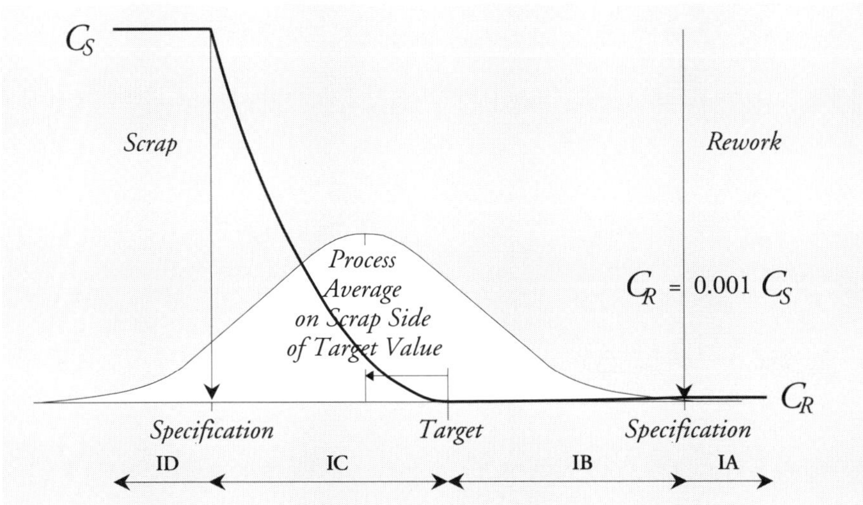

Table Four: Excess Cost Function when Average Is on Scrap Side of Target and Cost of Rework Is 0.1% of the Cost of Scrap

Corresponding to each pairing for C_p and C_{pk} there are three entries stacked in a column.

The topmost of these is the *Excess Costs of Production,* *ExCP*

The middle value is the *Excess Costs of Use,* *ExCU*

And the bottom value (in bold) is the *Effective Cost of Production and Use,* ***ECP&U***

The entries in Table Four assume that the Cost of Scrap, C_S, and the Nominal Cost of Production, *NC,* are equal, Table A.9 contains the normal theory values for **IA, IB, IC,** and **ID**. Those computing values should note the reversal of **IA, IB, IC,** and **ID** in the Figure above compared to the order shown in Figure 11.6. The general and specific formulas used for Table Four are:

$$ECP\&U = 1.0 + \frac{\frac{C_S}{NC}\{\mathbf{ID} + \frac{C_R}{C_S}\mathbf{IA}\}}{1.0 - \mathbf{ID}} + \frac{\frac{C_S}{NC}\{\mathbf{IC} + \frac{C_R}{C_S}\mathbf{IB}\}}{1.0 - \mathbf{ID}}$$

$$= 1.0 + \frac{\{\mathbf{ID} + 0.001\,\mathbf{IA}\}}{1.0 - \mathbf{ID}} + \frac{\{\mathbf{IC} + 0.001\,\mathbf{IB}\}}{1.0 - \mathbf{ID}}$$

$$= 1.0 + \textit{Excess Costs of Production} + \textit{Excess Costs of Use}$$

When the Cost of Rework is greater than 0.001 times the Cost of Scrap appropriate values may be obtained by interpolating between the entry in Table Four and the corresponding entry in Table Three. When the Cost of Scrap exceeds the Nominal Cost, the *Excess Costs* in Table Four, or any interpolated values, may be inflated as shown in the formula above.

Table Four: *ExCP, ExCU, and ECP&U* when Average Is on Scrap Side of Target and the Cost of Rework Is 0.1% of the Cost of Scrap

C_{pk} or P_{pk}	C_p or P_p values: 0.10	0.20	0.30	0.40	0.50	0.60	0.70	0.80	0.90	1.00	1.10	1.20	1.30	1.40
	13.97	13.97	13.97	13.97	13.97	13.97	13.97	13.97	13.97	13.97	13.97	13.97	13.97	13.97
	0.173	0.310	0.417	0.500	0.566	0.618	0.660	0.694	0.722	0.746	0.766	0.783	0.798	0.811
–0.50	15.14	15.28	15.39	15.47	15.53	15.59	15.63	15.66	15.69	15.71	15.73	15.75	15.77	15.78
	7.691	7.690	7.690	7.690	7.690	7.690	7.690	7.690	7.690	7.690	7.690	7.690	7.690	7.690
	0.154	0.280	0.383	0.466	0.532	0.586	0.630	0.666	0.696	0.721	0.743	0.761	0.778	0.792
–0.40	8.845	8.971	9.073	9.156	9.223	9.276	9.320	9.356	9.386	9.412	9.433	9.452	9.468	9.482
	4.433	4.433	4.433	4.433	4.433	4.433	4.433	4.433	4.433	4.433	4.433	4.433	4.433	4.433
	0.135	0.250	0.347	0.428	0.495	0.550	0.596	0.634	0.666	0.693	0.716	0.736	0.754	0.769
–0.30	5.568	5.683	5.780	5.861	5.928	5.983	6.029	6.067	6.099	6.126	6.149	6.169	6.187	6.202
	2.647	2.646	2.646	2.646	2.646	2.646	2.646	2.646	2.646	2.646	2.646	2.646	2.646	2.646
	0.116	0.219	0.310	0.387	0.454	0.510	0.557	0.597	0.631	0.660	0.685	0.707	0.726	0.743
–0.20	3.763	3.866	3.956	4.034	4.100	4.156	4.203	4.243	4.277	4.307	4.332	4.353	4.373	4.389
	1.618	1.617	1.617	1.617	1.617	1.617	1.617	1.617	1.617	1.617	1.617	1.617	1.617	1.617
	0.097	0.188	0.270	0.344	0.409	0.465	0.513	0.555	0.591	0.623	0.650	0.673	0.694	0.713
–0.10	2.715	2.805	2.888	2.961	3.026	3.082	3.131	3.172	3.209	3.240	3.267	3.291	3.311	3.330
	1.001	1.000	1.000	1.000	1.000	1.000	1.000	1.000	1.000	1.000	1.000	1.000	1.000	1.000
	0.080	0.157	0.230	0.298	0.360	0.416	0.465	0.508	0.546	0.579	0.608	0.634	0.657	0.677
0.00	2.080	2.157	2.230	2.298	2.360	2.416	2.465	2.508	2.546	2.579	2.608	2.634	2.657	2.677
	0.619	0.619	0.618	0.618	0.618	0.618	0.618	0.618	0.618	0.618	0.618	0.618	0.618	0.618
	0.063	0.127	0.190	0.252	0.310	0.363	0.412	0.456	0.495	0.530	0.561	0.589	0.614	0.636
0.10	1.682	1.746	1.809	1.870	1.928	1.981	2.030	2.074	2.114	2.149	2.180	2.207	2.232	2.254
		0.378	0.378	0.378	0.378	0.378	0.378	0.378	0.378	0.378	0.378	0.378	0.378	0.378
		0.099	0.152	0.205	0.257	0.308	0.355	0.399	0.439	0.475	0.508	0.538	0.564	0.588
0.20		1.477	1.530	1.583	1.635	1.686	1.733	1.777	1.817	1.853	1.886	1.916	1.942	1.966
			0.226	0.226	0.226	0.226	0.226	0.226	0.226	0.226	0.226	0.226	0.226	0.226
			0.116	0.160	0.206	0.252	0.296	0.339	0.379	0.416	0.450	0.481	0.510	0.535
0.30 »			1.342	1.386	1.432	1.477	1.522	1.564	1.604	1.641	1.676	1.707	1.735	1.761
				0.130	0.130	0.130	0.130	0.130	0.130	0.130	0.130	0.130	0.130	0.130
				0.119	0.157	0.197	0.237	0.277	0.316	0.353	0.388	0.420	0.450	0.477
0.40 »				1.250	1.287	1.327	1.367	1.407	1.446	1.483	1.518	1.550	1.580	1.607
					0.072	0.072	0.072	0.072	0.072	0.072	0.072	0.072	0.072	0.072
					0.114	0.147	0.181	0.217	0.253	0.289	0.323	0.356	0.387	0.416
0.50 »					1.186	1.218	1.253	1.289	1.325	1.361	1.395	1.428	1.458	1.487
						0.037	0.037	0.037	0.037	0.037	0.037	0.037	0.037	0.037
						0.103	0.132	0.162	0.195	0.227	0.260	0.292	0.323	0.352
0.60 »						1.140	1.169	1.200	1.232	1.265	1.297	1.330	1.360	1.390
							0.018	0.018	0.018	0.018	0.018	0.018	0.018	0.018
							0.090	0.115	0.142	0.171	0.201	0.231	0.261	0.290
0.70 »							1.108	1.133	1.160	1.189	1.219	1.250	1.279	1.309
								0.008	0.008	0.008	0.008	0.008	0.008	0.008
								0.077	0.098	0.123	0.149	0.176	0.204	0.232
0.80 »								1.085	1.107	1.131	1.157	1.184	1.212	1.240
									0.003	0.003	0.003	0.003	0.003	0.003
									0.065	0.084	0.105	0.129	0.154	0.180
0.90 »									1.068	1.087	1.109	1.132	1.157	1.183
										0.001	0.001	0.001	0.001	0.001
										0.054	0.071	0.090	0.111	0.134
1.00 »										1.055	1.072	1.091	1.113	1.136
											0.000	0.000	0.000	0.000
											0.045	0.060	0.078	0.097
1.10 »											1.046	1.061	1.078	1.097

C_{pk}	C_p: 1.20	1.30	1.40
	0.000	0.000	0.000
	0.038	0.052	0.067
1.20	1.039	1.052	1.067
		0.000	0.000
		0.033	0.045
1.30		1.033	1.045
			0.000
			0.028
1.40			1.028

Table Four: *ExCP, ExCU, and ECP&U* when Average Is on Scrap Side of Target and the Cost of Rework Is 0.1% of the Cost of Scrap

C_{pk} or P_{pk}	C_p or P_p value 1.50	1.60	1.70	1.80	1.90	2.00	2.10	2.20	2.30	2.40	2.50	2.60	2.70	2.80
	13.97	13.97	13.97	13.97	13.97	13.97	13.97	13.97	13.97	13.97	13.97	13.97	13.97	13.97
	0.822	0.833	0.842	0.850	0.857	0.864	0.870	0.876	0.881	0.886	0.890	0.894	0.898	0.901
–0.50	***15.79***	***15.80***	***15.81***	***15.82***	***15.83***	***15.83***	***15.84***	***15.84***	***15.85***	***15.85***	***15.86***	***15.86***	***15.87***	***15.87***
	7.690	7.690	7.690	7.690	7.690	7.690	7.690	7.690	7.690	7.690	7.690	7.690	7.690	7.690
	0.804	0.815	0.825	0.834	0.842	0.850	0.856	0.862	0.868	0.873	0.878	0.883	0.887	0.891
–0.40	***9.495***	***9.506***	***9.516***	***9.525***	***9.533***	***9.540***	***9.547***	***9.553***	***9.559***	***9.564***	***9.569***	***9.573***	***9.577***	***9.581***
	4.433	4.433	4.433	4.433	4.433	4.433	4.433	4.433	4.433	4.433	4.433	4.433	4.433	4.433
	0.783	0.795	0.806	0.816	0.825	0.833	0.840	0.847	0.853	0.859	0.864	0.869	0.874	0.878
–0.30	***6.216***	***6.228***	***6.239***	***6.249***	***6.258***	***6.266***	***6.273***	***6.280***	***6.286***	***6.292***	***6.297***	***6.302***	***6.307***	***6.311***
	2.646	2.646	2.646	2.646	2.646	2.646	2.646	2.646	2.646	2.646	2.646	2.646	2.646	2.646
	0.758	0.771	0.783	0.794	0.804	0.813	0.821	0.829	0.836	0.842	0.848	0.853	0.858	0.863
–0.20	***4.404***	***4.418***	***4.430***	***4.441***	***4.450***	***4.459***	***4.467***	***4.475***	***4.482***	***4.488***	***4.494***	***4.500***	***4.505***	***4.509***
	1.617	1.617	1.617	1.617	1.617	1.617	1.617	1.617	1.617	1.617	1.617	1.617	1.617	1.617
	0.729	0.744	0.757	0.769	0.780	0.790	0.799	0.807	0.815	0.822	0.828	0.834	0.840	0.845
–0.10	***3.346***	***3.361***	***3.374***	***3.386***	***3.397***	***3.407***	***3.416***	***3.424***	***3.432***	***3.439***	***3.446***	***3.452***	***3.457***	***3.463***
	1.000	1.000	1.000	1.000	1.000	1.000	1.000	1.000	1.000	1.000	1.000	1.000	1.000	1.000
	0.695	0.711	0.726	0.739	0.751	0.762	0.772	0.782	0.790	0.798	0.805	0.812	0.819	0.825
0.00	***2.695***	***2.711***	***2.726***	***2.739***	***2.751***	***2.762***	***2.772***	***2.782***	***2.790***	***2.798***	***2.805***	***2.812***	***2.819***	***2.825***
	0.618	0.618	0.618	0.618	0.618	0.618	0.618	0.618	0.618	0.618	0.618	0.618	0.618	0.618
	0.655	0.673	0.690	0.704	0.718	0.730	0.741	0.752	0.761	0.770	0.778	0.786	0.793	0.800
0.10	***2.274***	***2.292***	***2.308***	***2.323***	***2.336***	***2.348***	***2.360***	***2.370***	***2.380***	***2.388***	***2.397***	***2.404***	***2.412***	***2.418***
	0.378	0.378	0.378	0.378	0.378	0.378	0.378	0.378	0.378	0.378	0.378	0.378	0.378	0.378
	0.610	0.630	0.648	0.664	0.679	0.693	0.705	0.717	0.728	0.738	0.747	0.756	0.764	0.771
0.20	***1.988***	***2.008***	***2.026***	***2.042***	***2.057***	***2.070***	***2.083***	***2.095***	***2.105***	***2.115***	***2.125***	***2.133***	***2.142***	***2.149***
	0.226	0.226	0.226	0.226	0.226	0.226	0.226	0.226	0.226	0.226	0.226	0.226	0.226	0.226
	0.559	0.581	0.600	0.618	0.635	0.650	0.664	0.677	0.689	0.700	0.711	0.720	0.729	0.738
0.30	***1.785***	***1.806***	***1.826***	***1.844***	***1.860***	***1.875***	***1.889***	***1.902***	***1.915***	***1.926***	***1.936***	***1.946***	***1.955***	***1.964***
	0.130	0.130	0.130	0.130	0.130	0.130	0.130	0.130	0.130	0.130	0.130	0.130	0.130	0.130
	0.503	0.526	0.547	0.567	0.585	0.602	0.618	0.632	0.645	0.658	0.670	0.681	0.691	0.700
0.40	***1.633***	***1.656***	***1.677***	***1.697***	***1.715***	***1.732***	***1.748***	***1.762***	***1.775***	***1.788***	***1.800***	***1.811***	***1.821***	***1.830***
	0.072	0.072	0.072	0.072	0.072	0.072	0.072	0.072	0.072	0.072	0.072	0.072	0.072	0.072
	0.442	0.467	0.490	0.512	0.531	0.550	0.567	0.583	0.598	0.612	0.625	0.637	0.648	0.659
0.50	***1.514***	***1.539***	***1.562***	***1.583***	***1.603***	***1.621***	***1.639***	***1.654***	***1.669***	***1.683***	***1.696***	***1.708***	***1.720***	***1.730***
	0.037	0.037	0.037	0.037	0.037	0.037	0.037	0.037	0.037	0.037	0.037	0.037	0.037	0.037
	0.380	0.406	0.431	0.453	0.475	0.495	0.513	0.530	0.547	0.562	0.576	0.590	0.602	0.614
0.60	***1.418***	***1.444***	***1.468***	***1.491***	***1.512***	***1.532***	***1.550***	***1.568***	***1.584***	***1.599***	***1.613***	***1.627***	***1.640***	***1.651***
	0.018	0.018	0.018	0.018	0.018	0.018	0.018	0.018	0.018	0.018	0.018	0.018	0.018	0.018
	0.318	0.345	0.371	0.394	0.417	0.438	0.458	0.476	0.494	0.510	0.526	0.541	0.554	0.567
0.70	***1.337***	***1.363***	***1.389***	***1.413***	***1.435***	***1.456***	***1.476***	***1.495***	***1.512***	***1.529***	***1.544***	***1.559***	***1.573***	***1.586***
	0.008	0.008	0.008	0.008	0.008	0.008	0.008	0.008	0.008	0.008	0.008	0.008	0.008	0.008
	0.260	0.286	0.312	0.336	0.360	0.382	0.403	0.422	0.441	0.459	0.475	0.491	0.506	0.520
0.80	***1.268***	***1.295***	***1.320***	***1.345***	***1.368***	***1.390***	***1.411***	***1.431***	***1.449***	***1.467***	***1.483***	***1.499***	***1.514***	***1.528***
	0.003	0.003	0.003	0.003	0.003	0.003	0.003	0.003	0.003	0.003	0.003	0.003	0.003	0.003
	0.206	0.231	0.257	0.281	0.305	0.328	0.349	0.370	0.389	0.408	0.425	0.442	0.458	0.473
0.90	***1.209***	***1.235***	***1.260***	***1.285***	***1.308***	***1.331***	***1.353***	***1.373***	***1.393***	***1.411***	***1.429***	***1.445***	***1.461***	***1.476***
	0.001	0.001	0.001	0.001	0.001	0.001	0.001	0.001	0.001	0.001	0.001	0.001	0.001	0.001
	0.158	0.182	0.207	0.231	0.254	0.277	0.298	0.319	0.339	0.359	0.377	0.394	0.411	0.427
1.00	***1.159***	***1.184***	***1.208***	***1.232***	***1.255***	***1.278***	***1.300***	***1.321***	***1.341***	***1.360***	***1.378***	***1.396***	***1.412***	***1.428***
	0.000	0.000	0.000	0.000	0.000	0.000	0.000	0.000	0.000	0.000	0.000	0.000	0.000	0.000
	0.118	0.140	0.162	0.185	0.208	0.230	0.252	0.273	0.293	0.312	0.331	0.349	0.366	0.382
1.10	***1.118***	***1.140***	***1.163***	***1.185***	***1.208***	***1.230***	***1.252***	***1.273***	***1.293***	***1.313***	***1.332***	***1.349***	***1.367***	***1.383***

Table Four: *ExCP, ExCU, and ECP&U* when Average Is on Scrap Side of Target and the Cost of Rework Is 0.1% of the Cost of Scrap

C_p or P_p value

C_{pk} or P_{pk}	2.90	3.00	3.20	3.40	3.60	3.80	4.00	4.50	5.00	5.50	6.0	8.0	12.0	20.0
	13.97	13.97	13.97	13.97	13.97	13.97	13.97	13.97	13.97	13.97	13.97	13.97	13.97	13.97
	0.905	0.908	0.913	0.919	0.923	0.927	0.931	0.939	0.945	0.950	0.954	0.967	0.980	0.993
-0.50	***15.87***	***15.88***	***15.88***	***15.89***	***15.89***	***15.90***	***15.90***	***15.91***	***15.91***	***15.92***	***15.92***	***15.94***	***15.95***	***15.96***
	7.690	7.690	7.690	7.690	7.690	7.690	7.690	7.690	7.690	7.690	7.690	7.690	7.690	7.690
	0.894	0.898	0.904	0.909	0.914	0.919	0.923	0.931	0.938	0.944	0.949	0.963	0.977	0.990
-0.40	***9.585***	***9.588***	***9.594***	***9.600***	***9.605***	***9.609***	***9.613***	***9.622***	***9.629***	***9.635***	***9.639***	***9.653***	***9.667***	***9.681***
	4.433	4.433	4.433	4.433	4.433	4.433	4.433	4.433	4.433	4.433	4.433	4.433	4.433	4.433
	0.882	0.886	0.893	0.899	0.904	0.909	0.914	0.923	0.931	0.937	0.943	0.958	0.973	0.987
-0.30	***6.315***	***6.319***	***6.326***	***6.332***	***6.337***	***6.342***	***6.347***	***6.356***	***6.364***	***6.370***	***6.376***	***6.391***	***6.406***	***6.420***
	2.646	2.646	2.646	2.646	2.646	2.646	2.646	2.646	2.646	2.646	2.646	2.646	2.646	2.646
	0.868	0.872	0.879	0.886	0.892	0.898	0.903	0.913	0.922	0.929	0.935	0.952	0.969	0.984
-0.20	***4.514***	***4.518***	***4.526***	***4.533***	***4.539***	***4.544***	***4.549***	***4.560***	***4.568***	***4.575***	***4.581***	***4.598***	***4.615***	***4.631***
	1.617	1.617	1.617	1.617	1.617	1.617	1.617	1.617	1.617	1.617	1.617	1.617	1.617	1.617
	0.850	0.855	0.864	0.871	0.878	0.884	0.890	0.902	0.911	0.919	0.926	0.945	0.964	0.981
-0.10	***3.468***	***3.472***	***3.481***	***3.489***	***3.495***	***3.502***	***3.507***	***3.519***	***3.529***	***3.537***	***3.543***	***3.562***	***3.581***	***3.598***
	1.000	1.000	1.000	1.000	1.000	1.000	1.000	1.000	1.000	1.000	1.000	1.000	1.000	1.000
	0.830	0.835	0.845	0.854	0.861	0.868	0.875	0.888	0.899	0.908	0.915	0.936	0.958	0.977
0.00	***2.830***	***2.835***	***2.845***	***2.854***	***2.861***	***2.868***	***2.875***	***2.888***	***2.899***	***2.908***	***2.915***	***2.936***	***2.958***	***2.977***
	0.618	0.618	0.618	0.618	0.618	0.618	0.618	0.618	0.618	0.618	0.618	0.618	0.618	0.618
	0.806	0.812	0.823	0.833	0.841	0.849	0.856	0.872	0.884	0.894	0.903	0.927	0.951	0.972
0.10	***2.425***	***2.431***	***2.441***	***2.451***	***2.460***	***2.468***	***2.475***	***2.490***	***2.502***	***2.512***	***2.521***	***2.545***	***2.570***	***2.590***
	0.378	0.378	0.378	0.378	0.378	0.378	0.378	0.378	0.378	0.378	0.378	0.378	0.378	0.378
	0.778	0.785	0.797	0.808	0.818	0.827	0.835	0.852	0.866	0.878	0.888	0.915	0.943	0.967
0.20	***2.156***	***2.163***	***2.175***	***2.186***	***2.196***	***2.205***	***2.213***	***2.230***	***2.244***	***2.256***	***2.266***	***2.293***	***2.321***	***2.345***
	0.226	0.226	0.226	0.226	0.226	0.226	0.226	0.226	0.226	0.226	0.226	0.226	0.226	0.226
	0.746	0.754	0.768	0.780	0.791	0.801	0.811	0.830	0.846	0.859	0.871	0.902	0.934	0.961
0.30	***1.972***	***1.979***	***1.993***	***2.006***	***2.017***	***2.027***	***2.036***	***2.056***	***2.072***	***2.085***	***2.096***	***2.128***	***2.160***	***2.187***
	0.130	0.130	0.130	0.130	0.130	0.130	0.130	0.130	0.130	0.130	0.130	0.130	0.130	0.130
	0.710	0.718	0.734	0.748	0.760	0.772	0.782	0.805	0.823	0.838	0.851	0.887	0.924	0.954
0.40	***1.840***	***1.848***	***1.864***	***1.878***	***1.890***	***1.902***	***1.912***	***1.935***	***1.953***	***1.968***	***1.981***	***2.017***	***2.054***	***2.084***
	0.072	0.072	0.072	0.072	0.072	0.072	0.072	0.072	0.072	0.072	0.072	0.072	0.072	0.072
	0.669	0.679	0.696	0.712	0.726	0.739	0.751	0.776	0.797	0.814	0.829	0.870	0.912	0.947
0.50	***1.741***	***1.750***	***1.768***	***1.784***	***1.798***	***1.811***	***1.823***	***1.848***	***1.869***	***1.886***	***1.900***	***1.941***	***1.984***	***2.018***
	0.037	0.037	0.037	0.037	0.037	0.037	0.037	0.037	0.037	0.037	0.037	0.037	0.037	0.037
	0.625	0.636	0.656	0.673	0.689	0.704	0.717	0.745	0.769	0.788	0.805	0.851	0.899	0.939
0.60	***1.663***	***1.673***	***1.693***	***1.710***	***1.726***	***1.741***	***1.754***	***1.783***	***1.806***	***1.825***	***1.842***	***1.888***	***1.936***	***1.976***
	0.018	0.018	0.018	0.018	0.018	0.018	0.018	0.018	0.018	0.018	0.018	0.018	0.018	0.018
	0.580	0.591	0.613	0.632	0.650	0.666	0.681	0.712	0.739	0.760	0.779	0.831	0.885	0.930
0.70	***1.598***	***1.610***	***1.631***	***1.651***	***1.668***	***1.684***	***1.699***	***1.731***	***1.757***	***1.778***	***1.797***	***1.849***	***1.903***	***1.948***
	0.008	0.008	0.008	0.008	0.008	0.008	0.008	0.008	0.008	0.008	0.008	0.008	0.008	0.008
	0.533	0.546	0.569	0.590	0.610	0.627	0.644	0.679	0.707	0.731	0.752	0.810	0.871	0.921
0.80	***1.541***	***1.554***	***1.578***	***1.599***	***1.618***	***1.636***	***1.652***	***1.687***	***1.716***	***1.740***	***1.760***	***1.818***	***1.879***	***1.929***
	0.003	0.003	0.003	0.003	0.003	0.003	0.003	0.003	0.003	0.003	0.003	0.003	0.003	0.003
	0.487	0.500	0.526	0.548	0.569	0.588	0.606	0.644	0.676	0.702	0.725	0.789	0.856	0.912
0.90	***1.490***	***1.504***	***1.529***	***1.552***	***1.573***	***1.592***	***1.610***	***1.648***	***1.679***	***1.706***	***1.728***	***1.792***	***1.859***	***1.915***
	0.001	0.001	0.001	0.001	0.001	0.001	0.001	0.001	0.001	0.001	0.001	0.001	0.001	0.001
	0.442	0.456	0.483	0.507	0.529	0.550	0.569	0.610	0.644	0.673	0.697	0.767	0.841	0.903
1.00	***1.443***	***1.457***	***1.484***	***1.508***	***1.531***	***1.551***	***1.570***	***1.611***	***1.645***	***1.674***	***1.698***	***1.768***	***1.842***	***1.904***
	0.000	0.000	0.000	0.000	0.000	0.000	0.000	0.000	0.000	0.000	0.000	0.000	0.000	0.000
	0.398	0.413	0.441	0.467	0.491	0.512	0.532	0.576	0.613	0.643	0.670	0.746	0.826	0.893
1.10	***1.399***	***1.414***	***1.442***	***1.467***	***1.491***	***1.513***	***1.533***	***1.577***	***1.613***	***1.644***	***1.670***	***1.746***	***1.826***	***1.894***

Table Four: *ExCP, ExCU, and ECP&U* when Average Is on Scrap Side of Target and the Cost of Rework Is 0.1% of the Cost of Scrap

C_{pk} or P_{pk}	C_p or P_p value 1.50	1.60	1.70	1.80	1.90	2.00	2.10	2.20	2.30	2.40	2.50	2.60	2.70	2.80
	0.000	0.000	0.000	0.000	0.000	0.000	0.000	0.000	0.000	0.000	0.000	0.000	0.000	0.000
	0.085	0.104	0.124	0.145	0.166	0.188	0.209	0.229	0.250	0.269	0.288	0.306	0.324	0.341
1.20	***1.085***	***1.104***	***1.124***	***1.145***	***1.166***	***1.188***	***1.209***	***1.230***	***1.250***	***1.269***	***1.288***	***1.306***	***1.324***	***1.341***
	0.000	0.000	0.000	0.000	0.000	0.000	0.000	0.000	0.000	0.000	0.000	0.000	0.000	0.000
	0.059	0.074	0.092	0.111	0.130	0.150	0.170	0.190	0.210	0.229	0.248	0.266	0.284	0.301
1.30	***1.059***	***1.074***	***1.092***	***1.111***	***1.130***	***1.150***	***1.170***	***1.190***	***1.210***	***1.229***	***1.248***	***1.266***	***1.284***	***1.301***
	0.000	0.000	0.000	0.000	0.000	0.000	0.000	0.000	0.000	0.000	0.000	0.000	0.000	0.000
	0.039	0.052	0.066	0.082	0.099	0.117	0.136	0.155	0.174	0.193	0.211	0.229	0.247	0.264
1.40	***1.039***	***1.052***	***1.066***	***1.082***	***1.099***	***1.117***	***1.136***	***1.155***	***1.174***	***1.193***	***1.211***	***1.229***	***1.247***	***1.264***
	0.000	0.000	0.000	0.000	0.000	0.000	0.000	0.000	0.000	0.000	0.000	0.000	0.000	0.000
	0.025	0.034	0.046	0.059	0.074	0.090	0.107	0.124	0.142	0.160	0.178	0.195	0.213	0.230
1.50	***1.025***	***1.034***	***1.046***	***1.059***	***1.074***	***1.090***	***1.107***	***1.124***	***1.142***	***1.160***	***1.178***	***1.195***	***1.213***	***1.230***
		0.000	0.000	0.000	0.000	0.000	0.000	0.000	0.000	0.000	0.000	0.000	0.000	0.000
		0.022	0.030	0.041	0.053	0.066	0.081	0.097	0.114	0.130	0.147	0.164	0.181	0.198
1.60		***1.022***	***1.030***	***1.041***	***1.053***	***1.066***	***1.081***	***1.097***	***1.114***	***1.130***	***1.147***	***1.164***	***1.181***	***1.198***
			0.000	0.000	0.000	0.000	0.000	0.000	0.000	0.000	0.000	0.000	0.000	0.000
			0.019	0.027	0.037	0.048	0.060	0.074	0.089	0.104	0.120	0.136	0.152	0.169
1.70 »			***1.019***	***1.027***	***1.037***	***1.048***	***1.060***	***1.074***	***1.089***	***1.104***	***1.120***	***1.136***	***1.152***	***1.169***
				0.000	0.000	0.000	0.000	0.000	0.000	0.000	0.000	0.000	0.000	0.000
				0.017	0.024	0.033	0.043	0.055	0.068	0.082	0.096	0.111	0.126	0.142
1.80 »				***1.017***	***1.024***	***1.033***	***1.043***	***1.055***	***1.068***	***1.082***	***1.096***	***1.111***	***1.126***	***1.142***
					0.000	0.000	0.000	0.000	0.000	0.000	0.000	0.000	0.000	0.000
					0.015	0.022	0.030	0.039	0.050	0.062	0.075	0.089	0.103	0.117
1.90 »					***1.015***	***1.022***	***1.030***	***1.039***	***1.050***	***1.062***	***1.075***	***1.089***	***1.103***	***1.117***
						0.000	0.000	0.000	0.000	0.000	0.000	0.000	0.000	0.000
						0.014	0.020	0.027	0.036	0.046	0.057	0.070	0.082	0.096
2.00 »						***1.014***	***1.020***	***1.027***	***1.036***	***1.046***	***1.057***	***1.070***	***1.082***	***1.096***
							0.000	0.000	0.000	0.000	0.000	0.000	0.000	0.000
							0.013	0.018	0.025	0.033	0.043	0.053	0.064	0.077
2.10 »							***1.013***	***1.018***	***1.025***	***1.033***	***1.043***	***1.053***	***1.064***	***1.077***
								0.000	0.000	0.000	0.000	0.000	0.000	0.000
								0.011	0.017	0.023	0.031	0.039	0.049	0.060
2.20 »								***1.011***	***1.017***	***1.023***	***1.031***	***1.039***	***1.049***	***1.060***
									0.000	0.000	0.000	0.000	0.000	0.000
									0.011	0.015	0.021	0.028	0.036	0.046
2.30 »									***1.011***	***1.015***	***1.021***	***1.028***	***1.036***	***1.046***
										0.000	0.000	0.000	0.000	0.000
										0.010	0.014	0.020	0.026	0.034
2.40 »										***1.010***	***1.014***	***1.020***	***1.026***	***1.034***
											0.000	0.000	0.000	0.000
											0.009	0.013	0.018	0.024
2.50 »											***1.009***	***1.013***	***1.018***	***1.024***
												0.000	0.000	0.000
												0.008	0.012	0.017
2.60 »												***1.008***	***1.012***	***1.017***
													0.000	0.000
													0.008	0.011
2.70 »													***1.008***	***1.011***
														0.000
														0.007
2.80 »														***1.007***

Table Four: *ExCP, ExCU, and ECP&U* when Average Is on Scrap Side of Target and the Cost of Rework Is 0.1% of the Cost of Scrap

C_{pk} or P_{pk}	C_p or P_p value 2.90	3.00	3.20	3.40	3.60	3.80	4.00	4.50	5.00	5.50	6.0	8.0	12.0	20.0
	0.000	0.000	0.000	0.000	0.000	0.000	0.000	0.000	0.000	0.000	0.000	0.000	0.000	0.000
	0.357	0.372	0.401	0.428	0.453	0.476	0.497	0.543	0.582	0.615	0.643	0.724	0.811	0.884
1.20	***1.357***	***1.372***	***1.402***	***1.428***	***1.453***	***1.476***	***1.497***	***1.543***	***1.582***	***1.615***	***1.643***	***1.724***	***1.811***	***1.884***
	0.000	0.000	0.000	0.000	0.000	0.000	0.000	0.000	0.000	0.000	0.000	0.000	0.000	0.000
	0.318	0.333	0.363	0.391	0.417	0.440	0.463	0.511	0.552	0.587	0.617	0.703	0.796	0.874
1.30	***1.318***	***1.333***	***1.363***	***1.391***	***1.417***	***1.441***	***1.463***	***1.511***	***1.552***	***1.587***	***1.617***	***1.703***	***1.796***	***1.875***
	0.000	0.000	0.000	0.000	0.000	0.000	0.000	0.000	0.000	0.000	0.000	0.000	0.000	0.000
	0.281	0.297	0.327	0.356	0.382	0.407	0.429	0.480	0.523	0.559	0.591	0.682	0.781	0.865
1.40	***1.281***	***1.297***	***1.327***	***1.356***	***1.382***	***1.407***	***1.429***	***1.480***	***1.523***	***1.559***	***1.591***	***1.682***	***1.781***	***1.865***
	0.000	0.000	0.000	0.000	0.000	0.000	0.000	0.000	0.000	0.000	0.000	0.000	0.000	0.000
	0.246	0.262	0.293	0.322	0.349	0.374	0.398	0.450	0.494	0.533	0.566	0.662	0.766	0.856
1.50	***1.246***	***1.262***	***1.293***	***1.322***	***1.349***	***1.374***	***1.398***	***1.450***	***1.494***	***1.533***	***1.566***	***1.662***	***1.766***	***1.856***
	0.000	0.000	0.000	0.000	0.000	0.000	0.000	0.000	0.000	0.000	0.000	0.000	0.000	0.000
	0.214	0.230	0.261	0.290	0.317	0.343	0.367	0.421	0.467	0.506	0.541	0.642	0.752	0.847
1.60	***1.214***	***1.230***	***1.261***	***1.290***	***1.317***	***1.343***	***1.367***	***1.421***	***1.467***	***1.506***	***1.541***	***1.642***	***1.752***	***1.847***
	0.000	0.000	0.000	0.000	0.000	0.000	0.000	0.000	0.000	0.000	0.000	0.000	0.000	0.000
	0.184	0.200	0.231	0.260	0.287	0.313	0.338	0.393	0.440	0.481	0.517	0.622	0.738	0.838
1.70	***1.184***	***1.200***	***1.231***	***1.260***	***1.287***	***1.313***	***1.338***	***1.393***	***1.440***	***1.481***	***1.517***	***1.622***	***1.738***	***1.838***
	0.000	0.000	0.000	0.000	0.000	0.000	0.000	0.000	0.000	0.000	0.000	0.000	0.000	0.000
	0.157	0.172	0.202	0.231	0.259	0.285	0.309	0.365	0.414	0.456	0.493	0.602	0.723	0.828
1.80	***1.157***	***1.172***	***1.202***	***1.231***	***1.259***	***1.285***	***1.309***	***1.365***	***1.414***	***1.456***	***1.493***	***1.602***	***1.723***	***1.828***
	0.000	0.000	0.000	0.000	0.000	0.000	0.000	0.000	0.000	0.000	0.000	0.000	0.000	0.000
	0.132	0.147	0.176	0.204	0.232	0.258	0.283	0.339	0.389	0.432	0.470	0.583	0.709	0.819
1.90	***1.132***	***1.147***	***1.176***	***1.204***	***1.232***	***1.258***	***1.283***	***1.339***	***1.389***	***1.432***	***1.470***	***1.583***	***1.709***	***1.819***
	0.000	0.000	0.000	0.000	0.000	0.000	0.000	0.000	0.000	0.000	0.000	0.000	0.000	0.000
	0.110	0.123	0.151	0.179	0.206	0.232	0.257	0.314	0.364	0.409	0.448	0.564	0.695	0.810
2.00	***1.110***	***1.123***	***1.151***	***1.179***	***1.206***	***1.232***	***1.257***	***1.314***	***1.364***	***1.409***	***1.448***	***1.564***	***1.695***	***1.810***
	0.000	0.000	0.000	0.000	0.000	0.000	0.000	0.000	0.000	0.000	0.000	0.000	0.000	0.000
	0.089	0.102	0.129	0.156	0.182	0.208	0.233	0.290	0.341	0.386	0.426	0.546	0.681	0.801
2.10	***1.089***	***1.102***	***1.129***	***1.156***	***1.182***	***1.208***	***1.233***	***1.290***	***1.341***	***1.386***	***1.426***	***1.546***	***1.681***	***1.801***
	0.000	0.000	0.000	0.000	0.000	0.000	0.000	0.000	0.000	0.000	0.000	0.000	0.000	0.000
	0.071	0.083	0.109	0.134	0.160	0.185	0.209	0.267	0.318	0.364	0.404	0.527	0.668	0.792
2.20	***1.071***	***1.083***	***1.109***	***1.134***	***1.160***	***1.185***	***1.209***	***1.267***	***1.318***	***1.364***	***1.404***	***1.527***	***1.668***	***1.792***
	0.000	0.000	0.000	0.000	0.000	0.000	0.000	0.000	0.000	0.000	0.000	0.000	0.000	0.000
	0.056	0.067	0.090	0.114	0.139	0.164	0.188	0.244	0.296	0.342	0.383	0.509	0.654	0.784
2.30	***1.056***	***1.067***	***1.090***	***1.114***	***1.139***	***1.164***	***1.188***	***1.244***	***1.296***	***1.342***	***1.383***	***1.509***	***1.654***	***1.784***
	0.000	0.000	0.000	0.000	0.000	0.000	0.000	0.000	0.000	0.000	0.000	0.000	0.000	0.000
	0.043	0.052	0.073	0.096	0.120	0.143	0.167	0.223	0.275	0.321	0.363	0.492	0.641	0.775
2.40	***1.043***	***1.052***	***1.073***	***1.096***	***1.120***	***1.143***	***1.167***	***1.223***	***1.275***	***1.321***	***1.363***	***1.492***	***1.641***	***1.775***
	0.000	0.000	0.000	0.000	0.000	0.000	0.000	0.000	0.000	0.000	0.000	0.000	0.000	0.000
	0.032	0.040	0.059	0.080	0.102	0.125	0.148	0.203	0.254	0.301	0.343	0.474	0.628	0.766
2.50	***1.032***	***1.040***	***1.059***	***1.080***	***1.102***	***1.125***	***1.148***	***1.203***	***1.254***	***1.301***	***1.343***	***1.474***	***1.628***	***1.766***
	0.000	0.000	0.000	0.000	0.000	0.000	0.000	0.000	0.000	0.000	0.000	0.000	0.000	0.000
	0.023	0.030	0.046	0.065	0.086	0.107	0.129	0.184	0.235	0.282	0.324	0.457	0.614	0.757
2.60	***1.023***	***1.030***	***1.046***	***1.065***	***1.086***	***1.107***	***1.129***	***1.184***	***1.235***	***1.282***	***1.324***	***1.457***	***1.614***	***1.757***
	0.000	0.000	0.000	0.000	0.000	0.000	0.000	0.000	0.000	0.000	0.000	0.000	0.000	0.000
	0.016	0.021	0.035	0.052	0.071	0.091	0.113	0.165	0.216	0.263	0.306	0.441	0.601	0.749
2.70	***1.016***	***1.021***	***1.035***	***1.052***	***1.071***	***1.091***	***1.113***	***1.165***	***1.216***	***1.263***	***1.306***	***1.441***	***1.601***	***1.749***
	0.000	0.000	0.000	0.000	0.000	0.000	0.000	0.000	0.000	0.000	0.000	0.000	0.000	0.000
	0.010	0.015	0.026	0.041	0.058	0.077	0.097	0.148	0.198	0.245	0.288	0.424	0.589	0.740
2.80	***1.010***	***1.015***	***1.026***	***1.041***	***1.058***	***1.077***	***1.097***	***1.148***	***1.198***	***1.245***	***1.288***	***1.424***	***1.589***	***1.740***

Table Four: *ExCP, ExCU, and ECP&U* when Average Is on Scrap Side of Target and the Cost of Rework Is 0.1% of the Cost of Scrap

C_{pk} or P_{pk}	C_p or P_p value 2.90	3.00	3.20	3.40	3.60	3.80	4.00	4.50	5.00	5.50	6.0	8.0	12.0	20.0
	0.000	0.000	0.000	0.000	0.000	0.000	0.000	0.000	0.000	0.000	0.000	0.000	0.000	0.000
	0.007	0.010	0.019	0.031	0.046	0.064	0.083	0.132	0.181	0.227	0.270	0.408	0.576	0.731
2.90	***1.007***	***1.010***	***1.019***	***1.031***	***1.046***	***1.064***	***1.083***	***1.132***	***1.181***	***1.227***	***1.270***	***1.408***	***1.576***	***1.731***
		0.000	0.000	0.000	0.000	0.000	0.000	0.000	0.000	0.000	0.000	0.000	0.000	0.000
		0.006	0.013	0.023	0.036	0.052	0.069	0.117	0.164	0.210	0.253	0.392	0.563	0.723
3.00		***1.006***	***1.013***	***1.023***	***1.036***	***1.052***	***1.069***	***1.117***	***1.164***	***1.210***	***1.253***	***1.392***	***1.563***	***1.723***
			0.000	0.000	0.000	0.000	0.000	0.000	0.000	0.000	0.000	0.000	0.000	0.000
			0.005	0.011	0.021	0.033	0.047	0.089	0.134	0.179	0.221	0.362	0.539	0.706
3.20 »			***1.005***	***1.011***	***1.021***	***1.033***	***1.047***	***1.089***	***1.134***	***1.179***	***1.221***	***1.362***	***1.539***	***1.706***
				0.000	0.000	0.000	0.000	0.000	0.000	0.000	0.000	0.000	0.000	0.000
				0.005	0.010	0.018	0.029	0.065	0.107	0.149	0.191	0.332	0.514	0.689
3.40 »				***1.005***	***1.010***	***1.018***	***1.029***	***1.065***	***1.107***	***1.149***	***1.191***	***1.332***	***1.514***	***1.689***
					0.000	0.000	0.000	0.000	0.000	0.000	0.000	0.000	0.000	0.000
					0.004	0.009	0.017	0.045	0.083	0.123	0.163	0.304	0.491	0.673
3.60 »					***1.004***	***1.009***	***1.017***	***1.045***	***1.083***	***1.123***	***1.163***	***1.304***	***1.491***	***1.673***
						0.000	0.000	0.000	0.000	0.000	0.000	0.000	0.000	0.000
						0.004	0.008	0.030	0.062	0.099	0.138	0.277	0.468	0.656
3.80 »						***1.004***	***1.008***	***1.030***	***1.062***	***1.099***	***1.138***	***1.277***	***1.468***	***1.656***
							0.000	0.000	0.000	0.000	0.000	0.000	0.000	0.000
							0.003	0.018	0.044	0.078	0.114	0.252	0.445	0.640
4.00 »							***1.003***	***1.018***	***1.044***	***1.078***	***1.114***	***1.252***	***1.445***	***1.640***
								0.000	0.000	0.000	0.000	0.000	0.000	0.000
								0.003	0.014	0.037	0.066	0.193	0.391	0.601
4.50 »								***1.003***	***1.014***	***1.037***	***1.066***	***1.193***	***1.391***	***1.601***
									0.000	0.000	0.000	0.000	0.000	0.000
									0.002	0.012	0.031	0.142	0.341	0.563
5.00 »									***1.002***	***1.012***	***1.031***	***1.142***	***1.341***	***1.563***
										0.000	0.000	0.000	0.000	0.000
										0.002	0.010	0.099	0.294	0.526
5.50 »										***1.002***	***1.010***	***1.099***	***1.294***	***1.526***
											0.000	0.000	0.000	0.000
											0.002	0.064	0.251	0.490
6.00 »											***1.002***	***1.064***	***1.251***	***1.490***
												0.000	0.000	0.000
												0.001	0.112	0.360
8.00 »												***1.001***	***1.112***	***1.360***
													0.000	0.000
													0.000	0.160
12.0 »													***1.000***	***1.160***
														0.000
														0.000
20.0 »														***1.000***

Table Five: The *Excess Costs of Production, Excess Costs of Use,* and *Effective Cost of Production and Use* when Average Is on the Rework Side of Target and the Cost of Rework Is Equal to the Cost of Scrap

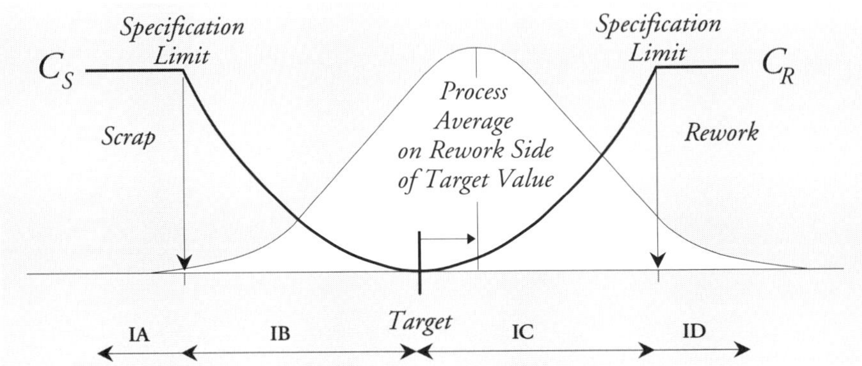

Table Five: Excess Cost Function when Average Is on Rework Side of Target and Cost of Rework Is Equal to Cost of Scrap

Corresponding to each pairing for C_p and C_{pk} there are three entries stacked in a column.

The topmost of these is the *Excess Costs of Production,* *ExCP*

The middle value is the *Excess Costs of Use,* *ExCU*

And the bottom value (in bold) is the *Effective Cost of Production and Use,* ***ECP&U***

The entries in Table Five assume that the Cost of Rework, C_R, the Cost of Scrap, C_S, and the Nominal Cost of Production, *NC*, are all equal. Table A.9 contains the normal theory values for **IA, IB, IC,** and **ID**. The general and specific formulas used for Table Five are:

$$ECP\&U = 1.0 + \frac{\frac{C_S}{NC}\left\{\mathbf{IA} + \frac{C_R}{C_S}\mathbf{ID}\right\}}{1.0 - \mathbf{IA}} + \frac{\frac{C_S}{NC}\left\{\mathbf{IB} + \frac{C_R}{C_S}\mathbf{IC}\right\}}{1.0 - \mathbf{IA}}$$

$$= 1.0 + \frac{\{\mathbf{IA} + \mathbf{ID}\}}{1.0 - \mathbf{IA}} + \frac{\{\mathbf{IB} + \mathbf{IC}\}}{1.0 - \mathbf{IA}}$$

$$= 1.0 + \textit{Excess Costs of Production} + \textit{Excess Costs of Use}$$

When the Cost of Rework is less than the Cost of Scrap appropriate values may be obtained by interpolating between the entry in Table Five and the corresponding entry in Table Six.

When the Cost of Scrap exceeds the Nominal Cost, the *Excess Costs* in Table Five, or any interpolated values, may be inflated as shown in the general formula above.

Table Five: *ExCP, ExCU, and ECP&U* when Average Is on Rework Side of Target and the Cost of Rework Is Equal to the Cost of Scrap

C_{pk} or P_{pk}	C_p or P_p values 0.10	0.20	0.30	0.40	0.50	0.60	0.70	0.80	0.90	1.00	1.10	1.20	1.30	1.40
	0.968	0.940	0.934	0.933	0.933	0.933	0.933	0.933	0.933	0.933	0.933	0.933	0.933	0.933
	0.017	0.024	0.029	0.034	0.038	0.041	0.044	0.046	0.048	0.050	0.051	0.052	0.053	0.054
–0.50	***1.985***	***1.964***	***1.963***	***1.967***	***1.971***	***1.975***	***1.977***	***1.980***	***1.981***	***1.983***	***1.984***	***1.986***	***1.986***	***1.987***
	0.955	0.901	0.887	0.885	0.885	0.885	0.885	0.885	0.885	0.885	0.885	0.885	0.885	0.885
	0.028	0.039	0.047	0.055	0.061	0.067	0.072	0.077	0.080	0.083	0.085	0.088	0.089	0.091
–0.40	***1.983***	***1.940***	***1.934***	***1.940***	***1.946***	***1.952***	***1.957***	***1.962***	***1.965***	***1.968***	***1.970***	***1.973***	***1.974***	***1.976***
	0.946	0.849	0.822	0.817	0.816	0.816	0.816	0.816	0.816	0.816	0.816	0.816	0.816	0.816
	0.042	0.059	0.070	0.081	0.092	0.101	0.110	0.117	0.123	0.128	0.132	0.136	0.139	0.142
–0.30	***1.988***	***1.908***	***1.892***	***1.898***	***1.908***	***1.917***	***1.926***	***1.933***	***1.938***	***1.943***	***1.948***	***1.951***	***1.955***	***1.958***
	0.950	0.790	0.740	0.728	0.726	0.726	0.726	0.726	0.726	0.726	0.726	0.726	0.726	0.726
	0.060	0.084	0.098	0.111	0.126	0.140	0.153	0.164	0.173	0.181	0.188	0.194	0.199	0.204
–0.20	***2.010***	***1.874***	***1.838***	***1.839***	***1.852***	***1.866***	***1.879***	***1.890***	***1.899***	***1.907***	***1.914***	***1.920***	***1.925***	***1.930***
	0.983	0.734	0.647	0.624	0.619	0.618	0.618	0.618	0.618	0.618	0.618	0.618	0.618	0.618
	0.080	0.111	0.127	0.142	0.160	0.179	0.197	0.212	0.226	0.238	0.248	0.257	0.265	0.272
–0.10	***2.063***	***1.845***	***1.774***	***1.766***	***1.779***	***1.797***	***1.814***	***1.830***	***1.844***	***1.856***	***1.866***	***1.875***	***1.883***	***1.890***
	1.067	0.695	0.556	0.512	0.502	0.500	0.500	0.500	0.500	0.500	0.500	0.500	0.500	0.500
	0.103	0.141	0.156	0.170	0.189	0.211	0.233	0.254	0.273	0.290	0.304	0.317	0.328	0.338
0.00	***2.169***	***1.836***	***1.712***	***1.682***	***1.691***	***1.711***	***1.733***	***1.754***	***1.773***	***1.790***	***1.804***	***1.817***	***1.828***	***1.838***
	1.237	0.694	0.481	0.407	0.387	0.383	0.382	0.382	0.382	0.382	0.382	0.382	0.382	0.382
	0.126	0.170	0.183	0.192	0.208	0.231	0.257	0.282	0.306	0.328	0.347	0.364	0.379	0.393
0.10	***2.362***	***1.864***	***1.664***	***1.600***	***1.595***	***1.614***	***1.639***	***1.665***	***1.688***	***1.710***	***1.729***	***1.746***	***1.761***	***1.775***
		0.756	0.440	0.322	0.285	0.276	0.274	0.274	0.274	0.274	0.274	0.274	0.274	0.274
		0.198	0.208	0.210	0.218	0.237	0.263	0.291	0.319	0.345	0.369	0.390	0.410	0.427
0.20		***1.953***	***1.648***	***1.531***	***1.502***	***1.512***	***1.537***	***1.566***	***1.593***	***1.619***	***1.643***	***1.665***	***1.684***	***1.701***
			0.451	0.269	0.206	0.188	0.185	0.184	0.184	0.184	0.184	0.184	0.184	0.184
			0.231	0.224	0.220	0.230	0.252	0.280	0.310	0.340	0.367	0.393	0.416	0.437
0.30 »			***1.683***	***1.493***	***1.426***	***1.418***	***1.437***	***1.464***	***1.494***	***1.524***	***1.551***	***1.577***	***1.600***	***1.621***
				0.260	0.157	0.124	0.117	0.115	0.115	0.115	0.115	0.115	0.115	0.115
				0.238	0.222	0.218	0.230	0.253	0.283	0.313	0.343	0.372	0.398	0.422
0.40 »				***1.499***	***1.378***	***1.342***	***1.346***	***1.369***	***1.398***	***1.428***	***1.458***	***1.487***	***1.513***	***1.538***
					0.143	0.086	0.071	0.067	0.067	0.067	0.067	0.067	0.067	0.067
					0.228	0.207	0.205	0.219	0.243	0.272	0.303	0.333	0.361	0.388
0.50 »					***1.371***	***1.294***	***1.275***	***1.286***	***1.310***	***1.339***	***1.370***	***1.399***	***1.428***	***1.455***
						0.075	0.044	0.037	0.036	0.036	0.036	0.036	0.036	0.036
						0.206	0.186	0.185	0.200	0.225	0.253	0.283	0.312	0.340
0.60 »						***1.281***	***1.230***	***1.222***	***1.236***	***1.260***	***1.289***	***1.318***	***1.348***	***1.376***
							0.036	0.021	0.018	0.018	0.018	0.018	0.018	0.018
							0.180	0.161	0.163	0.178	0.202	0.229	0.257	0.286
0.70 »							***1.216***	***1.183***	***1.181***	***1.196***	***1.220***	***1.247***	***1.275***	***1.303***
								0.017	0.010	0.008	0.008	0.008	0.008	0.008
								0.153	0.138	0.141	0.156	0.178	0.204	0.231
0.80 »								***1.170***	***1.147***	***1.149***	***1.164***	***1.187***	***1.212***	***1.239***
									0.007	0.004	0.004	0.003	0.003	0.003
									0.129	0.116	0.121	0.136	0.156	0.180
0.90 »									***1.136***	***1.120***	***1.124***	***1.139***	***1.160***	***1.184***
										0.003	0.002	0.001	0.001	0.001
										0.108	0.098	0.103	0.117	0.137
1.00 »										***1.111***	***1.100***	***1.105***	***1.119***	***1.138***
											0.001	0.001	0.000	0.000
											0.091	0.083	0.089	0.102
1.10 »											***1.092***	***1.084***	***1.089***	***1.103***

C_{pk}	C_p 1.20	1.30	1.40
	0.000	0.000	0.000
	0.077	0.071	0.077
1.20	***1.077***	***1.072***	***1.077***
		0.000	0.000
		0.066	0.062
1.30		***1.066***	***1.062***
			0.000
			0.057
1.40			***1.057***

Table Five: *ExCP, ExCU, and ECP&U when* Average Is on Rework Side of Target and the Cost of Rework Is Equal to the Cost of Scrap

C_{pk} or P_{pk}	C_p or P_p value 1.50	1.60	1.70	1.80	1.90	2.00	2.10	2.20	2.30	2.40	2.50	2.60	2.70	2.80
	0.933	0.933	0.933	0.933	0.933	0.933	0.933	0.933	0.933	0.933	0.933	0.933	0.933	0.933
	0.055	0.056	0.056	0.057	0.057	0.058	0.058	0.059	0.059	0.059	0.059	0.060	0.060	0.060
–0.50	***1.988***	***1.989***	***1.989***	***1.990***	***1.990***	***1.991***	***1.991***	***1.992***	***1.992***	***1.992***	***1.993***	***1.993***	***1.993***	***1.993***
	0.885	0.885	0.885	0.885	0.885	0.885	0.885	0.885	0.885	0.885	0.885	0.885	0.885	0.885
	0.093	0.094	0.095	0.096	0.097	0.098	0.099	0.099	0.100	0.100	0.101	0.102	0.102	0.102
–0.40	***1.977***	***1.979***	***1.980***	***1.981***	***1.982***	***1.983***	***1.983***	***1.984***	***1.985***	***1.985***	***1.986***	***1.986***	***1.987***	***1.987***
	0.816	0.816	0.816	0.816	0.816	0.816	0.816	0.816	0.816	0.816	0.816	0.816	0.816	0.816
	0.144	0.146	0.148	0.150	0.152	0.153	0.155	0.156	0.157	0.158	0.159	0.160	0.161	0.162
–0.30	***1.960***	***1.962***	***1.964***	***1.966***	***1.968***	***1.969***	***1.971***	***1.972***	***1.973***	***1.974***	***1.975***	***1.976***	***1.977***	***1.978***
	0.726	0.726	0.726	0.726	0.726	0.726	0.726	0.726	0.726	0.726	0.726	0.726	0.726	0.726
	0.208	0.212	0.215	0.218	0.221	0.223	0.225	0.227	0.229	0.231	0.233	0.234	0.235	0.237
–0.20	***1.934***	***1.937***	***1.941***	***1.944***	***1.946***	***1.949***	***1.951***	***1.953***	***1.955***	***1.957***	***1.958***	***1.960***	***1.961***	***1.962***
	0.618	0.618	0.618	0.618	0.618	0.618	0.618	0.618	0.618	0.618	0.618	0.618	0.618	0.618
	0.279	0.284	0.289	0.294	0.298	0.302	0.305	0.308	0.311	0.314	0.317	0.319	0.321	0.323
–0.10	***1.896***	***1.902***	***1.907***	***1.912***	***1.916***	***1.920***	***1.923***	***1.926***	***1.929***	***1.932***	***1.934***	***1.937***	***1.939***	***1.941***
	0.500	0.500	0.500	0.500	0.500	0.500	0.500	0.500	0.500	0.500	0.500	0.500	0.500	0.500
	0.347	0.356	0.363	0.370	0.376	0.381	0.386	0.391	0.395	0.399	0.403	0.406	0.409	0.412
0.00	***1.847***	***1.856***	***1.863***	***1.870***	***1.876***	***1.881***	***1.886***	***1.891***	***1.895***	***1.899***	***1.903***	***1.906***	***1.909***	***1.912***
	0.382	0.382	0.382	0.382	0.382	0.382	0.382	0.382	0.382	0.382	0.382	0.382	0.382	0.382
	0.405	0.416	0.426	0.435	0.443	0.451	0.458	0.464	0.470	0.476	0.481	0.486	0.490	0.494
0.10	***1.787***	***1.798***	***1.808***	***1.817***	***1.826***	***1.833***	***1.840***	***1.847***	***1.852***	***1.858***	***1.863***	***1.868***	***1.872***	***1.876***
	0.274	0.274	0.274	0.274	0.274	0.274	0.274	0.274	0.274	0.274	0.274	0.274	0.274	0.274
	0.443	0.457	0.470	0.482	0.493	0.503	0.512	0.520	0.528	0.535	0.542	0.548	0.554	0.560
0.20	***1.717***	***1.731***	***1.744***	***1.756***	***1.767***	***1.777***	***1.786***	***1.794***	***1.802***	***1.810***	***1.816***	***1.823***	***1.828***	***1.834***
	0.184	0.184	0.184	0.184	0.184	0.184	0.184	0.184	0.184	0.184	0.184	0.184	0.184	0.184
	0.456	0.474	0.490	0.504	0.518	0.530	0.542	0.552	0.562	0.571	0.580	0.588	0.595	0.602
0.30	***1.640***	***1.658***	***1.674***	***1.688***	***1.702***	***1.714***	***1.726***	***1.736***	***1.746***	***1.755***	***1.764***	***1.772***	***1.779***	***1.786***
	0.115	0.115	0.115	0.115	0.115	0.115	0.115	0.115	0.115	0.115	0.115	0.115	0.115	0.115
	0.445	0.465	0.484	0.502	0.518	0.533	0.547	0.559	0.571	0.582	0.593	0.602	0.611	0.620
0.40	***1.560***	***1.581***	***1.599***	***1.617***	***1.633***	***1.648***	***1.662***	***1.674***	***1.686***	***1.697***	***1.708***	***1.717***	***1.726***	***1.735***
	0.067	0.067	0.067	0.067	0.067	0.067	0.067	0.067	0.067	0.067	0.067	0.067	0.067	0.067
	0.413	0.436	0.458	0.477	0.496	0.513	0.529	0.544	0.558	0.571	0.583	0.594	0.605	0.615
0.50	***1.480***	***1.503***	***1.524***	***1.544***	***1.563***	***1.580***	***1.596***	***1.611***	***1.625***	***1.638***	***1.650***	***1.661***	***1.672***	***1.682***
	0.036	0.036	0.036	0.036	0.036	0.036	0.036	0.036	0.036	0.036	0.036	0.036	0.036	0.036
	0.367	0.392	0.415	0.437	0.458	0.477	0.495	0.511	0.527	0.542	0.555	0.568	0.581	0.592
0.60	***1.403***	***1.428***	***1.451***	***1.473***	***1.494***	***1.513***	***1.531***	***1.547***	***1.563***	***1.578***	***1.591***	***1.604***	***1.617***	***1.628***
	0.018	0.018	0.018	0.018	0.018	0.018	0.018	0.018	0.018	0.018	0.018	0.018	0.018	0.018
	0.313	0.339	0.364	0.387	0.409	0.430	0.450	0.468	0.485	0.501	0.517	0.531	0.544	0.557
0.70	***1.331***	***1.357***	***1.382***	***1.405***	***1.427***	***1.448***	***1.467***	***1.486***	***1.503***	***1.519***	***1.534***	***1.549***	***1.562***	***1.575***
	0.008	0.008	0.008	0.008	0.008	0.008	0.008	0.008	0.008	0.008	0.008	0.008	0.008	0.008
	0.258	0.284	0.309	0.334	0.357	0.379	0.399	0.419	0.437	0.455	0.471	0.487	0.502	0.516
0.80	***1.266***	***1.292***	***1.318***	***1.342***	***1.365***	***1.387***	***1.408***	***1.427***	***1.446***	***1.463***	***1.479***	***1.495***	***1.510***	***1.524***
	0.003	0.003	0.003	0.003	0.003	0.003	0.003	0.003	0.003	0.003	0.003	0.003	0.003	0.003
	0.205	0.231	0.256	0.280	0.304	0.326	0.348	0.368	0.388	0.406	0.424	0.440	0.456	0.471
0.90	***1.209***	***1.234***	***1.259***	***1.284***	***1.307***	***1.330***	***1.351***	***1.372***	***1.391***	***1.410***	***1.427***	***1.444***	***1.459***	***1.474***
	0.001	0.001	0.001	0.001	0.001	0.001	0.001	0.001	0.001	0.001	0.001	0.001	0.001	0.001
	0.159	0.183	0.206	0.230	0.254	0.276	0.298	0.319	0.339	0.358	0.376	0.394	0.410	0.426
1.00	***1.160***	***1.184***	***1.208***	***1.232***	***1.255***	***1.278***	***1.299***	***1.320***	***1.340***	***1.359***	***1.378***	***1.395***	***1.412***	***1.427***
	0.000	0.000	0.000	0.000	0.000	0.000	0.000	0.000	0.000	0.000	0.000	0.000	0.000	0.000
	0.120	0.141	0.162	0.185	0.208	0.230	0.251	0.272	0.293	0.312	0.331	0.349	0.366	0.382
1.10	***1.120***	***1.141***	***1.163***	***1.185***	***1.208***	***1.230***	***1.252***	***1.273***	***1.293***	***1.313***	***1.331***	***1.349***	***1.366***	***1.383***

Table Five: *ExCP, ExCU, and ECP&U* when Average Is on Rework Side of Target and the Cost of Rework Is Equal to the Cost of Scrap

C_{pk} or P_{pk}	C_p or P_p value 2.90	3.00	3.20	3.40	3.60	3.80	4.00	4.50	5.00	5.50	6.0	8.0	12.0	20.0
	0.933	0.933	0.933	0.933	0.933	0.933	0.933	0.933	0.933	0.933	0.933	0.933	0.933	0.933
	0.060	0.061	0.061	0.061	0.062	0.062	0.062	0.063	0.063	0.063	0.064	0.065	0.065	0.066
-0.50	***1.994***	***1.994***	***1.994***	***1.995***	***1.995***	***1.995***	***1.995***	***1.996***	***1.996***	***1.997***	***1.997***	***1.998***	***1.999***	***2.000***
	0.885	0.885	0.885	0.885	0.885	0.885	0.885	0.885	0.885	0.885	0.885	0.885	0.885	0.885
	0.103	0.103	0.104	0.105	0.105	0.106	0.106	0.107	0.108	0.109	0.109	0.111	0.112	0.114
-0.40	***1.988***	***1.988***	***1.989***	***1.990***	***1.990***	***1.991***	***1.991***	***1.992***	***1.993***	***1.994***	***1.994***	***1.996***	***1.997***	***1.999***
	0.816	0.816	0.816	0.816	0.816	0.816	0.816	0.816	0.816	0.816	0.816	0.816	0.816	0.816
	0.162	0.163	0.164	0.165	0.166	0.167	0.168	0.170	0.171	0.173	0.173	0.176	0.179	0.182
-0.30	***1.978***	***1.979***	***1.980***	***1.981***	***1.982***	***1.983***	***1.984***	***1.986***	***1.987***	***1.988***	***1.989***	***1.992***	***1.995***	***1.998***
	0.726	0.726	0.726	0.726	0.726	0.726	0.726	0.726	0.726	0.726	0.726	0.726	0.726	0.726
	0.238	0.239	0.241	0.243	0.245	0.246	0.248	0.250	0.253	0.255	0.256	0.261	0.266	0.270
-0.20	***1.964***	***1.965***	***1.967***	***1.969***	***1.970***	***1.972***	***1.973***	***1.976***	***1.979***	***1.981***	***1.982***	***1.987***	***1.992***	***1.996***
	0.618	0.618	0.618	0.618	0.618	0.618	0.618	0.618	0.618	0.618	0.618	0.618	0.618	0.618
	0.325	0.327	0.330	0.333	0.336	0.338	0.340	0.345	0.348	0.351	0.354	0.361	0.368	0.375
-0.10	***1.943***	***1.945***	***1.948***	***1.951***	***1.953***	***1.956***	***1.958***	***1.962***	***1.966***	***1.969***	***1.972***	***1.979***	***1.986***	***1.993***
	0.500	0.500	0.500	0.500	0.500	0.500	0.500	0.500	0.500	0.500	0.500	0.500	0.500	0.500
	0.415	0.418	0.423	0.427	0.431	0.434	0.437	0.444	0.449	0.454	0.458	0.468	0.479	0.488
0.00	***1.915***	***1.918***	***1.923***	***1.927***	***1.931***	***1.934***	***1.937***	***1.944***	***1.949***	***1.954***	***1.958***	***1.968***	***1.979***	***1.988***
	0.382	0.382	0.382	0.382	0.382	0.382	0.382	0.382	0.382	0.382	0.382	0.382	0.382	0.382
	0.498	0.502	0.509	0.515	0.520	0.525	0.529	0.539	0.546	0.553	0.558	0.573	0.588	0.601
0.10	***1.880***	***1.884***	***1.891***	***1.897***	***1.902***	***1.907***	***1.911***	***1.921***	***1.928***	***1.935***	***1.940***	***1.955***	***1.970***	***1.983***
	0.274	0.274	0.274	0.274	0.274	0.274	0.274	0.274	0.274	0.274	0.274	0.274	0.274	0.274
	0.565	0.570	0.579	0.587	0.594	0.600	0.606	0.619	0.629	0.637	0.644	0.664	0.685	0.702
0.20	***1.839***	***1.844***	***1.853***	***1.861***	***1.868***	***1.875***	***1.880***	***1.893***	***1.903***	***1.912***	***1.919***	***1.939***	***1.959***	***1.976***
	0.184	0.184	0.184	0.184	0.184	0.184	0.184	0.184	0.184	0.184	0.184	0.184	0.184	0.184
	0.609	0.615	0.626	0.636	0.646	0.654	0.661	0.677	0.690	0.701	0.710	0.736	0.762	0.784
0.30	***1.793***	***1.799***	***1.810***	***1.821***	***1.830***	***1.838***	***1.845***	***1.861***	***1.874***	***1.885***	***1.894***	***1.920***	***1.946***	***1.968***
	0.115	0.115	0.115	0.115	0.115	0.115	0.115	0.115	0.115	0.115	0.115	0.115	0.115	0.115
	0.628	0.635	0.649	0.662	0.673	0.683	0.692	0.712	0.728	0.742	0.753	0.785	0.817	0.844
0.40	***1.743***	***1.751***	***1.764***	***1.777***	***1.788***	***1.798***	***1.807***	***1.827***	***1.843***	***1.857***	***1.868***	***1.900***	***1.932***	***1.960***
	0.067	0.067	0.067	0.067	0.067	0.067	0.067	0.067	0.067	0.067	0.067	0.067	0.067	0.067
	0.624	0.633	0.650	0.664	0.678	0.690	0.701	0.724	0.744	0.760	0.773	0.812	0.851	0.884
0.50	***1.691***	***1.700***	***1.716***	***1.731***	***1.745***	***1.757***	***1.768***	***1.791***	***1.811***	***1.827***	***1.840***	***1.878***	***1.918***	***1.950***
	0.036	0.036	0.036	0.036	0.036	0.036	0.036	0.036	0.036	0.036	0.036	0.036	0.036	0.036
	0.603	0.613	0.632	0.649	0.664	0.678	0.691	0.719	0.741	0.760	0.776	0.820	0.867	0.905
0.60	***1.639***	***1.649***	***1.668***	***1.685***	***1.700***	***1.714***	***1.727***	***1.754***	***1.777***	***1.796***	***1.812***	***1.856***	***1.903***	***1.941***
	0.018	0.018	0.018	0.018	0.018	0.018	0.018	0.018	0.018	0.018	0.018	0.018	0.018	0.018
	0.569	0.581	0.602	0.621	0.638	0.654	0.669	0.700	0.725	0.747	0.765	0.816	0.869	0.914
0.70	***1.587***	***1.599***	***1.620***	***1.639***	***1.656***	***1.672***	***1.686***	***1.718***	***1.743***	***1.765***	***1.783***	***1.834***	***1.887***	***1.931***
	0.008	0.008	0.008	0.008	0.008	0.008	0.008	0.008	0.008	0.008	0.008	0.008	0.008	0.008
	0.529	0.541	0.565	0.586	0.605	0.622	0.638	0.673	0.701	0.725	0.746	0.803	0.864	0.914
0.80	***1.537***	***1.550***	***1.573***	***1.594***	***1.613***	***1.630***	***1.646***	***1.681***	***1.710***	***1.734***	***1.754***	***1.812***	***1.872***	***1.922***
	0.003	0.003	0.003	0.003	0.003	0.003	0.003	0.003	0.003	0.003	0.003	0.003	0.003	0.003
	0.485	0.499	0.524	0.547	0.567	0.586	0.604	0.642	0.673	0.700	0.722	0.786	0.853	0.909
0.90	***1.489***	***1.502***	***1.527***	***1.550***	***1.571***	***1.590***	***1.607***	***1.645***	***1.677***	***1.703***	***1.725***	***1.789***	***1.856***	***1.912***
	0.001	0.001	0.001	0.001	0.001	0.001	0.001	0.001	0.001	0.001	0.001	0.001	0.001	0.001
	0.441	0.455	0.482	0.506	0.529	0.549	0.568	0.609	0.643	0.672	0.696	0.766	0.840	0.901
1.00	***1.442***	***1.457***	***1.483***	***1.508***	***1.530***	***1.551***	***1.569***	***1.610***	***1.644***	***1.673***	***1.697***	***1.767***	***1.841***	***1.903***
	0.000	0.000	0.000	0.000	0.000	0.000	0.000	0.000	0.000	0.000	0.000	0.000	0.000	0.000
	0.398	0.413	0.441	0.467	0.490	0.512	0.532	0.576	0.612	0.643	0.670	0.745	0.825	0.893
1.10	***1.398***	***1.413***	***1.441***	***1.467***	***1.491***	***1.513***	***1.533***	***1.576***	***1.613***	***1.644***	***1.670***	***1.746***	***1.826***	***1.893***

Table Five: *ExCP, ExCU, and ECP&U* when Average Is on Rework Side of Target and the Cost of Rework Is Equal to the Cost of Scrap

C_{pk} or P_{pk}	C_p or P_p value 1.50	1.60	1.70	1.80	1.90	2.00	2.10	2.20	2.30	2.40	2.50	2.60	2.70	2.80
	0.000	0.000	0.000	0.000	0.000	0.000	0.000	0.000	0.000	0.000	0.000	0.000	0.000	0.000
	0.089	0.106	0.125	0.145	0.166	0.188	0.209	0.229	0.250	0.269	0.288	0.306	0.324	0.341
1.20	***1.089***	***1.106***	***1.125***	***1.145***	***1.166***	***1.188***	***1.209***	***1.230***	***1.250***	***1.269***	***1.288***	***1.306***	***1.324***	***1.341***
	0.000	0.000	0.000	0.000	0.000	0.000	0.000	0.000	0.000	0.000	0.000	0.000	0.000	0.000
	0.067	0.079	0.094	0.111	0.130	0.150	0.170	0.190	0.210	0.229	0.248	0.266	0.284	0.301
1.30	***1.067***	***1.079***	***1.094***	***1.111***	***1.130***	***1.150***	***1.170***	***1.190***	***1.210***	***1.229***	***1.248***	***1.266***	***1.284***	***1.301***
	0.000	0.000	0.000	0.000	0.000	0.000	0.000	0.000	0.000	0.000	0.000	0.000	0.000	0.000
	0.054	0.059	0.070	0.084	0.100	0.118	0.136	0.155	0.174	0.193	0.211	0.229	0.247	0.264
1.40	***1.054***	***1.059***	***1.070***	***1.084***	***1.100***	***1.118***	***1.136***	***1.155***	***1.174***	***1.193***	***1.211***	***1.229***	***1.247***	***1.264***
	0.000	0.000	0.000	0.000	0.000	0.000	0.000	0.000	0.000	0.000	0.000	0.000	0.000	0.000
	0.049	0.047	0.052	0.062	0.075	0.090	0.107	0.124	0.142	0.160	0.178	0.195	0.213	0.230
1.50	***1.049***	***1.047***	***1.052***	***1.062***	***1.075***	***1.090***	***1.107***	***1.124***	***1.142***	***1.160***	***1.178***	***1.195***	***1.213***	***1.230***
		0.000	0.000	0.000	0.000	0.000	0.000	0.000	0.000	0.000	0.000	0.000	0.000	0.000
		0.043	0.042	0.047	0.056	0.068	0.082	0.097	0.114	0.130	0.147	0.164	0.181	0.198
1.60		***1.043***	***1.042***	***1.047***	***1.056***	***1.068***	***1.082***	***1.097***	***1.114***	***1.130***	***1.147***	***1.164***	***1.181***	***1.198***
			0.000	0.000	0.000	0.000	0.000	0.000	0.000	0.000	0.000	0.000	0.000	0.000
			0.038	0.037	0.042	0.050	0.061	0.075	0.089	0.104	0.120	0.136	0.152	0.169
1.70 »			***1.038***	***1.037***	***1.042***	***1.050***	***1.061***	***1.075***	***1.089***	***1.104***	***1.120***	***1.136***	***1.152***	***1.169***
				0.000	0.000	0.000	0.000	0.000	0.000	0.000	0.000	0.000	0.000	0.000
				0.034	0.034	0.038	0.046	0.056	0.068	0.082	0.096	0.111	0.126	0.142
1.80 »				***1.034***	***1.034***	***1.038***	***1.046***	***1.056***	***1.068***	***1.082***	***1.096***	***1.111***	***1.126***	***1.142***
					0.000	0.000	0.000	0.000	0.000	0.000	0.000	0.000	0.000	0.000
					0.031	0.030	0.034	0.042	0.051	0.063	0.075	0.089	0.103	0.117
1.90 »					***1.031***	***1.030***	***1.034***	***1.042***	***1.051***	***1.063***	***1.075***	***1.089***	***1.103***	***1.117***
						0.000	0.000	0.000	0.000	0.000	0.000	0.000	0.000	0.000
						0.028	0.027	0.031	0.038	0.047	0.058	0.070	0.082	0.096
2.00 »						***1.028***	***1.027***	***1.031***	***1.038***	***1.047***	***1.058***	***1.070***	***1.082***	***1.096***
							0.000	0.000	0.000	0.000	0.000	0.000	0.000	0.000
							0.025	0.025	0.029	0.035	0.043	0.053	0.065	0.077
2.10 »							***1.025***	***1.025***	***1.029***	***1.035***	***1.043***	***1.053***	***1.065***	***1.077***
								0.000	0.000	0.000	0.000	0.000	0.000	0.000
								0.023	0.023	0.026	0.032	0.040	0.050	0.060
2.20 »								***1.023***	***1.023***	***1.026***	***1.032***	***1.040***	***1.050***	***1.060***
									0.000	0.000	0.000	0.000	0.000	0.000
									0.021	0.021	0.024	0.030	0.037	0.046
2.30 »									***1.021***	***1.021***	***1.024***	***1.030***	***1.037***	***1.046***
										0.000	0.000	0.000	0.000	0.000
										0.019	0.019	0.022	0.028	0.035
2.40 »										***1.019***	***1.019***	***1.022***	***1.028***	***1.035***
											0.000	0.000	0.000	0.000
											0.018	0.018	0.021	0.026
2.50 »											***1.018***	***1.018***	***1.021***	***1.026***
												0.000	0.000	0.000
												0.016	0.017	0.019
2.60 »												***1.016***	***1.017***	***1.019***
													0.000	0.000
													0.015	0.015
2.70 »													***1.015***	***1.015***
														0.000
														0.014
2.80 »														***1.014***

Table Five: *ExCP, ExCU, and ECP&U* when Average Is on Rework Side of Target and the Cost of Rework Is Equal to the Cost of Scrap

C_{pk} or P_{pk}	C_p or P_p value 2.90	3.00	3.20	3.40	3.60	3.80	4.00	4.50	5.00	5.50	6.0	8.0	12.0	20.0
	0.000	0.000	0.000	0.000	0.000	0.000	0.000	0.000	0.000	0.000	0.000	0.000	0.000	0.000
	0.357	0.372	0.401	0.428	0.453	0.476	0.497	0.543	0.582	0.615	0.643	0.724	0.811	0.884
1.20	***1.357***	***1.372***	***1.402***	***1.428***	***1.453***	***1.476***	***1.497***	***1.543***	***1.582***	***1.615***	***1.643***	***1.724***	***1.811***	***1.884***
	0.000	0.000	0.000	0.000	0.000	0.000	0.000	0.000	0.000	0.000	0.000	0.000	0.000	0.000
	0.318	0.333	0.363	0.391	0.417	0.440	0.463	0.511	0.552	0.587	0.617	0.703	0.796	0.874
1.30	***1.318***	***1.333***	***1.363***	***1.391***	***1.417***	***1.441***	***1.463***	***1.511***	***1.552***	***1.587***	***1.617***	***1.703***	***1.796***	***1.875***
	0.000	0.000	0.000	0.000	0.000	0.000	0.000	0.000	0.000	0.000	0.000	0.000	0.000	0.000
	0.281	0.297	0.327	0.356	0.382	0.407	0.429	0.480	0.523	0.559	0.591	0.682	0.781	0.865
1.40	***1.281***	***1.297***	***1.327***	***1.356***	***1.382***	***1.407***	***1.429***	***1.480***	***1.523***	***1.559***	***1.591***	***1.682***	***1.781***	***1.865***
	0.000	0.000	0.000	0.000	0.000	0.000	0.000	0.000	0.000	0.000	0.000	0.000	0.000	0.000
	0.246	0.262	0.293	0.322	0.349	0.374	0.398	0.450	0.494	0.533	0.566	0.662	0.766	0.856
1.50	***1.246***	***1.262***	***1.293***	***1.322***	***1.349***	***1.374***	***1.398***	***1.450***	***1.494***	***1.533***	***1.566***	***1.662***	***1.766***	***1.856***
	0.000	0.000	0.000	0.000	0.000	0.000	0.000	0.000	0.000	0.000	0.000	0.000	0.000	0.000
	0.214	0.230	0.261	0.290	0.317	0.343	0.367	0.421	0.467	0.506	0.541	0.642	0.752	0.847
1.60	***1.214***	***1.230***	***1.261***	***1.290***	***1.317***	***1.343***	***1.367***	***1.421***	***1.467***	***1.506***	***1.541***	***1.642***	***1.752***	***1.847***
	0.000	0.000	0.000	0.000	0.000	0.000	0.000	0.000	0.000	0.000	0.000	0.000	0.000	0.000
	0.184	0.200	0.231	0.260	0.287	0.313	0.338	0.393	0.440	0.481	0.517	0.622	0.738	0.838
1.70	***1.184***	***1.200***	***1.231***	***1.260***	***1.287***	***1.313***	***1.338***	***1.393***	***1.440***	***1.481***	***1.517***	***1.622***	***1.738***	***1.838***
	0.000	0.000	0.000	0.000	0.000	0.000	0.000	0.000	0.000	0.000	0.000	0.000	0.000	0.000
	0.157	0.172	0.202	0.231	0.259	0.285	0.309	0.365	0.414	0.456	0.493	0.602	0.723	0.828
1.80	***1.157***	***1.172***	***1.202***	***1.231***	***1.259***	***1.285***	***1.309***	***1.365***	***1.414***	***1.456***	***1.493***	***1.602***	***1.723***	***1.828***
	0.000	0.000	0.000	0.000	0.000	0.000	0.000	0.000	0.000	0.000	0.000	0.000	0.000	0.000
	0.132	0.147	0.176	0.204	0.232	0.258	0.283	0.339	0.389	0.432	0.470	0.583	0.709	0.819
1.90	***1.132***	***1.147***	***1.176***	***1.204***	***1.232***	***1.258***	***1.283***	***1.339***	***1.389***	***1.432***	***1.470***	***1.583***	***1.709***	***1.819***
	0.000	0.000	0.000	0.000	0.000	0.000	0.000	0.000	0.000	0.000	0.000	0.000	0.000	0.000
	0.110	0.123	0.151	0.179	0.206	0.232	0.257	0.314	0.364	0.409	0.448	0.564	0.695	0.810
2.00	***1.110***	***1.123***	***1.151***	***1.179***	***1.206***	***1.232***	***1.257***	***1.314***	***1.364***	***1.409***	***1.448***	***1.564***	***1.695***	***1.810***
	0.000	0.000	0.000	0.000	0.000	0.000	0.000	0.000	0.000	0.000	0.000	0.000	0.000	0.000
	0.089	0.102	0.129	0.156	0.182	0.208	0.233	0.290	0.341	0.386	0.426	0.546	0.681	0.801
2.10	***1.089***	***1.102***	***1.129***	***1.156***	***1.182***	***1.208***	***1.233***	***1.290***	***1.341***	***1.386***	***1.426***	***1.546***	***1.681***	***1.801***
	0.000	0.000	0.000	0.000	0.000	0.000	0.000	0.000	0.000	0.000	0.000	0.000	0.000	0.000
	0.071	0.083	0.109	0.134	0.160	0.185	0.209	0.267	0.318	0.364	0.404	0.527	0.668	0.792
2.20	***1.071***	***1.083***	***1.109***	***1.134***	***1.160***	***1.185***	***1.209***	***1.267***	***1.318***	***1.364***	***1.404***	***1.527***	***1.668***	***1.792***
	0.000	0.000	0.000	0.000	0.000	0.000	0.000	0.000	0.000	0.000	0.000	0.000	0.000	0.000
	0.056	0.067	0.090	0.114	0.139	0.164	0.188	0.244	0.296	0.342	0.383	0.509	0.654	0.784
2.30	***1.056***	***1.067***	***1.090***	***1.114***	***1.139***	***1.164***	***1.188***	***1.244***	***1.296***	***1.342***	***1.383***	***1.509***	***1.654***	***1.784***
	0.000	0.000	0.000	0.000	0.000	0.000	0.000	0.000	0.000	0.000	0.000	0.000	0.000	0.000
	0.043	0.052	0.073	0.096	0.120	0.143	0.167	0.223	0.275	0.321	0.363	0.492	0.641	0.775
2.40	***1.043***	***1.052***	***1.073***	***1.096***	***1.120***	***1.143***	***1.167***	***1.223***	***1.275***	***1.321***	***1.363***	***1.492***	***1.641***	***1.775***
	0.000	0.000	0.000	0.000	0.000	0.000	0.000	0.000	0.000	0.000	0.000	0.000	0.000	0.000
	0.032	0.040	0.059	0.080	0.102	0.125	0.148	0.203	0.254	0.301	0.343	0.474	0.628	0.766
2.50	***1.032***	***1.040***	***1.059***	***1.080***	***1.102***	***1.125***	***1.148***	***1.203***	***1.254***	***1.301***	***1.343***	***1.474***	***1.628***	***1.766***
	0.000	0.000	0.000	0.000	0.000	0.000	0.000	0.000	0.000	0.000	0.000	0.000	0.000	0.000
	0.024	0.030	0.046	0.065	0.086	0.107	0.129	0.184	0.235	0.282	0.324	0.457	0.614	0.757
2.60	***1.024***	***1.030***	***1.046***	***1.065***	***1.086***	***1.107***	***1.129***	***1.184***	***1.235***	***1.282***	***1.324***	***1.457***	***1.614***	***1.757***
	0.000	0.000	0.000	0.000	0.000	0.000	0.000	0.000	0.000	0.000	0.000	0.000	0.000	0.000
	0.018	0.022	0.035	0.052	0.071	0.091	0.113	0.165	0.216	0.263	0.306	0.441	0.601	0.749
2.70	***1.018***	***1.022***	***1.035***	***1.052***	***1.071***	***1.091***	***1.113***	***1.165***	***1.216***	***1.263***	***1.306***	***1.441***	***1.601***	***1.749***
	0.000	0.000	0.000	0.000	0.000	0.000	0.000	0.000	0.000	0.000	0.000	0.000	0.000	0.000
	0.014	0.017	0.026	0.041	0.058	0.077	0.097	0.148	0.198	0.245	0.288	0.424	0.589	0.740
2.80	***1.014***	***1.017***	***1.026***	***1.041***	***1.058***	***1.077***	***1.097***	***1.148***	***1.198***	***1.245***	***1.288***	***1.424***	***1.589***	***1.740***

Table Five: *ExCP, ExCU, and ECP&U* when Average Is on Rework Side of Target and the Cost of Rework Is Equal to the Cost of Scrap

C_{pk} or P_{pk}	*C_p or P_p value* 2.90	3.00	3.20	3.40	3.60	3.80	4.00	4.50	5.00	5.50	6.0	8.0	12.0	20.0
	0.000	0.000	0.000	0.000	0.000	0.000	0.000	0.000	0.000	0.000	0.000	0.000	0.000	0.000
	0.013	0.013	0.020	0.031	0.046	0.064	0.083	0.132	0.181	0.227	0.270	0.408	0.576	0.731
2.90	***1.013***	***1.013***	***1.020***	***1.031***	***1.046***	***1.064***	***1.083***	***1.132***	***1.181***	***1.227***	***1.270***	***1.408***	***1.576***	***1.731***
		0.000	0.000	0.000	0.000	0.000	0.000	0.000	0.000	0.000	0.000	0.000	0.000	0.000
		0.012	0.015	0.023	0.036	0.052	0.069	0.117	0.164	0.210	0.253	0.392	0.563	0.723
3.00		***1.012***	***1.015***	***1.023***	***1.036***	***1.052***	***1.069***	***1.117***	***1.164***	***1.210***	***1.253***	***1.392***	***1.563***	***1.723***
			0.000	0.000	0.000	0.000	0.000	0.000	0.000	0.000	0.000	0.000	0.000	0.000
			0.011	0.013	0.021	0.033	0.047	0.089	0.134	0.179	0.221	0.362	0.539	0.706
3.20 »			***1.011***	***1.013***	***1.021***	***1.033***	***1.047***	***1.089***	***1.134***	***1.179***	***1.221***	***1.362***	***1.539***	***1.706***
				0.000	0.000	0.000	0.000	0.000	0.000	0.000	0.000	0.000	0.000	0.000
				0.010	0.012	0.019	0.029	0.065	0.107	0.149	0.191	0.332	0.514	0.689
3.40 »				***1.010***	***1.012***	***1.019***	***1.029***	***1.065***	***1.107***	***1.149***	***1.191***	***1.332***	***1.514***	***1.689***
					0.000	0.000	0.000	0.000	0.000	0.000	0.000	0.000	0.000	0.000
					0.009	0.010	0.017	0.045	0.083	0.123	0.163	0.304	0.491	0.673
3.60 »					***1.009***	***1.010***	***1.017***	***1.045***	***1.083***	***1.123***	***1.163***	***1.304***	***1.491***	***1.673***
						0.000	0.000	0.000	0.000	0.000	0.000	0.000	0.000	0.000
						0.008	0.009	0.030	0.062	0.099	0.138	0.277	0.468	0.656
3.80 »						***1.008***	***1.009***	***1.030***	***1.062***	***1.099***	***1.138***	***1.277***	***1.468***	***1.656***
							0.000	0.000	0.000	0.000	0.000	0.000	0.000	0.000
							0.007	0.018	0.044	0.078	0.114	0.252	0.445	0.640
4.00 »							***1.007***	***1.018***	***1.044***	***1.078***	***1.114***	***1.252***	***1.445***	***1.640***
								0.000	0.000	0.000	0.000	0.000	0.000	0.000
								0.005	0.014	0.037	0.066	0.193	0.391	0.601
4.50 »								***1.005***	***1.014***	***1.037***	***1.066***	***1.193***	***1.391***	***1.601***
									0.000	0.000	0.000	0.000	0.000	0.000
									0.004	0.012	0.031	0.142	0.341	0.563
5.00 »									***1.004***	***1.012***	***1.031***	***1.142***	***1.341***	***1.563***
										0.000	0.000	0.000	0.000	0.000
										0.004	0.010	0.099	0.294	0.526
5.50 »										***1.004***	***1.010***	***1.099***	***1.294***	***1.526***
											0.000	0.000	0.000	0.000
											0.003	0.064	0.251	0.490
6.00 »											***1.003***	***1.064***	***1.251***	***1.490***
												0.000	0.000	0.000
												0.002	0.112	0.360
8.00 »												***1.002***	***1.112***	***1.360***
													0.000	0.000
													0.001	0.160
12.0 »													***1.001***	***1.160***
														0.000
														0.000
20.0 »														***1.000***

Table Six: The *Excess Costs of Production, Excess Costs of Use,* and *Effective Cost of Production and Use* when Average Is on the Rework Side of Target and the Cost of Rework Is 0.1% of the Cost of Scrap

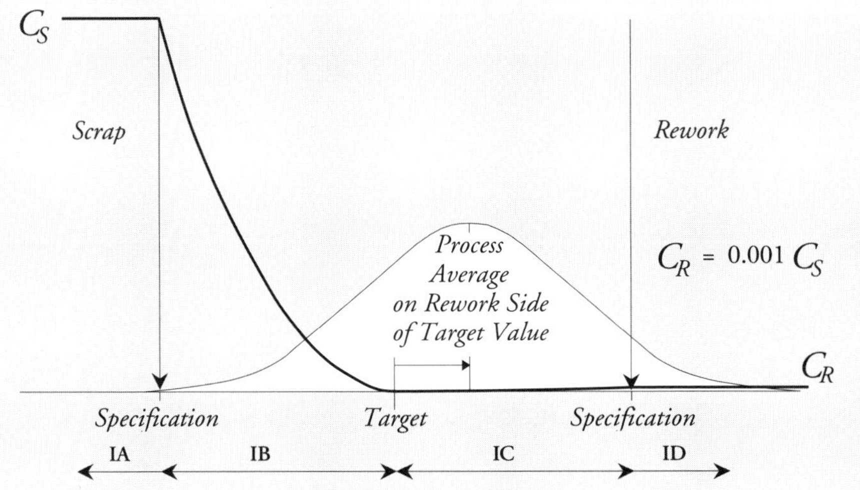

Table Six: Excess Cost Function when Average Is on Rework Side of Target and Cost of Rework Is 0.1% of the Cost of Scrap

Corresponding to each pairing for C_p and C_{pk} there are three entries stacked in a column.

The topmost of these is the *Excess Costs of Production,* *ExCP*

The middle value is the *Excess Costs of Use,* *ExCU*

And the bottom value (in bold) is the *Effective Cost of Production and Use,* ***ECP&U***

The entries in Table Six assume that the Cost of Scrap, C_S, and the Nominal Cost of Production, *NC,* are equal, Table A.9 contains the normal theory values for **IA, IB, IC,** and **ID**. The general and specific formulas used for Table Six are:

$$ECP\&U \quad = \quad 1.0 \; + \; \frac{\frac{C_S}{NC}\,\{\,\mathbf{IA} + \frac{C_R}{C_S}\,\mathbf{ID}\,\}}{1.0 \; - \; \mathbf{IA}} \; + \; \frac{\frac{C_S}{NC}\,\{\,\mathbf{IB} + \frac{C_R}{C_S}\,\mathbf{IC}\,\}}{1.0 \; - \; \mathbf{IA}}$$

$$= \quad 1.0 \; + \; \frac{\{\,\mathbf{IA} + 0.001\,\mathbf{ID}\,\}}{1.0 \; - \; \mathbf{IA}} \; + \; \frac{\{\,\mathbf{IB} + 0.001\,\mathbf{IC}\,\}}{1.0 \; - \; \mathbf{IA}}$$

$$= \quad 1.0 \; + \; \textit{Excess Costs of Production} \; + \; \textit{Excess Costs of Use}$$

When the Cost of Rework is greater than 0.001 times the Cost of Scrap appropriate values may be obtained by interpolating between the entry in Table Six and the corresponding entry in Table Five.

When the Cost of Scrap exceeds the Nominal Cost, the *Excess Costs* in Table Six, or any interpolated values, may be inflated as shown in the general formula above.

Table Six: *ExCP, ExCU, and ECP&U* when Average Is on Rework Side of Target and the Cost of Rework Is 0.1% of the Cost of Scrap

C_{pk} or P_{pk}	C_p or P_p values: 0.10	0.20	0.30	0.40	0.50	0.60	0.70	0.80	0.90	1.00	1.10	1.20	1.30	1.40
	0.019	0.004	0.001	0.001	0.001	0.001	0.001	0.001	0.001	0.001	0.001	0.001	0.001	0.001
	0.005	0.003	0.001	0.000	0.000	0.000	0.000	0.000	0.000	0.000	0.000	0.000	0.000	0.000
–0.50	***1.024***	***1.008***	***1.003***	***1.001***	***1.001***	***1.001***	***1.001***	***1.001***	***1.001***	***1.001***	***1.001***	***1.001***	***1.001***	***1.001***
	0.038	0.009	0.002	0.001	0.001	0.001	0.001	0.001	0.001	0.001	0.001	0.001	0.001	0.001
	0.009	0.007	0.003	0.001	0.000	0.000	0.000	0.000	0.000	0.000	0.000	0.000	0.000	0.000
–0.40	***1.048***	***1.016***	***1.005***	***1.002***	***1.001***	***1.001***	***1.001***	***1.001***	***1.001***	***1.001***	***1.001***	***1.001***	***1.001***	***1.001***
	0.072	0.019	0.004	0.001	0.001	0.001	0.001	0.001	0.001	0.001	0.001	0.001	0.001	0.001
	0.016	0.012	0.006	0.002	0.001	0.000	0.000	0.000	0.000	0.000	0.000	0.000	0.000	0.000
–0.30	***1.088***	***1.031***	***1.010***	***1.004***	***1.002***	***1.001***	***1.001***	***1.001***	***1.001***	***1.001***	***1.001***	***1.001***	***1.001***	***1.001***
	0.131	0.038	0.009	0.002	0.001	0.001	0.001	0.001	0.001	0.001	0.001	0.001	0.001	0.001
	0.024	0.022	0.012	0.005	0.002	0.001	0.000	0.000	0.000	0.000	0.000	0.000	0.000	0.000
–0.20	***1.155***	***1.060***	***1.021***	***1.007***	***1.003***	***1.001***	***1.001***	***1.001***	***1.001***	***1.001***	***1.001***	***1.001***	***1.001***	***1.001***
	0.226	0.072	0.019	0.004	0.001	0.001	0.001	0.001	0.001	0.001	0.001	0.001	0.001	0.001
	0.035	0.035	0.022	0.010	0.004	0.001	0.001	0.000	0.000	0.000	0.000	0.000	0.000	0.000
–0.10	***1.261***	***1.107***	***1.041***	***1.014***	***1.005***	***1.002***	***1.001***	***1.001***	***1.001***	***1.001***	***1.001***	***1.001***	***1.001***	***1.001***
	0.379	0.131	0.038	0.009	0.002	0.001	0.001	0.001	0.001	0.001	0.001	0.001	0.001	0.001
	0.048	0.052	0.037	0.020	0.008	0.003	0.001	0.001	0.000	0.000	0.000	0.000	0.000	0.000
0.00	***1.426***	***1.183***	***1.075***	***1.028***	***1.010***	***1.004***	***1.002***	***1.001***	***1.001***	***1.001***	***1.001***	***1.001***	***1.001***	***1.001***
	0.619	0.226	0.072	0.019	0.004	0.001	0.000	0.000	0.000	0.000	0.000	0.000	0.000	0.000
	0.063	0.074	0.057	0.034	0.016	0.007	0.003	0.001	0.001	0.000	0.000	0.000	0.000	0.000
0.10	***1.682***	***1.300***	***1.129***	***1.053***	***1.020***	***1.008***	***1.003***	***1.001***	***1.001***	***1.001***	***1.001***	***1.001***	***1.001***	***1.001***
		0.378	0.130	0.038	0.009	0.002	0.000	0.000	0.000	0.000	0.000	0.000	0.000	0.000
		0.099	0.084	0.055	0.029	0.013	0.005	0.002	0.001	0.001	0.000	0.000	0.000	0.000
0.20		***1.477***	***1.214***	***1.093***	***1.038***	***1.015***	***1.006***	***1.002***	***1.001***	***1.001***	***1.001***	***1.001***	***1.001***	***1.001***
			0.226	0.072	0.018	0.004	0.001	0.000	0.000	0.000	0.000	0.000	0.000	0.000
			0.116	0.084	0.050	0.024	0.010	0.004	0.002	0.001	0.001	0.000	0.000	0.000
	0.30 »		***1.342***	***1.156***	***1.068***	***1.028***	***1.011***	***1.004***	***1.002***	***1.001***	***1.001***	***1.001***	***1.001***	***1.001***
				0.130	0.037	0.008	0.001	0.000	0.000	0.000	0.000	0.000	0.000	0.000
				0.119	0.078	0.042	0.020	0.008	0.003	0.001	0.001	0.001	0.000	0.000
		0.40 »		***1.250***	***1.115***	***1.051***	***1.021***	***1.009***	***1.004***	***1.002***	***1.001***	***1.001***	***1.001***	***1.001***
					0.072	0.018	0.004	0.001	0.000	0.000	0.000	0.000	0.000	0.000
					0.114	0.068	0.035	0.016	0.007	0.003	0.001	0.001	0.000	0.000
			0.50 »		***1.186***	***1.086***	***1.038***	***1.016***	***1.007***	***1.003***	***1.001***	***1.001***	***1.001***	***1.000***
						0.037	0.008	0.001	0.000	0.000	0.000	0.000	0.000	0.000
						0.103	0.058	0.029	0.013	0.006	0.002	0.001	0.001	0.000
				0.60 »		***1.140***	***1.066***	***1.030***	***1.013***	***1.006***	***1.002***	***1.001***	***1.001***	***1.000***
							0.018	0.003	0.001	0.000	0.000	0.000	0.000	0.000
							0.090	0.048	0.023	0.011	0.005	0.002	0.001	0.001
					0.70 »		***1.108***	***1.052***	***1.024***	***1.011***	***1.005***	***1.002***	***1.001***	***1.001***
								0.008	0.001	0.000	0.000	0.000	0.000	0.000
								0.077	0.040	0.019	0.009	0.004	0.002	0.001
						0.80 »		***1.085***	***1.041***	***1.019***	***1.009***	***1.004***	***1.002***	***1.001***
									0.003	0.000	0.000	0.000	0.000	0.000
									0.065	0.033	0.016	0.007	0.003	0.001
							0.90 »		***1.068***	***1.034***	***1.016***	***1.007***	***1.003***	***1.001***
										0.001	0.000	0.000	0.000	0.000
										0.054	0.028	0.013	0.006	0.003
								1.00 »		***1.055***	***1.028***	***1.013***	***1.006***	***1.003***
											0.000	0.000	0.000	0.000
											0.045	0.023	0.011	0.005
									1.10 »		***1.046***	***1.023***	***1.011***	***1.005***

C_{pk}	C_p: 1.20	1.30	1.40
	0.000	0.000	0.000
	0.038	0.020	0.010
1.20	***1.039***	***1.020***	***1.010***
		0.000	0.000
		0.033	0.017
1.30		***1.033***	***1.017***
			0.000
			0.028
1.40			***1.028***

Table Six: *ExCP, ExCU, and ECP&U* when Average Is on Rework Side of Target and the Cost of Rework Is 0.1% of the Cost of Scrap

C_{pk} or P_{pk}	*C_p or P_p value* 1.50	1.60	1.70	1.80	1.90	2.00	2.10	2.20	2.30	2.40	2.50	2.60	2.70	2.80
	0.001	0.001	0.001	0.001	0.001	0.001	0.001	0.001	0.001	0.001	0.001	0.001	0.001	0.001
	0.000	0.000	0.000	0.000	0.000	0.000	0.000	0.000	0.000	0.000	0.000	0.000	0.000	0.000
–0.50	***1.001***	***1.001***	***1.001***	***1.001***	***1.001***	***1.001***	***1.001***	***1.001***	***1.001***	***1.001***	***1.001***	***1.001***	***1.001***	***1.001***
	0.001	0.001	0.001	0.001	0.001	0.001	0.001	0.001	0.001	0.001	0.001	0.001	0.001	0.001
	0.000	0.000	0.000	0.000	0.000	0.000	0.000	0.000	0.000	0.000	0.000	0.000	0.000	0.000
–0.40	***1.001***	***1.001***	***1.001***	***1.001***	***1.001***	***1.001***	***1.001***	***1.001***	***1.001***	***1.001***	***1.001***	***1.001***	***1.001***	***1.001***
	0.001	0.001	0.001	0.001	0.001	0.001	0.001	0.001	0.001	0.001	0.001	0.001	0.001	0.001
	0.000	0.000	0.000	0.000	0.000	0.000	0.000	0.000	0.000	0.000	0.000	0.000	0.000	0.000
–0.30	***1.001***	***1.001***	***1.001***	***1.001***	***1.001***	***1.001***	***1.001***	***1.001***	***1.001***	***1.001***	***1.001***	***1.001***	***1.001***	***1.001***
	0.001	0.001	0.001	0.001	0.001	0.001	0.001	0.001	0.001	0.001	0.001	0.001	0.001	0.001
	0.000	0.000	0.000	0.000	0.000	0.000	0.000	0.000	0.000	0.000	0.000	0.000	0.000	0.000
–0.20	***1.001***	***1.001***	***1.001***	***1.001***	***1.001***	***1.001***	***1.001***	***1.001***	***1.001***	***1.001***	***1.001***	***1.001***	***1.001***	***1.001***
	0.001	0.001	0.001	0.001	0.001	0.001	0.001	0.001	0.001	0.001	0.001	0.001	0.001	0.001
	0.000	0.000	0.000	0.000	0.000	0.000	0.000	0.000	0.000	0.000	0.000	0.000	0.000	0.000
–0.10	***1.001***	***1.001***	***1.001***	***1.001***	***1.001***	***1.001***	***1.001***	***1.001***	***1.001***	***1.001***	***1.001***	***1.001***	***1.001***	***1.001***
	0.001	0.001	0.001	0.001	0.001	0.001	0.001	0.001	0.001	0.001	0.001	0.001	0.001	0.001
	0.000	0.000	0.000	0.000	0.000	0.000	0.000	0.000	0.000	0.000	0.000	0.000	0.000	0.000
0.00	***1.001***	***1.001***	***1.001***	***1.001***	***1.001***	***1.001***	***1.001***	***1.001***	***1.001***	***1.001***	***1.001***	***1.001***	***1.001***	***1.001***
	0.000	0.000	0.000	0.000	0.000	0.000	0.000	0.000	0.000	0.000	0.000	0.000	0.000	0.000
	0.000	0.000	0.000	0.000	0.000	0.000	0.000	0.000	0.000	0.000	0.000	0.000	0.000	0.000
0.10	***1.001***	***1.001***	***1.001***	***1.001***	***1.001***	***1.001***	***1.001***	***1.001***	***1.001***	***1.001***	***1.001***	***1.001***	***1.001***	***1.001***
	0.000	0.000	0.000	0.000	0.000	0.000	0.000	0.000	0.000	0.000	0.000	0.000	0.000	0.000
	0.000	0.000	0.000	0.000	0.000	0.001	0.001	0.001	0.001	0.001	0.001	0.001	0.001	0.001
0.20	***1.001***	***1.001***	***1.001***	***1.001***	***1.001***	***1.001***	***1.001***	***1.001***	***1.001***	***1.001***	***1.001***	***1.001***	***1.001***	***1.001***
	0.000	0.000	0.000	0.000	0.000	0.000	0.000	0.000	0.000	0.000	0.000	0.000	0.000	0.000
	0.000	0.000	0.000	0.001	0.001	0.001	0.001	0.001	0.001	0.001	0.001	0.001	0.001	0.001
0.30	***1.001***	***1.001***	***1.001***	***1.001***	***1.001***	***1.001***	***1.001***	***1.001***	***1.001***	***1.001***	***1.001***	***1.001***	***1.001***	***1.001***
	0.000	0.000	0.000	0.000	0.000	0.000	0.000	0.000	0.000	0.000	0.000	0.000	0.000	0.000
	0.000	0.000	0.000	0.001	0.001	0.001	0.001	0.001	0.001	0.001	0.001	0.001	0.001	0.001
0.40	***1.001***	***1.001***	***1.001***	***1.001***	***1.001***	***1.001***	***1.001***	***1.001***	***1.001***	***1.001***	***1.001***	***1.001***	***1.001***	***1.001***
	0.000	0.000	0.000	0.000	0.000	0.000	0.000	0.000	0.000	0.000	0.000	0.000	0.000	0.000
	0.000	0.000	0.000	0.000	0.000	0.001	0.001	0.001	0.001	0.001	0.001	0.001	0.001	0.001
0.50	***1.000***	***1.001***	***1.001***	***1.001***	***1.001***	***1.001***	***1.001***	***1.001***	***1.001***	***1.001***	***1.001***	***1.001***	***1.001***	***1.001***
	0.000	0.000	0.000	0.000	0.000	0.000	0.000	0.000	0.000	0.000	0.000	0.000	0.000	0.000
	0.000	0.000	0.000	0.000	0.000	0.000	0.000	0.001	0.001	0.001	0.001	0.001	0.001	0.001
0.60	***1.000***	***1.000***	***1.000***	***1.000***	***1.000***	***1.001***	***1.001***	***1.001***	***1.001***	***1.001***	***1.001***	***1.001***	***1.001***	***1.001***
	0.000	0.000	0.000	0.000	0.000	0.000	0.000	0.000	0.000	0.000	0.000	0.000	0.000	0.000
	0.000	0.000	0.000	0.000	0.000	0.000	0.000	0.000	0.000	0.001	0.001	0.001	0.001	0.001
0.70	***1.000***	***1.000***	***1.000***	***1.000***	***1.000***	***1.000***	***1.000***	***1.000***	***1.001***	***1.001***	***1.001***	***1.001***	***1.001***	***1.001***
	0.000	0.000	0.000	0.000	0.000	0.000	0.000	0.000	0.000	0.000	0.000	0.000	0.000	0.000
	0.000	0.000	0.000	0.000	0.000	0.000	0.000	0.000	0.000	0.000	0.000	0.000	0.001	0.001
0.80	***1.000***	***1.000***	***1.000***	***1.000***	***1.000***	***1.000***	***1.000***	***1.000***	***1.000***	***1.000***	***1.000***	***1.000***	***1.001***	***1.001***
	0.000	0.000	0.000	0.000	0.000	0.000	0.000	0.000	0.000	0.000	0.000	0.000	0.000	0.000
	0.001	0.000	0.000	0.000	0.000	0.000	0.000	0.000	0.000	0.000	0.000	0.000	0.000	0.000
0.90	***1.001***	***1.000***	***1.000***	***1.000***	***1.000***	***1.000***	***1.000***	***1.000***	***1.000***	***1.000***	***1.000***	***1.000***	***1.000***	***1.000***
	0.000	0.000	0.000	0.000	0.000	0.000	0.000	0.000	0.000	0.000	0.000	0.000	0.000	0.000
	0.001	0.001	0.000	0.000	0.000	0.000	0.000	0.000	0.000	0.000	0.000	0.000	0.000	0.000
1.00	***1.001***	***1.001***	***1.000***	***1.000***	***1.000***	***1.000***	***1.000***	***1.000***	***1.000***	***1.000***	***1.000***	***1.000***	***1.000***	***1.000***
	0.000	0.000	0.000	0.000	0.000	0.000	0.000	0.000	0.000	0.000	0.000	0.000	0.000	0.000
	0.002	0.001	0.001	0.000	0.000	0.000	0.000	0.000	0.000	0.000	0.000	0.000	0.000	0.000
1.10	***1.002***	***1.001***	***1.001***	***1.000***	***1.000***	***1.000***	***1.000***	***1.000***	***1.000***	***1.000***	***1.000***	***1.000***	***1.000***	***1.000***

Table Six: *ExCP, ExCU, and ECP&U* when Average Is on Rework Side of Target and the Cost of Rework Is 0.1% of the Cost of Scrap

C_{pk} or P_{pk}	C_p or P_p value 2.90	3.00	3.20	3.40	3.60	3.80	4.00	4.50	5.00	5.50	6.0	8.0	12.0	20.0
	0.001	0.001	0.001	0.001	0.001	0.001	0.001	0.001	0.001	0.001	0.001	0.001	0.001	0.001
	0.000	0.000	0.000	0.000	0.000	0.000	0.000	0.000	0.000	0.000	0.000	0.000	0.000	0.000
-0.50	***1.001***	***1.001***	***1.001***	***1.001***	***1.001***	***1.001***	***1.001***	***1.001***	***1.001***	***1.001***	***1.001***	***1.001***	***1.001***	***1.001***
	0.001	0.001	0.001	0.001	0.001	0.001	0.001	0.001	0.001	0.001	0.001	0.001	0.001	0.001
	0.000	0.000	0.000	0.000	0.000	0.000	0.000	0.000	0.000	0.000	0.000	0.000	0.000	0.000
-0.40	***1.001***	***1.001***	***1.001***	***1.001***	***1.001***	***1.001***	***1.001***	***1.001***	***1.001***	***1.001***	***1.001***	***1.001***	***1.001***	***1.001***
	0.001	0.001	0.001	0.001	0.001	0.001	0.001	0.001	0.001	0.001	0.001	0.001	0.001	0.001
	0.000	0.000	0.000	0.000	0.000	0.000	0.000	0.000	0.000	0.000	0.000	0.000	0.000	0.000
-0.30	***1.001***	***1.001***	***1.001***	***1.001***	***1.001***	***1.001***	***1.001***	***1.001***	***1.001***	***1.001***	***1.001***	***1.001***	***1.001***	***1.001***
	0.001	0.001	0.001	0.001	0.001	0.001	0.001	0.001	0.001	0.001	0.001	0.001	0.001	0.001
	0.000	0.000	0.000	0.000	0.000	0.000	0.000	0.000	0.000	0.000	0.000	0.000	0.000	0.000
-0.20	***1.001***	***1.001***	***1.001***	***1.001***	***1.001***	***1.001***	***1.001***	***1.001***	***1.001***	***1.001***	***1.001***	***1.001***	***1.001***	***1.001***
	0.001	0.001	0.001	0.001	0.001	0.001	0.001	0.001	0.001	0.001	0.001	0.001	0.001	0.001
	0.000	0.000	0.000	0.000	0.000	0.000	0.000	0.000	0.000	0.000	0.000	0.000	0.000	0.000
-0.10	***1.001***	***1.001***	***1.001***	***1.001***	***1.001***	***1.001***	***1.001***	***1.001***	***1.001***	***1.001***	***1.001***	***1.001***	***1.001***	***1.001***
	0.001	0.001	0.001	0.001	0.001	0.001	0.000	0.000	0.000	0.000	0.000	0.000	0.001	0.001
	0.000	0.000	0.000	0.000	0.000	0.000	0.000	0.000	0.000	0.000	0.000	0.000	0.000	0.000
0.00	***1.001***	***1.001***	***1.001***	***1.001***	***1.001***	***1.001***	***1.001***	***1.001***	***1.001***	***1.001***	***1.001***	***1.001***	***1.001***	***1.001***
	0.000	0.000	0.000	0.000	0.000	0.000	0.000	0.000	0.000	0.000	0.000	0.000	0.000	0.000
	0.000	0.001	0.001	0.001	0.001	0.001	0.001	0.001	0.001	0.001	0.001	0.001	0.001	0.001
0.10	***1.001***	***1.001***	***1.001***	***1.001***	***1.001***	***1.001***	***1.001***	***1.001***	***1.001***	***1.001***	***1.001***	***1.001***	***1.001***	***1.001***
	0.000	0.000	0.000	0.000	0.000	0.000	0.000	0.000	0.000	0.000	0.000	0.000	0.000	0.000
	0.001	0.001	0.001	0.001	0.001	0.001	0.001	0.001	0.001	0.001	0.001	0.001	0.001	0.001
0.20	***1.001***	***1.001***	***1.001***	***1.001***	***1.001***	***1.001***	***1.001***	***1.001***	***1.001***	***1.001***	***1.001***	***1.001***	***1.001***	***1.001***
	0.000	0.000	0.000	0.000	0.000	0.000	0.000	0.000	0.000	0.000	0.000	0.000	0.000	0.000
	0.001	0.001	0.001	0.001	0.001	0.001	0.001	0.001	0.001	0.001	0.001	0.001	0.001	0.001
0.30	***1.001***	***1.001***	***1.001***	***1.001***	***1.001***	***1.001***	***1.001***	***1.001***	***1.001***	***1.001***	***1.001***	***1.001***	***1.001***	***1.001***
	0.000	0.000	0.000	0.000	0.000	0.000	0.000	0.000	0.000	0.000	0.000	0.000	0.000	0.000
	0.001	0.001	0.001	0.001	0.001	0.001	0.001	0.001	0.001	0.001	0.001	0.001	0.001	0.001
0.40	***1.001***	***1.001***	***1.001***	***1.001***	***1.001***	***1.001***	***1.001***	***1.001***	***1.001***	***1.001***	***1.001***	***1.001***	***1.001***	***1.001***
	0.000	0.000	0.000	0.000	0.000	0.000	0.000	0.000	0.000	0.000	0.000	0.000	0.000	0.000
	0.001	0.001	0.001	0.001	0.001	0.001	0.001	0.001	0.001	0.001	0.001	0.001	0.001	0.001
0.50	***1.001***	***1.001***	***1.001***	***1.001***	***1.001***	***1.001***	***1.001***	***1.001***	***1.001***	***1.001***	***1.001***	***1.001***	***1.001***	***1.001***
	0.000	0.000	0.000	0.000	0.000	0.000	0.000	0.000	0.000	0.000	0.000	0.000	0.000	0.000
	0.001	0.001	0.001	0.001	0.001	0.001	0.001	0.001	0.001	0.001	0.001	0.001	0.001	0.001
0.60	***1.001***	***1.001***	***1.001***	***1.001***	***1.001***	***1.001***	***1.001***	***1.001***	***1.001***	***1.001***	***1.001***	***1.001***	***1.001***	***1.001***
	0.000	0.000	0.000	0.000	0.000	0.000	0.000	0.000	0.000	0.000	0.000	0.000	0.000	0.000
	0.001	0.001	0.001	0.001	0.001	0.001	0.001	0.001	0.001	0.001	0.001	0.001	0.001	0.001
0.70	***1.001***	***1.001***	***1.001***	***1.001***	***1.001***	***1.001***	***1.001***	***1.001***	***1.001***	***1.001***	***1.001***	***1.001***	***1.001***	***1.001***
	0.000	0.000	0.000	0.000	0.000	0.000	0.000	0.000	0.000	0.000	0.000	0.000	0.000	0.000
	0.001	0.001	0.001	0.001	0.001	0.001	0.001	0.001	0.001	0.001	0.001	0.001	0.001	0.001
0.80	***1.001***	***1.001***	***1.001***	***1.001***	***1.001***	***1.001***	***1.001***	***1.001***	***1.001***	***1.001***	***1.001***	***1.001***	***1.001***	***1.001***
	0.000	0.000	0.000	0.000	0.000	0.000	0.000	0.000	0.000	0.000	0.000	0.000	0.000	0.000
	0.000	0.000	0.001	0.001	0.001	0.001	0.001	0.001	0.001	0.001	0.001	0.001	0.001	0.001
0.90	***1.000***	***1.001***	***1.001***	***1.001***	***1.001***	***1.001***	***1.001***	***1.001***	***1.001***	***1.001***	***1.001***	***1.001***	***1.001***	***1.001***
	0.000	0.000	0.000	0.000	0.000	0.000	0.000	0.000	0.000	0.000	0.000	0.000	0.000	0.000
	0.000	0.000	0.000	0.001	0.001	0.001	0.001	0.001	0.001	0.001	0.001	0.001	0.001	0.001
1.00	***1.000***	***1.000***	***1.000***	***1.001***	***1.001***	***1.001***	***1.001***	***1.001***	***1.001***	***1.001***	***1.001***	***1.001***	***1.001***	***1.001***
	0.000	0.000	0.000	0.000	0.000	0.000	0.000	0.000	0.000	0.000	0.000	0.000	0.000	0.000
	0.000	0.000	0.000	0.000	0.000	0.001	0.001	0.001	0.001	0.001	0.001	0.001	0.001	0.001
1.10	***1.000***	***1.000***	***1.000***	***1.000***	***1.000***	***1.001***	***1.001***	***1.001***	***1.001***	***1.001***	***1.001***	***1.001***	***1.001***	***1.001***

Table Six: *ExCP, ExCU, and ECP&U* when Average Is on Rework Side of Target and the Cost of Rework Is 0.1% of the Cost of Scrap

C_{pk} or P_{pk}	C_p or P_p value 1.50	1.60	1.70	1.80	1.90	2.00	2.10	2.20	2.30	2.40	2.50	2.60	2.70	2.80
	0.000	0.000	0.000	0.000	0.000	0.000	0.000	0.000	0.000	0.000	0.000	0.000	0.000	0.000
	0.005	0.002	0.001	0.000	0.000	0.000	0.000	0.000	0.000	0.000	0.000	0.000	0.000	0.000
1.20	***1.005***	***1.002***	***1.001***	***1.000***	***1.000***	***1.000***	***1.000***	***1.000***	***1.000***	***1.000***	***1.000***	***1.000***	***1.000***	***1.000***
	0.000	0.000	0.000	0.000	0.000	0.000	0.000	0.000	0.000	0.000	0.000	0.000	0.000	0.000
	0.009	0.004	0.002	0.001	0.000	0.000	0.000	0.000	0.000	0.000	0.000	0.000	0.000	0.000
1.30	***1.009***	***1.004***	***1.002***	***1.001***	***1.000***	***1.000***	***1.000***	***1.000***	***1.000***	***1.000***	***1.000***	***1.000***	***1.000***	***1.000***
	0.000	0.000	0.000	0.000	0.000	0.000	0.000	0.000	0.000	0.000	0.000	0.000	0.000	0.000
	0.015	0.008	0.004	0.002	0.001	0.000	0.000	0.000	0.000	0.000	0.000	0.000	0.000	0.000
1.40	***1.015***	***1.008***	***1.004***	***1.002***	***1.001***	***1.000***	***1.000***	***1.000***	***1.000***	***1.000***	***1.000***	***1.000***	***1.000***	***1.000***
	0.000	0.000	0.000	0.000	0.000	0.000	0.000	0.000	0.000	0.000	0.000	0.000	0.000	0.000
	0.025	0.013	0.007	0.003	0.002	0.001	0.000	0.000	0.000	0.000	0.000	0.000	0.000	0.000
1.50	***1.025***	***1.013***	***1.007***	***1.003***	***1.002***	***1.001***	***1.000***	***1.000***	***1.000***	***1.000***	***1.000***	***1.000***	***1.000***	***1.000***
		0.000	0.000	0.000	0.000	0.000	0.000	0.000	0.000	0.000	0.000	0.000	0.000	0.000
		0.022	0.012	0.006	0.003	0.001	0.001	0.000	0.000	0.000	0.000	0.000	0.000	0.000
1.60		***1.022***	***1.012***	***1.006***	***1.003***	***1.001***	***1.001***	***1.000***	***1.000***	***1.000***	***1.000***	***1.000***	***1.000***	***1.000***
			0.000	0.000	0.000	0.000	0.000	0.000	0.000	0.000	0.000	0.000	0.000	0.000
			0.019	0.010	0.005	0.003	0.001	0.001	0.000	0.000	0.000	0.000	0.000	0.000
1.70 »			***1.019***	***1.010***	***1.005***	***1.003***	***1.001***	***1.001***	***1.000***	***1.000***	***1.000***	***1.000***	***1.000***	***1.000***
				0.000	0.000	0.000	0.000	0.000	0.000	0.000	0.000	0.000	0.000	0.000
				0.017	0.009	0.005	0.002	0.001	0.001	0.000	0.000	0.000	0.000	0.000
1.80 »				***1.017***	***1.009***	***1.005***	***1.002***	***1.001***	***1.001***	***1.000***	***1.000***	***1.000***	***1.000***	***1.000***
					0.000	0.000	0.000	0.000	0.000	0.000	0.000	0.000	0.000	0.000
					0.015	0.008	0.004	0.002	0.001	0.001	0.000	0.000	0.000	0.000
1.90 »					***1.015***	***1.008***	***1.004***	***1.002***	***1.001***	***1.001***	***1.000***	***1.000***	***1.000***	***1.000***
						0.000	0.000	0.000	0.000	0.000	0.000	0.000	0.000	0.000
						0.014	0.008	0.004	0.002	0.001	0.000	0.000	0.000	0.000
2.00 »						***1.014***	***1.008***	***1.004***	***1.002***	***1.001***	***1.000***	***1.000***	***1.000***	***1.000***
							0.000	0.000	0.000	0.000	0.000	0.000	0.000	0.000
							0.013	0.007	0.004	0.002	0.001	0.000	0.000	0.000
2.10 »							***1.013***	***1.007***	***1.004***	***1.002***	***1.001***	***1.000***	***1.000***	***1.000***
								0.000	0.000	0.000	0.000	0.000	0.000	0.000
								0.011	0.006	0.003	0.002	0.001	0.000	0.000
2.20 »								***1.011***	***1.006***	***1.003***	***1.002***	***1.001***	***1.000***	***1.000***
									0.000	0.000	0.000	0.000	0.000	0.000
									0.011	0.006	0.003	0.002	0.001	0.000
2.30 »									***1.011***	***1.006***	***1.003***	***1.002***	***1.001***	***1.000***
										0.000	0.000	0.000	0.000	0.000
										0.010	0.005	0.003	0.001	0.001
2.40 »										***1.010***	***1.005***	***1.003***	***1.001***	***1.001***
											0.000	0.000	0.000	0.000
											0.009	0.005	0.003	0.001
2.50 »											***1.009***	***1.005***	***1.003***	***1.001***
												0.000	0.000	0.000
												0.008	0.005	0.002
2.60 »												***1.008***	***1.005***	***1.002***
													0.000	0.000
													0.008	0.004
2.70 »													***1.008***	***1.004***
														0.000
														0.007
2.80 »														***1.007***

Table Six: *ExCP, ExCU, and ECP&U* when Average Is on Rework Side of Target and the Cost of Rework Is 0.1% of the Cost of Scrap

C_{pk} or P_{pk}	C_p or P_p value 2.90	3.00	3.20	3.40	3.60	3.80	4.00	4.50	5.00	5.50	6.0	8.0	12.0	20.0
	0.000	0.000	0.000	0.000	0.000	0.000	0.000	0.000	0.000	0.000	0.000	0.000	0.000	0.000
	0.000	0.000	0.000	0.000	0.000	0.000	0.000	0.001	0.001	0.001	0.001	0.001	0.001	0.001
1.20	**1.000**	**1.000**	**1.000**	**1.000**	**1.000**	**1.000**	**1.000**	**1.001**	**1.001**	**1.001**	**1.001**	**1.001**	**1.001**	**1.001**
	0.000	0.000	0.000	0.000	0.000	0.000	0.000	0.000	0.000	0.000	0.000	0.000	0.000	0.000
	0.000	0.000	0.000	0.000	0.000	0.000	0.000	0.001	0.001	0.001	0.001	0.001	0.001	0.001
1.30	**1.000**	**1.000**	**1.000**	**1.000**	**1.000**	**1.000**	**1.000**	**1.001**	**1.001**	**1.001**	**1.001**	**1.001**	**1.001**	**1.001**
	0.000	0.000	0.000	0.000	0.000	0.000	0.000	0.000	0.000	0.000	0.000	0.000	0.000	0.000
	0.000	0.000	0.000	0.000	0.000	0.000	0.000	0.000	0.001	0.001	0.001	0.001	0.001	0.001
1.40	**1.000**	**1.000**	**1.000**	**1.000**	**1.000**	**1.000**	**1.000**	**1.000**	**1.001**	**1.001**	**1.001**	**1.001**	**1.001**	**1.001**
	0.000	0.000	0.000	0.000	0.000	0.000	0.000	0.000	0.000	0.000	0.000	0.000	0.000	0.000
	0.000	0.000	0.000	0.000	0.000	0.000	0.000	0.000	0.000	0.001	0.001	0.001	0.001	0.001
1.50	**1.000**	**1.000**	**1.000**	**1.000**	**1.000**	**1.000**	**1.000**	**1.000**	**1.000**	**1.001**	**1.001**	**1.001**	**1.001**	**1.001**
	0.000	0.000	0.000	0.000	0.000	0.000	0.000	0.000	0.000	0.000	0.000	0.000	0.000	0.000
	0.000	0.000	0.000	0.000	0.000	0.000	0.000	0.000	0.000	0.001	0.001	0.001	0.001	0.001
1.60	**1.000**	**1.000**	**1.000**	**1.000**	**1.000**	**1.000**	**1.000**	**1.000**	**1.000**	**1.001**	**1.001**	**1.001**	**1.001**	**1.001**
	0.000	0.000	0.000	0.000	0.000	0.000	0.000	0.000	0.000	0.000	0.000	0.000	0.000	0.000
	0.000	0.000	0.000	0.000	0.000	0.000	0.000	0.000	0.000	0.000	0.001	0.001	0.001	0.001
1.70	**1.000**	**1.000**	**1.000**	**1.000**	**1.000**	**1.000**	**1.000**	**1.000**	**1.000**	**1.000**	**1.001**	**1.001**	**1.001**	**1.001**
	0.000	0.000	0.000	0.000	0.000	0.000	0.000	0.000	0.000	0.000	0.000	0.000	0.000	0.000
	0.000	0.000	0.000	0.000	0.000	0.000	0.000	0.000	0.000	0.000	0.000	0.001	0.001	0.001
1.80	**1.000**	**1.000**	**1.000**	**1.000**	**1.000**	**1.000**	**1.000**	**1.000**	**1.000**	**1.000**	**1.000**	**1.001**	**1.001**	**1.001**
	0.000	0.000	0.000	0.000	0.000	0.000	0.000	0.000	0.000	0.000	0.000	0.000	0.000	0.000
	0.000	0.000	0.000	0.000	0.000	0.000	0.000	0.000	0.000	0.000	0.000	0.001	0.001	0.001
1.90	**1.000**	**1.000**	**1.000**	**1.000**	**1.000**	**1.000**	**1.000**	**1.000**	**1.000**	**1.000**	**1.000**	**1.001**	**1.001**	**1.001**
	0.000	0.000	0.000	0.000	0.000	0.000	0.000	0.000	0.000	0.000	0.000	0.000	0.000	0.000
	0.000	0.000	0.000	0.000	0.000	0.000	0.000	0.000	0.000	0.000	0.000	0.001	0.001	0.001
2.00	**1.000**	**1.000**	**1.000**	**1.000**	**1.000**	**1.000**	**1.000**	**1.000**	**1.000**	**1.000**	**1.000**	**1.001**	**1.001**	**1.001**
	0.000	0.000	0.000	0.000	0.000	0.000	0.000	0.000	0.000	0.000	0.000	0.000	0.000	0.000
	0.000	0.000	0.000	0.000	0.000	0.000	0.000	0.000	0.000	0.000	0.000	0.001	0.001	0.001
2.10	**1.000**	**1.000**	**1.000**	**1.000**	**1.000**	**1.000**	**1.000**	**1.000**	**1.000**	**1.000**	**1.000**	**1.001**	**1.001**	**1.001**
	0.000	0.000	0.000	0.000	0.000	0.000	0.000	0.000	0.000	0.000	0.000	0.000	0.000	0.000
	0.000	0.000	0.000	0.000	0.000	0.000	0.000	0.000	0.000	0.000	0.000	0.001	0.001	0.001
2.20	**1.000**	**1.000**	**1.000**	**1.000**	**1.000**	**1.000**	**1.000**	**1.000**	**1.000**	**1.000**	**1.000**	**1.001**	**1.001**	**1.001**
	0.000	0.000	0.000	0.000	0.000	0.000	0.000	0.000	0.000	0.000	0.000	0.000	0.000	0.000
	0.000	0.000	0.000	0.000	0.000	0.000	0.000	0.000	0.000	0.000	0.000	0.001	0.001	0.001
2.30	**1.000**	**1.000**	**1.000**	**1.000**	**1.000**	**1.000**	**1.000**	**1.000**	**1.000**	**1.000**	**1.000**	**1.001**	**1.001**	**1.001**
	0.000	0.000	0.000	0.000	0.000	0.000	0.000	0.000	0.000	0.000	0.000	0.000	0.000	0.000
	0.000	0.000	0.000	0.000	0.000	0.000	0.000	0.000	0.000	0.000	0.000	0.000	0.001	0.001
2.40	**1.000**	**1.000**	**1.000**	**1.000**	**1.000**	**1.000**	**1.000**	**1.000**	**1.000**	**1.000**	**1.000**	**1.000**	**1.001**	**1.001**
	0.000	0.000	0.000	0.000	0.000	0.000	0.000	0.000	0.000	0.000	0.000	0.000	0.000	0.000
	0.001	0.000	0.000	0.000	0.000	0.000	0.000	0.000	0.000	0.000	0.000	0.000	0.001	0.001
2.50	**1.001**	**1.000**	**1.000**	**1.000**	**1.000**	**1.000**	**1.000**	**1.000**	**1.000**	**1.000**	**1.000**	**1.000**	**1.001**	**1.001**
	0.000	0.000	0.000	0.000	0.000	0.000	0.000	0.000	0.000	0.000	0.000	0.000	0.000	0.000
	0.001	0.001	0.000	0.000	0.000	0.000	0.000	0.000	0.000	0.000	0.000	0.000	0.001	0.001
2.60	**1.001**	**1.001**	**1.000**	**1.000**	**1.000**	**1.000**	**1.000**	**1.000**	**1.000**	**1.000**	**1.000**	**1.000**	**1.001**	**1.001**
	0.000	0.000	0.000	0.000	0.000	0.000	0.000	0.000	0.000	0.000	0.000	0.000	0.000	0.000
	0.002	0.001	0.000	0.000	0.000	0.000	0.000	0.000	0.000	0.000	0.000	0.000	0.001	0.001
2.70	**1.002**	**1.001**	**1.000**	**1.000**	**1.000**	**1.000**	**1.000**	**1.000**	**1.000**	**1.000**	**1.000**	**1.000**	**1.001**	**1.001**
	0.000	0.000	0.000	0.000	0.000	0.000	0.000	0.000	0.000	0.000	0.000	0.000	0.000	0.000
	0.004	0.002	0.001	0.000	0.000	0.000	0.000	0.000	0.000	0.000	0.000	0.000	0.001	0.001
2.80	**1.004**	**1.002**	**1.001**	**1.000**	**1.000**	**1.000**	**1.000**	**1.000**	**1.000**	**1.000**	**1.000**	**1.000**	**1.001**	**1.001**

Table Six: *ExCP, ExCU, and ECP&U* when Average Is on Rework Side of Target and the Cost of Rework Is 0.1% of the Cost of Scrap

C_{pk} or P_{pk}	C_p or P_p value 2.90	3.00	3.20	3.40	3.60	3.80	4.00	4.50	5.00	5.50	6.0	8.0	12.0	20.0
	0.000	0.000	0.000	0.000	0.000	0.000	0.000	0.000	0.000	0.000	0.000	0.000	0.000	0.000
	0.007	0.004	0.001	0.000	0.000	0.000	0.000	0.000	0.000	0.000	0.000	0.000	0.001	0.001
2.90	***1.007***	***1.004***	***1.001***	***1.000***	***1.000***	***1.000***	***1.000***	***1.000***	***1.000***	***1.000***	***1.000***	***1.000***	***1.001***	***1.001***
		0.000	0.000	0.000	0.000	0.000	0.000	0.000	0.000	0.000	0.000	0.000	0.000	0.000
		0.006	0.002	0.000	0.000	0.000	0.000	0.000	0.000	0.000	0.000	0.000	0.001	0.001
3.00		***1.006***	***1.002***	***1.000***	***1.000***	***1.000***	***1.000***	***1.000***	***1.000***	***1.000***	***1.000***	***1.000***	***1.001***	***1.001***
			0.000	0.000	0.000	0.000	0.000	0.000	0.000	0.000	0.000	0.000	0.000	0.000
			0.005	0.002	0.000	0.000	0.000	0.000	0.000	0.000	0.000	0.000	0.001	0.001
3.20 »			***1.005***	***1.002***	***1.000***	***1.000***	***1.000***	***1.000***	***1.000***	***1.000***	***1.000***	***1.000***	***1.001***	***1.001***
				0.000	0.000	0.000	0.000	0.000	0.000	0.000	0.000	0.000	0.000	0.000
				0.005	0.001	0.000	0.000	0.000	0.000	0.000	0.000	0.000	0.001	0.001
3.40 »				***1.005***	***1.001***	***1.000***	***1.000***	***1.000***	***1.000***	***1.000***	***1.000***	***1.000***	***1.001***	***1.001***
					0.000	0.000	0.000	0.000	0.000	0.000	0.000	0.000	0.000	0.000
					0.004	0.001	0.000	0.000	0.000	0.000	0.000	0.000	0.000	0.001
3.60 »					***1.004***	***1.001***	***1.000***	***1.000***	***1.000***	***1.000***	***1.000***	***1.000***	***1.000***	***1.001***
						0.000	0.000	0.000	0.000	0.000	0.000	0.000	0.000	0.000
						0.004	0.001	0.000	0.000	0.000	0.000	0.000	0.000	0.001
3.80 »						***1.004***	***1.001***	***1.000***	***1.000***	***1.000***	***1.000***	***1.000***	***1.000***	***1.001***
							0.000	0.000	0.000	0.000	0.000	0.000	0.000	0.000
							0.003	0.000	0.000	0.000	0.000	0.000	0.000	0.001
4.00 »							***1.003***	***1.000***	***1.000***	***1.000***	***1.000***	***1.000***	***1.000***	***1.001***
								0.000	0.000	0.000	0.000	0.000	0.000	0.000
								0.003	0.000	0.000	0.000	0.000	0.000	0.001
4.50 »								***1.003***	***1.000***	***1.000***	***1.000***	***1.000***	***1.000***	***1.001***
									0.000	0.000	0.000	0.000	0.000	0.000
									0.002	0.000	0.000	0.000	0.000	0.001
5.00 »									***1.002***	***1.000***	***1.000***	***1.000***	***1.000***	***1.001***
										0.000	0.000	0.000	0.000	0.000
										0.002	0.000	0.000	0.000	0.001
5.50 »										***1.002***	***1.000***	***1.000***	***1.000***	***1.001***
											0.000	0.000	0.000	0.000
											0.002	0.000	0.000	0.000
6.00 »											***1.002***	***1.000***	***1.000***	***1.000***
												0.000	0.000	0.000
												0.001	0.000	0.000
8.00 »												***1.001***	***1.000***	***1.000***
													0.000	0.000
													0.000	0.000
12.0 »													***1.000***	***1.000***
														0.000
														0.000
20.0 »														***1.000***

Table Seven: *Excess Costs of Production* (*ExCP*) and *Excess Costs of Use* (*ExCU*) for One-Sided Product Specifications when Nonconforming Product Is Scrapped

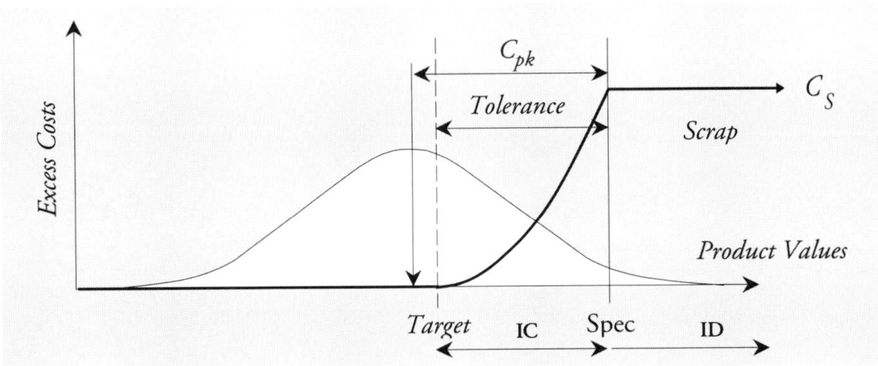

Table Seven: The Excess Cost Function for One-Sided Product Specifications with a Target Value

The *Excess Costs of Production* (*ExCP*) will depend only upon the value for C_{pk} (or P_{pk}). These values were obtained from:

$$ExCP = \frac{\mathbf{ID}}{1 - \mathbf{ID}}$$

The *Excess Cost of Use* (*ExCU*) will depend upon both C_{pk} and *Tolerance*. These values were obtained from Table A.9 using $C_p = 0.333$ *Tolerance* and the formula:

$$ExCU = \frac{\mathbf{IC}}{1 - \mathbf{ID}}$$

Use the values of C_{pk} and *Tolerance* to find tabled values to compute the following as needed:

$$\textit{Effective Cost of Production} = \left\{ 1.0 + \frac{C_S}{NC} \, ExCP \right\}$$

$$\textit{Actual Cost of Production} = \left\{ 1.0 + \frac{C_S}{NC} \, ExCP \right\} \times \textit{Nominal Cost}$$

$$\textit{Effective Cost of Production and Use} = 1.0 + \frac{C_S}{NC} \left\{ ExCP + ExCU \right\}$$

$$\textit{Actual Cost of Production and Use} = ECP\&U \times \textit{Nominal Cost}$$

Table Seven: *Excess Costs of Production* (*ExCP*) and *Excess Costs of Use* (*ExCU*) for One-Sided Product Specifications when Nonconforming Product Is Scrapped

		Tolerance in Std. Dev. Units										
	C_{pk} or	***0.3***	***0.6***	***0.9***	***1.2***	***1.5***	***1.8***	***2.1***	***2.4***	***2.7***	***3.0***	***3.3***
ExCP	P_{pk}	***ExCU***	***ExCU***	***ExCU***	***ExCU***	***ExCU***	***ExCU***	***ExCU***	***ExCU***	***ExCU***	***ExCU***	***ExCU***
13.97	**–0.5**	0.173	0.310	0.417	0.500	0.566	0.618	0.660	0.694	0.722	0.746	0.766
7.690	**–0.4**	0.154	0.280	0.383	0.466	0.532	0.586	0.630	0.666	0.696	0.721	0.743
4.433	**–0.3**	0.135	0.250	0.347	0.428	0.495	0.550	0.596	0.634	0.666	0.693	0.716
2.656	**–0.2**	0.116	0.219	0.309	0.387	0.454	0.510	0.557	0.597	0.631	0.660	0.685
1.617	**–0.1**	0.097	0.188	0.270	0.344	0.409	0.465	0.513	0.555	0.591	0.623	0.650
1.000	**0.0**	0.079	0.157	0.230	0.298	0.360	0.416	0.465	0.508	0.546	0.579	0.608
0.618	**0.1**	0.063	0.127	0.190	0.252	0.309	0.363	0.412	0.456	0.495	0.530	0.561
0.378	**0.2**	0.048	0.099	0.152	0.205	0.257	0.308	0.355	0.399	0.439	0.475	0.508
0.226	**0.3**	0.035	0.074	0.116	0.160	0.206	0.251	0.296	0.339	0.379	0.416	0.450
0.130	**0.4**	0.024	0.052	0.084	0.119	0.157	0.197	0.237	0.277	0.316	0.353	0.388
0.072	**0.5**	0.016	0.035	0.057	0.084	0.114	0.147	0.181	0.217	0.253	0.289	0.323
0.037	**0.6**	0.009	0.021	0.037	0.055	0.078	0.103	0.132	0.162	0.195	0.227	0.260
0.018	**0.7**	0.005	0.012	0.022	0.034	0.049	0.068	0.090	0.115	0.142	0.171	0.201
0.008	**0.8**	0.003	0.007	0.012	0.019	0.029	0.042	0.058	0.077	0.098	0.123	0.149
0.003	**0.9**	0.001	0.003	0.006	0.010	0.016	0.024	0.035	0.048	0.064	0.084	0.105
0.001	**1.0**	0.001	0.001	0.003	0.005	0.008	0.013	0.020	0.028	0.040	0.054	0.071
0.000	**1.1**	0.000	0.001	0.001	0.002	0.004	0.006	0.010	0.016	0.023	0.033	0.045
	1.2	.	0.000	0.000	0.001	0.002	0.003	0.005	0.008	0.013	0.019	0.028
	1.3	.	.	0.000	0.000	0.001	0.001	0.002	0.004	0.006	0.010	0.016
	1.4	.	.	.	.	0.000	0.000	0.001	0.002	0.003	0.005	0.009
	1.5	.	.	.	.	.	.	0.000	0.001	0.001	0.003	0.004
	1.6	.	.	.	.	.	.	.	0.000	0.001	0.001	0.002
	1.7	.	.	.	.	.	.	.	.	0.000	0.000	0.001
	1.8	.	.	.	.	.	.	.	.	.	.	0.000

Continued on the following pages »

Table Seven: *Excess Costs of Use (ExCU)* for One-Sided Product Specifications when Nonconforming Product Is Scrapped

	Tolerance in Std. Dev. Units											
C_{pk} *or*	***3.6***	***3.9***	***4.2***	***4.5***	***4.8***	***5.1***	***5.4***	***5.7***	***6.0***	***6.3***	***6.6***	***6.9***
P_{pk}	***ExCU***	***ExCU***	***ExCU***	***ExCU***	***ExCU***	***ExCU***	***ExCU***	***ExCU***	***ExCU***	***ExCU***	***ExCU***	***ExCU***
–0.5	0.783	0.798	0.811	0.822	0.833	0.842	0.850	0.857	0.864	0.870	0.876	0.881
–0.4	0.762	0.778	0.792	0.804	0.815	0.825	0.834	0.842	0.850	0.856	0.862	0.868
–0.3	0.736	0.754	0.769	0.783	0.795	0.806	0.816	0.825	0.833	0.840	0.847	0.853
–0.2	0.707	0.726	0.743	0.758	0.771	0.783	0.794	0.804	0.813	0.821	0.829	0.836
–0.1	0.673	0.694	0.713	0.729	0.744	0.757	0.769	0.780	0.790	0.799	0.807	0.815
0.0	0.634	0.657	0.677	0.695	0.711	0.726	0.739	0.751	0.762	0.772	0.782	0.790
0.1	0.589	0.614	0.636	0.655	0.673	0.690	0.704	0.718	0.730	0.741	0.752	0.761
0.2	0.538	0.564	0.588	0.610	0.630	0.648	0.664	0.679	0.693	0.705	0.717	0.728
0.3	0.481	0.510	0.535	0.559	0.581	0.600	0.618	0.635	0.650	0.664	0.677	0.689
0.4	0.420	0.450	0.477	0.503	0.526	0.547	0.567	0.585	0.602	0.618	0.632	0.645
0.5	0.356	0.387	0.416	0.442	0.467	0.490	0.512	0.531	0.550	0.567	0.583	0.598
0.6	0.292	0.323	0.353	0.380	0.406	0.431	0.453	0.475	0.495	0.513	0.530	0.547
0.7	0.231	0.261	0.290	0.318	0.345	0.371	0.394	0.417	0.438	0.458	0.476	0.494
0.8	0.176	0.204	0.232	0.260	0.286	0.312	0.336	0.360	0.382	0.403	0.422	0.441
0.9	0.129	0.154	0.180	0.206	0.231	0.257	0.281	0.305	0.328	0.349	0.370	0.389
1.0	0.090	0.111	0.134	0.158	0.182	0.207	0.231	0.254	0.277	0.298	0.319	0.339
1.1	0.060	0.078	0.097	0.118	0.140	0.162	0.185	0.208	0.230	0.252	0.273	0.293
1.2	0.038	0.052	0.067	0.085	0.104	0.124	0.145	0.166	0.188	0.209	0.229	0.250
1.3	0.023	0.033	0.045	0.059	0.074	0.092	0.111	0.130	0.150	0.170	0.190	0.210
1.4	0.013	0.020	0.028	0.039	0.052	0.066	0.082	0.099	0.117	0.136	0.155	0.174
1.5	0.007	0.011	0.017	0.025	0.034	0.046	0.059	0.074	0.090	0.107	0.124	0.142
1.6	0.004	0.006	0.010	0.015	0.022	0.030	0.041	0.053	0.066	0.081	0.097	0.114
1.7	0.002	0.003	0.005	0.009	0.013	0.019	0.027	0.037	0.048	0.060	0.074	0.089
1.8	0.001	0.002	0.003	0.005	0.008	0.012	0.017	0.024	0.033	0.043	0.055	0.068
1.9	0.000	0.001	0.001	0.002	0.004	0.007	0.010	0.015	0.022	0.030	0.039	0.050
2.0	.	0.000	0.001	0.001	0.002	0.004	0.006	0.009	0.014	0.020	0.027	0.036
2.1	.	.	0.000	0.001	0.001	0.002	0.003	0.005	0.008	0.013	0.018	0.025
2.2	.	.	.	0.000	0.000	0.001	0.002	0.003	0.005	0.008	0.011	0.017
2.3	.	.	.	.	.	0.000	0.001	0.001	0.003	0.004	0.007	0.011
2.4	.	.	.	.	.	.	0.000	0.001	0.001	0.000	0.004	0.006
2.5	.	.	.	.	.	.	.	0.000	0.001	0.001	0.002	0.004
2.6	.	.	.	.	.	.	.	.	0.000	0.001	0.001	0.002
2.7	.	.	.	.	.	.	.	.	.	0.000	0.000	0.000

Table Seven: *Excess Costs of Use* (*ExCU*) for One-Sided Product Specifications when Nonconforming Product Is Scrapped

	Tolerance in Std. Dev. Units											
C_{pk} or	*7.2*	*7.5*	*7.8*	*8.1*	*8.4*	*8.7*	*9.0*	*9.6*	*10.2*	*10.8*	*11.4*	*12.0*
P_{pk}	*ExCU*	*ExCU*	*ExCU*	*ExCU*	*ExCU*	*ExCU*	*ExCU*	*ExCU*	*ExCU*	*ExCU*	*ExCU*	*ExCU*
–0.5	0.886	0.890	0.894	0.898	0.901	0.905	0.908	0.913	0.919	0.923	0.927	0.931
–0.4	0.873	0.878	0.883	0.887	0.891	0.894	0.898	0.904	0.909	0.914	0.919	0.923
–0.3	0.859	0.864	0.869	0.874	0.878	0.882	0.886	0.893	0.899	0.904	0.909	0.914
–0.2	0.842	0.848	0.853	0.858	0.863	0.868	0.872	0.879	0.886	0.892	0.898	0.903
–0.1	0.822	0.828	0.834	0.840	0.845	0.850	0.855	0.864	0.871	0.878	0.884	0.890
0.0	0.798	0.805	0.812	0.819	0.825	0.830	0.835	0.845	0.854	0.861	0.868	0.875
0.1	0.770	0.778	0.786	0.793	0.800	0.806	0.812	0.823	0.833	0.841	0.849	0.856
0.2	0.738	0.747	0.756	0.764	0.771	0.778	0.785	0.797	0.808	0.818	0.827	0.835
0.3	0.700	0.711	0.720	0.729	0.738	0.746	0.754	0.768	0.780	0.791	0.801	0.811
0.4	0.658	0.670	0.681	0.691	0.700	0.710	0.718	0.734	0.748	0.760	0.772	0.782
0.5	0.612	0.625	0.637	0.648	0.659	0.669	0.679	0.696	0.712	0.726	0.739	0.751
0.6	0.562	0.576	0.590	0.602	0.614	0.625	0.636	0.656	0.673	0.689	0.704	0.717
0.7	0.510	0.526	0.541	0.554	0.567	0.580	0.591	0.613	0.632	0.650	0.666	0.681
0.8	0.459	0.475	0.491	0.506	0.520	0.533	0.546	0.569	0.590	0.610	0.627	0.644
0.9	0.408	0.425	0.442	0.458	0.473	0.487	0.500	0.526	0.548	0.569	0.588	0.606
1.0	0.359	0.377	0.394	0.411	0.427	0.442	0.456	0.483	0.507	0.529	0.550	0.569
1.1	0.312	0.331	0.349	0.366	0.382	0.398	0.413	0.441	0.467	0.491	0.512	0.532
1.2	0.269	0.288	0.306	0.324	0.341	0.357	0.372	0.401	0.428	0.453	0.476	0.497
1.3	0.229	0.248	0.266	0.284	0.301	0.318	0.333	0.363	0.391	0.417	0.440	0.463
1.4	0.193	0.211	0.229	0.247	0.264	0.281	0.297	0.327	0.356	0.382	0.407	0.429
1.5	0.160	0.178	0.195	0.213	0.230	0.246	0.262	0.293	0.322	0.349	0.374	0.398
1.6	0.130	0.147	0.164	0.181	0.198	0.214	0.230	0.261	0.290	0.317	0.343	0.367
1.7	0.104	0.120	0.136	0.152	0.169	0.184	0.200	0.231	0.260	0.287	0.313	0.338
1.8	0.082	0.096	0.111	0.126	0.142	0.157	0.172	0.202	0.231	0.259	0.285	0.309
1.9	0.062	0.075	0.089	0.103	0.117	0.132	0.147	0.176	0.204	0.232	0.258	0.283
2.0	0.046	0.057	0.070	0.082	0.096	0.110	0.123	0.151	0.179	0.206	0.232	0.257
2.1	0.033	0.043	0.053	0.064	0.077	0.089	0.102	0.129	0.156	0.182	0.208	0.233
2.2	0.023	0.031	0.039	0.049	0.060	0.071	0.083	0.109	0.134	0.160	0.185	0.209
2.3	0.015	0.021	0.028	0.036	0.046	0.056	0.067	0.090	0.114	0.139	0.164	0.188
2.4	0.010	0.014	0.020	0.026	0.034	0.043	0.052	0.073	0.096	0.120	0.143	0.167
2.5	0.006	0.009	0.013	0.018	0.024	0.032	0.040	0.059	0.080	0.102	0.125	0.148
2.6	0.003	0.005	0.008	0.012	0.017	0.023	0.030	0.046	0.065	0.086	0.107	0.129
2.7	0.002	0.003	0.005	0.008	0.011	0.016	0.021	0.035	0.052	0.071	0.091	0.113
2.8	0.001	0.002	0.003	0.005	0.007	0.010	0.015	0.026	0.041	0.058	0.077	0.097
2.9	0.000	0.001	0.002	0.003	0.004	0.007	0.010	0.019	0.031	0.046	0.064	0.083
3.0	.	0.000	0.001	0.001	0.002	0.004	0.006	0.013	0.023	0.036	0.052	0.069
3.2	.	.	0.000	0.000	0.001	0.001	0.002	0.005	0.011	0.021	0.033	0.047
3.4	.	.	.	.	0.000	0.000	0.001	0.002	0.005	0.010	0.018	0.029
3.6	.	.	.	.	.	.	0.000	0.001	0.002	0.004	0.009	0.017
3.8	.	.	.	.	.	.	.	0.000	0.000	0.001	0.004	0.008
4.0	.	.	.	.	.	.	.	.	.	0.000	0.001	0.003

Table Eight: *Excess Costs of Production* (*ExCP*) and *Excess Costs of Use* (*ExCU*) for One-Sided Product Specifications when Nonconforming Product Is Reworked

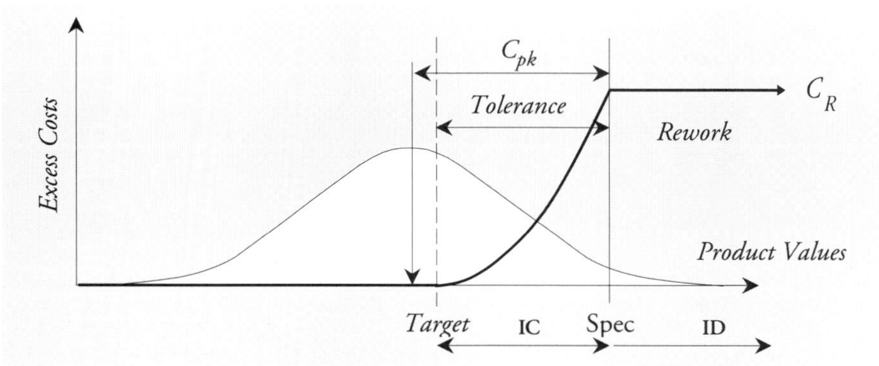

Table Eight: The Excess Cost Function for One-Sided Product Specifications with a Target Value

The *Excess Costs of Production* (*ExCP*) will depend only upon the value for C_{pk} (or P_{pk}). These values were obtained from the formula:

$$ExCP = \mathbf{ID}$$

The *Excess Cost of Use* (*ExCU*) will depend upon both C_{pk} and *Tolerance*. These values were obtained from Table A.9 using $C_p = 0.333$ *Tolerance* and the formula:

$$ExCU = \mathbf{IC}$$

Use the values of C_{pk} and *Tolerance* to find tabled values to compute the following as needed:

$$\textit{Effective Cost of Production} = \left\{ 1.0 + \frac{C_R}{NC} ExCP \right\}$$

$$\textit{Actual Cost of Production} = \left\{ 1.0 + \frac{C_R}{NC} ExCP \right\} \times \textit{Nominal Cost}$$

$$\textit{Effective Cost of Production and Use} = 1.0 + \frac{C_R}{NC} \left\{ ExCP + ExCU \right\}$$

$$\textit{Actual Cost of Production and Use} = ECP\&U \times \textit{Nominal Cost}$$

Table Eight: *Excess Costs of Production* (*ExCP*) and *Excess Costs of Use* (*ExCU*) for One-Sided Product Specifications when Nonconforming Product Is Reworked

		Tolerance in Std. Dev. Units										
	C_{pk} or	***0.3***	***0.6***	***0.9***	***1.2***	***1.5***	***1.8***	***2.1***	***2.4***	***2.7***	***3.0***	***3.3***
ExCP	P_{pk}	***ExCU***	***ExCU***	***ExCU***	***ExCU***	***ExCU***	***ExCU***	***ExCU***	***ExCU***	***ExCU***	***ExCU***	***ExCU***
0.933	**–0.5**	0.012	0.021	0.028	0.033	0.038	0.041	0.044	0.046	0.048	0.050	0.051
0.885	**–0.4**	0.018	0.032	0.044	0.054	0.061	0.067	0.072	0.077	0.080	0.083	0.085
0.816	**–0.3**	0.025	0.046	0.064	0.079	0.091	0.101	0.110	0.117	0.123	0.128	0.132
0.726	**–0.2**	0.032	0.060	0.085	0.106	0.124	0.140	0.153	0.164	0.173	0.181	0.188
0.618	**–0.1**	0.037	0.072	0.103	0.131	0.156	0.178	0.196	0.212	0.226	0.238	0.248
0.500	**0**	0.040	0.078	0.115	0.149	0.180	0.208	0.233	0.254	0.273	0.290	0.304
0.382	**0.1**	0.039	0.078	0.118	0.155	0.191	0.224	0.255	0.282	0.306	0.328	0.347
0.274	**0.2**	0.035	0.072	0.110	0.149	0.187	0.223	0.258	0.290	0.319	0.345	0.369
0.184	**0.3**	0.028	0.060	0.094	0.131	0.168	0.205	0.242	0.276	0.309	0.339	0.367
0.115	**0.4**	0.021	0.046	0.074	0.106	0.139	0.174	0.210	0.245	0.280	0.312	0.343
0.067	**0.5**	0.014	0.032	0.054	0.078	0.106	0.137	0.169	0.203	0.237	0.270	0.302
0.036	**0.6**	0.009	0.021	0.035	0.053	0.075	0.099	0.127	0.156	0.188	0.219	0.251
0.018	**0.7**	0.005	0.012	0.021	0.033	0.049	0.067	0.088	0.113	0.140	0.168	0.198
0.008	**0.8**	0.003	0.007	0.012	0.019	0.029	0.042	0.057	0.076	0.098	0.122	0.148
0.004	**0.9**	0.001	0.003	0.006	0.010	0.016	0.024	0.035	0.048	0.064	0.083	0.105
0.001	**1.0**	0.001	0.001	0.003	0.005	0.008	0.013	0.019	0.028	0.040	0.054	0.071
0.000	**1.1**	0.000	0.001	0.001	0.002	0.004	0.006	0.010	0.016	0.023	0.033	0.045
	1.2	.	0.000	0.000	0.001	0.002	0.003	0.005	0.008	0.013	0.019	0.028
	1.3	.	.	.	0.000	0.001	0.001	0.002	0.004	0.006	0.010	0.016
	1.4	.	.	.	.	0.000	0.000	0.001	0.002	0.003	0.005	0.009
	1.5	.	.	.	.	.	.	0.000	0.001	0.001	0.003	0.004
	1.6	.	.	.	.	.	.	.	0.000	0.001	0.001	0.002
	1.7	.	.	.	.	.	.	.		0.000	0.000	0.001
	1.8	.	.	.	.	.	.	.	.	.	.	0.000

Continued on the following pages »

Table Eight: *Excess Costs of Use (ExCU)* for One-Sided Product Specifications when Nonconforming Product Is Reworked

	Tolerance in Std. Dev. Units											
C_{pk} or	***3.6***	***3.9***	***4.2***	***4.5***	***4.8***	***5.1***	***5.4***	***5.7***	***6.0***	***6.3***	***6.6***	***6.9***
P_{pk}	***ExCU***	***ExCU***	***ExCU***	***ExCU***	***ExCU***	***ExCU***	***ExCU***	***ExCU***	***ExCU***	***ExCU***	***ExCU***	***ExCU***
–0.5	0.052	0.053	0.054	0.055	0.056	0.056	0.057	0.057	0.058	0.058	0.059	0.059
–0.4	0.088	0.089	0.091	0.093	0.094	0.095	0.096	0.097	0.098	0.099	0.099	0.100
–0.3	0.136	0.139	0.142	0.144	0.146	0.148	0.150	0.152	0.153	0.155	0.156	0.157
–0.2	0.194	0.199	0.204	0.208	0.212	0.215	0.218	0.221	0.223	0.225	0.227	0.229
–0.1	0.257	0.265	0.272	0.279	0.284	0.289	0.294	0.298	0.302	0.305	0.308	0.311
0.0	0.317	0.328	0.338	0.347	0.356	0.363	0.370	0.376	0.381	0.386	0.391	0.395
0.1	0.364	0.379	0.393	0.405	0.416	0.426	0.435	0.443	0.451	0.458	0.464	0.470
0.2	0.390	0.410	0.427	0.443	0.457	0.470	0.482	0.493	0.503	0.512	0.520	0.528
0.3	0.393	0.416	0.437	0.456	0.474	0.490	0.504	0.518	0.530	0.542	0.552	0.562
0.4	0.372	0.398	0.422	0.445	0.465	0.484	0.502	0.518	0.533	0.547	0.559	0.571
0.5	0.332	0.361	0.388	0.413	0.436	0.458	0.477	0.496	0.513	0.529	0.544	0.558
0.6	0.282	0.312	0.340	0.367	0.392	0.415	0.437	0.458	0.477	0.495	0.511	0.527
0.7	0.227	0.257	0.285	0.313	0.339	0.364	0.387	0.409	0.430	0.450	0.468	0.485
0.8	0.175	0.202	0.230	0.257	0.284	0.309	0.334	0.357	0.379	0.399	0.419	0.437
0.9	0.128	0.153	0.179	0.205	0.231	0.256	0.280	0.304	0.326	0.348	0.368	0.388
1.0	0.090	0.111	0.134	0.158	0.182	0.206	0.230	0.254	0.276	0.298	0.319	0.339
1.1	0.060	0.077	0.097	0.118	0.140	0.162	0.185	0.207	0.230	0.251	0.272	0.293
1.2	0.038	0.052	0.067	0.085	0.104	0.124	0.145	0.166	0.188	0.209	0.229	0.250
1.3	0.023	0.033	0.045	0.059	0.074	0.092	0.111	0.130	0.150	0.170	0.190	0.210
1.4	0.013	0.020	0.028	0.039	0.052	0.066	0.082	0.099	0.117	0.136	0.155	0.174
1.5	0.007	0.011	0.017	0.025	0.034	0.046	0.059	0.074	0.090	0.107	0.124	0.142
1.6	0.004	0.006	0.010	0.015	0.022	0.030	0.041	0.053	0.066	0.081	0.097	0.114
1.7	0.002	0.003	0.005	0.009	0.013	0.019	0.027	0.037	0.048	0.060	0.074	0.089
1.8	0.001	0.002	0.003	0.005	0.008	0.012	0.017	0.024	0.033	0.043	0.055	0.068
1.9	0.000	0.001	0.001	0.002	0.004	0.007	0.010	0.015	0.022	0.030	0.039	0.050
2.0	.	0.000	0.001	0.001	0.002	0.004	0.006	0.009	0.014	0.020	0.027	0.036
2.1	.	.	0.000	0.001	0.001	0.002	0.003	0.005	0.008	0.013	0.018	0.025
2.2	.	.	.	0.000	0.000	0.001	0.002	0.003	0.005	0.008	0.011	0.017
2.3	.	.	.	.	.	0.000	0.001	0.001	0.003	0.004	0.007	0.011
2.4	.	.	.	.	.	.	0.000	0.001	0.001	0.000	0.004	0.006
2.5	.	.	.	.	.	.	.	0.000	0.001	0.001	0.002	0.004
2.6	.	.	.	.	.	.	.	.	0.000	0.001	0.001	0.002
2.7	.	.	.	.	.	.	.	.	.	0.000	0.000	0.000

Table Eight: *Excess Costs of Use* (*ExCU*) for One-Sided Product Specifications when Nonconforming Product Is Reworked

Tolerance in Std. Dev. Units												
C_{pk} *or*	*7.2*	*7.5*	*7.8*	*8.1*	*8.4*	*8.7*	*9.0*	*9.6*	*10.2*	*10.8*	*11.4*	*12.0*
P_{pk}	*ExCU*	*ExCU*	*ExCU*	*ExCU*	*ExCU*	*ExCU*	*ExCU*	*ExCU*	*ExCU*	*ExCU*	*ExCU*	*ExCU*
–0.5	0.059	0.059	0.060	0.060	0.060	0.060	0.061	0.061	0.061	0.062	0.062	0.062
–0.4	0.100	0.101	0.102	0.102	0.102	0.103	0.103	0.104	0.105	0.105	0.106	0.106
–0.3	0.158	0.159	0.160	0.161	0.162	0.162	0.163	0.164	0.165	0.166	0.167	0.168
–0.2	0.231	0.233	0.234	0.235	0.237	0.238	0.239	0.241	0.243	0.245	0.246	0.248
–0.1	0.314	0.317	0.319	0.321	0.323	0.325	0.327	0.330	0.333	0.336	0.338	0.340
0.0	0.399	0.403	0.406	0.409	0.412	0.415	0.418	0.423	0.427	0.431	0.434	0.437
0.1	0.476	0.481	0.486	0.490	0.494	0.498	0.502	0.509	0.515	0.520	0.525	0.529
0.2	0.535	0.542	0.548	0.554	0.560	0.565	0.570	0.579	0.587	0.594	0.600	0.606
0.3	0.571	0.580	0.588	0.595	0.602	0.609	0.615	0.626	0.636	0.646	0.654	0.661
0.4	0.582	0.593	0.602	0.611	0.620	0.628	0.635	0.649	0.662	0.673	0.683	0.692
0.5	0.571	0.583	0.594	0.605	0.615	0.624	0.633	0.650	0.664	0.678	0.690	0.701
0.6	0.542	0.555	0.568	0.581	0.592	0.603	0.613	0.632	0.649	0.664	0.678	0.691
0.7	0.501	0.517	0.531	0.544	0.557	0.569	0.581	0.602	0.621	0.638	0.654	0.669
0.8	0.455	0.471	0.487	0.502	0.516	0.529	0.541	0.565	0.586	0.605	0.622	0.638
0.9	0.406	0.424	0.440	0.456	0.471	0.485	0.499	0.524	0.547	0.567	0.586	0.604
1.0	0.358	0.376	0.394	0.410	0.426	0.441	0.455	0.482	0.506	0.529	0.549	0.568
1.1	0.312	0.331	0.349	0.366	0.382	0.398	0.413	0.441	0.467	0.490	0.512	0.532
1.2	0.269	0.288	0.306	0.324	0.341	0.357	0.372	0.401	0.428	0.453	0.476	0.497
1.3	0.229	0.248	0.266	0.284	0.301	0.318	0.333	0.363	0.391	0.417	0.440	0.463
1.4	0.193	0.211	0.229	0.247	0.264	0.281	0.297	0.327	0.356	0.382	0.407	0.429
1.5	0.160	0.178	0.195	0.213	0.230	0.246	0.262	0.293	0.322	0.349	0.374	0.398
1.6	0.130	0.147	0.164	0.181	0.198	0.214	0.230	0.261	0.290	0.317	0.343	0.367
1.7	0.104	0.120	0.136	0.152	0.169	0.184	0.200	0.231	0.260	0.287	0.313	0.338
1.8	0.082	0.096	0.111	0.126	0.142	0.157	0.172	0.202	0.231	0.259	0.285	0.309
1.9	0.062	0.075	0.089	0.103	0.117	0.132	0.147	0.176	0.204	0.232	0.258	0.283
2.0	0.046	0.057	0.070	0.082	0.096	0.110	0.123	0.151	0.179	0.206	0.232	0.257
2.1	0.033	0.043	0.053	0.064	0.077	0.089	0.102	0.129	0.156	0.182	0.208	0.233
2.2	0.023	0.031	0.039	0.049	0.060	0.071	0.083	0.109	0.134	0.160	0.185	0.209
2.3	0.015	0.021	0.028	0.036	0.046	0.056	0.067	0.090	0.114	0.139	0.164	0.188
2.4	0.010	0.014	0.020	0.026	0.034	0.043	0.052	0.073	0.096	0.120	0.143	0.167
2.5	0.006	0.009	0.013	0.018	0.024	0.032	0.040	0.059	0.080	0.102	0.125	0.148
2.6	0.003	0.005	0.008	0.012	0.017	0.023	0.030	0.046	0.065	0.086	0.107	0.129
2.7	0.002	0.003	0.005	0.008	0.011	0.016	0.021	0.035	0.052	0.071	0.091	0.113
2.8	0.001	0.002	0.003	0.005	0.007	0.010	0.015	0.026	0.041	0.058	0.077	0.097
2.9	0.000	0.001	0.002	0.003	0.004	0.007	0.010	0.019	0.031	0.046	0.064	0.083
3.0	0.000	0.001	0.001	0.002	0.004	0.006	0.013	0.023	0.036	0.052	0.069	
3.2	.	.	0.000	0.000	0.001	0.001	0.002	0.005	0.011	0.021	0.033	0.047
3.4	.	.	.	.	0.000	0.000	0.001	0.002	0.005	0.010	0.018	0.029
3.6	.	.	.	.	.	.	0.000	0.001	0.002	0.004	0.009	0.017
3.8	.	.	.	.	.	.	.	0.000	0.000	0.001	0.004	0.008
4.0	.	.	.	.	.	.	.	.	.	0.000	0.001	0.003

Table Nine: Target at Boundary: The *Excess Costs of Production, Excess Costs of Use,* and *Effective Cost of Production and Use* when Nonconforming Units Are Scrapped

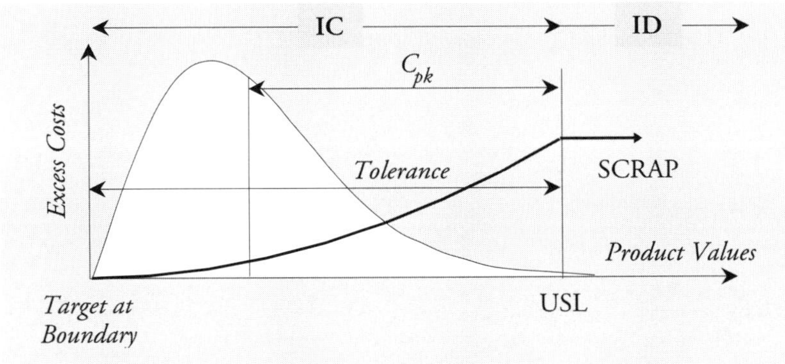

Table Nine: Excess Cost Function when Target Is at a Boundary Value

Corresponding to each pairing for *Tolerance* and C_{pk} there are three entries:

The topmost of these is the *Excess Costs of Production,* *ExCP*

The middle value is the *Excess Costs of Use,* *ExCU*

And the bottom value (in bold) is the *Effective Cost of Production and Use,* ***ECP&U***

The entries in Table Nine are for the case where nonconforming product is scrapped and the Cost of Scrap, C_S, is assumed to be equal to the Nominal Cost of Production, *NC*. The values for **IC** and **ID** are given in Table A.10. The proportion scrapped is given by the integral **ID**, making the general and specific formulas for Table Nine:

$$ECP\&U = 1.0 + \frac{C_S}{NC}\frac{\mathbf{ID}}{1-\mathbf{ID}} + \frac{C_S}{NC}\frac{\mathbf{IC}}{1-\mathbf{ID}}$$

$$= 1.0 + \frac{\mathbf{ID}}{1-\mathbf{ID}} + \frac{\mathbf{IC}}{1-\mathbf{ID}}$$

$$= 1.0 + \textit{Excess Costs of Production} + \textit{Excess Costs of Use}$$

Since the Cost of Scrap can be appreciably greater than *NC*, the entries in Table Nine are essentially *lower bounds* for all three quantities. If greater precision is required, the entries for the *Excess Costs of Production* and the *Excess Costs of Use* may be inflated by the ratio of C_S to *NC*, as shown in the general formula above, and added to 1.00 to get a customized value for the *Effective Cost of Production and Use*.

For example, with *Tolerance* = 2.7 and C_{pk} = 0.5, and with C_S = 1.5 *NC* we would convert the tabled value of 1.287 for *ECP&U* into the value 1.431 (*ExCP* = 0.135 and *ExCU* = 0.293).

Table Nine: Target at Boundary: The *Excess Costs of Production, Excess Costs of Use,* and *Effective Cost of Production and Use* when Nonconforming Units Are Scrapped

C_{pk} or P_{pk}	*Tolerance in Standard Deviation Units* 0.3	0.6	0.9	1.2	1.5	1.8	2.1	2.4	2.7	3.0	3.3	3.6	3.9	4.2
	36.09	20.98	16.75	14.88	13.79	13.93	13.93	13.93	13.93	13.93	13.93	13.93	13.93	13.93
	0.484	0.526	0.562	0.594	0.621	0.612	0.657	0.687	0.716	0.746	0.761	0.776	0.791	0.806
–0.50	***37.57***	***22.51***	***18.31***	***16.48***	***15.41***	***15.54***	***15.58***	***15.61***	***15.64***	***15.67***	***15.69***	***15.70***	***15.72***	***15.73***
	13.60	9.39	8.18	7.68	7.44	7.30	7.70	7.70	7.70	7.70	7.70	7.70	7.70	7.70
	0.431	0.478	0.520	0.557	0.589	0.617	0.626	0.670	0.696	0.722	0.739	0.765	0.774	0.791
–0.40	***15.03***	***10.86***	***9.70***	***9.23***	***9.03***	***8.91***	***9.32***	***9.37***	***9.39***	***9.42***	***9.43***	***9.46***	***9.47***	***9.49***
	5.324	4.335	4.112	4.072	4.093	4.137	4.181	4.435	4.435	4.435	4.435	4.435	4.435	4.435
	0.365	0.419	0.468	0.511	0.549	0.582	0.610	0.636	0.668	0.696	0.717	0.739	0.755	0.772
–0.30	***6.689***	***5.754***	***5.581***	***5.583***	***5.642***	***5.719***	***5.791***	***6.071***	***6.103***	***6.130***	***6.152***	***6.174***	***6.190***	***6.207***
	2.158	2.031	2.106	2.202	2.295	2.380	2.454	2.520	2.650	2.650	2.650	2.650	2.650	2.650
	0.286	0.345	0.404	0.455	0.499	0.538	0.572	0.602	0.631	0.661	0.686	0.708	0.726	0.745
–0.20	***3.444***	***3.376***	***3.510***	***3.657***	***3.795***	***3.918***	***4.027***	***4.121***	***4.281***	***4.310***	***4.336***	***4.358***	***4.376***	***4.394***
		0.945	1.080	1.201	1.302	1.386	1.456	1.513	1.565	1.618	1.618	1.618	1.618	1.618
		0.259	0.324	0.385	0.438	0.484	0.524	0.560	0.590	0.623	0.649	0.673	0.694	0.712
–0.10		***2.204***	***2.404***	***2.586***	***2.740***	***2.870***	***2.980***	***3.073***	***3.154***	***3.241***	***3.267***	***3.291***	***3.312***	***3.330***
			0.540	0.653	0.741	0.812	0.869	0.915	0.952	0.985	1.000	1.000	1.000	1.000
			0.234	0.302	0.364	0.419	0.466	0.507	0.543	0.574	0.608	0.634	0.656	0.676
	0.00 »		***1.774***	***1.954***	***2.105***	***2.230***	***2.335***	***2.422***	***2.495***	***2.559***	***2.608***	***2.634***	***2.656***	***2.676***
				0.344	0.420	0.477	0.520	0.554	0.581	0.601	0.620	0.618	0.618	0.618
				0.212	0.279	0.342	0.397	0.445	0.487	0.523	0.555	0.589	0.613	0.636
		0.10 »		***1.556***	***1.700***	***1.819***	***1.917***	***1.999***	***2.067***	***2.125***	***2.175***	***2.207***	***2.231***	***2.254***
					0.232	0.281	0.314	0.338	0.355	0.368	0.377	0.385	0.377	0.377
					0.191	0.257	0.318	0.373	0.420	0.462	0.499	0.532	0.565	0.588
			0.20 »		***1.423***	***1.538***	***1.632***	***1.710***	***1.776***	***1.830***	***1.876***	***1.917***	***1.942***	***1.966***
						0.162	0.192	0.209	0.220	0.226	0.230	0.232	0.233	0.225
						0.173	0.236	0.294	0.347	0.394	0.435	0.472	0.505	0.536
				0.30 »		***1.335***	***1.427***	***1.503***	***1.567***	***1.620***	***1.665***	***1.704***	***1.738***	***1.761***
							0.116	0.132	0.139	0.142	0.142	0.140	0.138	0.136
							0.156	0.215	0.270	0.320	0.366	0.406	0.442	0.474
					0.40 »		***1.272***	***1.347***	***1.409***	***1.462***	***1.507***	***1.546***	***1.580***	***1.610***
								0.085	0.092	0.093	0.090	0.087	0.083	0.079
								0.141	0.195	0.247	0.294	0.336	0.375	0.409
						0.50 »		***1.225***	***1.287***	***1.339***	***1.384***	***1.423***	***1.458***	***1.488***
									0.062	0.064	0.061	0.057	0.052	0.047
									0.127	0.177	0.224	0.267	0.307	0.343
							0.60 »		***1.189***	***1.241***	***1.285***	***1.324***	***1.359***	***1.390***
										0.046	0.045	0.040	0.035	0.030
										0.115	0.160	0.203	0.242	0.278
								0.70 »		***1.161***	***1.205***	***1.243***	***1.277***	***1.308***
											0.035	0.032	0.026	0.021
											0.104	0.145	0.183	0.218
									0.80 »		***1.139***	***1.176***	***1.209***	***1.239***
												0.026	0.022	0.017
												0.094	0.131	0.165
										0.90 »		***1.120***	***1.153***	***1.181***
													0.020	0.015
													0.085	0.118
											1.00 »		***1.105***	***1.133***
														0.015
														0.078
												1.10 »		***1.093***

Table Nine: Target at Boundary: The *Excess Costs of Production, Excess Costs of Use,* and *Effective Cost of Production and Use* when Nonconforming Units Are Scrapped

C_{pk} or P_{pk}	*Tolerance in Standard Deviation Units* 4.5	4.8	5.1	5.4	5.7	6.0	6.3	6.6	6.9	7.2	7.5	7.8	8.1	8.4
	13.93	13.93	13.93	13.93	13.93	13.93	13.93	13.93	13.93	13.93	13.93	13.93	13.93	13.93
	0.821	0.836	0.836	0.851	0.851	0.866	0.866	0.881	0.881	0.881	0.881	0.896	0.896	0.896
–0.50	***15.75***	***15.76***	***15.76***	***15.78***	***15.78***	***15.79***	***15.79***	***15.81***	***15.81***	***15.81***	***15.81***	***15.82***	***15.82***	***15.82***
	7.70	7.70	7.70	7.70	7.70	7.70	7.70	7.70	7.70	7.70	7.70	7.70	7.70	7.70
	0.809	0.817	0.826	0.835	0.843	0.852	0.861	0.861	0.870	0.870	0.878	0.887	0.887	0.887
–0.40	***9.50***	***9.51***	***9.52***	***9.53***	***9.54***	***9.55***	***9.56***	***9.56***	***9.57***	***9.57***	***9.57***	***9.58***	***9.58***	***9.58***
	4.435	4.435	4.435	4.435	4.435	4.435	4.435	4.435	4.435	4.435	4.435	4.435	4.435	4.435
	0.783	0.793	0.804	0.815	0.826	0.832	0.842	0.848	0.853	0.859	0.864	0.870	0.875	0.880
–0.30	***6.217***	***6.228***	***6.239***	***6.250***	***6.261***	***6.266***	***6.277***	***6.283***	***6.288***	***6.293***	***6.299***	***6.304***	***6.310***	***6.315***
	2.650	2.650	2.650	2.650	2.650	2.650	2.650	2.650	2.650	2.650	2.650	2.650	2.650	2.650
	0.759	0.774	0.785	0.796	0.807	0.814	0.821	0.828	0.836	0.843	0.850	0.854	0.858	0.865
–0.20	***4.409***	***4.423***	***4.434***	***4.445***	***4.456***	***4.464***	***4.471***	***4.478***	***4.485***	***4.493***	***4.500***	***4.504***	***4.507***	***4.515***
	1.618	1.618	1.618	1.618	1.618	1.618	1.618	1.618	1.618	1.618	1.618	1.618	1.618	1.618
	0.730	0.743	0.757	0.770	0.780	0.791	0.798	0.806	0.814	0.822	0.830	0.835	0.840	0.846
–0.10	***3.348***	***3.361***	***3.374***	***3.387***	***3.398***	***3.408***	***3.416***	***3.424***	***3.432***	***3.440***	***3.448***	***3.453***	***3.458***	***3.463***
	1.000	1.000	1.000	1.000	1.000	1.000	1.000	1.000	1.000	1.000	1.000	1.000	1.000	1.000
	0.694	0.712	0.726	0.740	0.752	0.762	0.772	0.782	0.790	0.798	0.806	0.812	0.818	0.824
0.00	***2.694***	***2.712***	***2.726***	***2.740***	***2.752***	***2.762***	***2.772***	***2.782***	***2.790***	***2.798***	***2.806***	***2.812***	***2.818***	***2.824***
	0.618	0.618	0.618	0.618	0.618	0.618	0.618	0.618	0.618	0.618	0.618	0.618	0.618	0.618
	0.655	0.673	0.689	0.704	0.717	0.730	0.741	0.751	0.761	0.770	0.778	0.786	0.793	0.799
0.10	***2.273***	***2.291***	***2.307***	***2.322***	***2.335***	***2.348***	***2.359***	***2.369***	***2.379***	***2.388***	***2.396***	***2.405***	***2.411***	***2.417***
	0.377	0.377	0.377	0.377	0.377	0.377	0.377	0.377	0.377	0.377	0.377	0.377	0.377	0.377
	0.610	0.629	0.647	0.664	0.679	0.693	0.705	0.716	0.727	0.737	0.747	0.755	0.763	0.771
0.20	***1.988***	***2.007***	***2.025***	***2.041***	***2.056***	***2.070***	***2.083***	***2.094***	***2.105***	***2.114***	***2.124***	***2.132***	***2.140***	***2.149***
	0.225	0.225	0.225	0.225	0.225	0.225	0.225	0.225	0.225	0.225	0.225	0.225	0.225	0.225
	0.559	0.581	0.600	0.618	0.635	0.650	0.664	0.676	0.689	0.700	0.711	0.721	0.729	0.738
0.30	***1.784***	***1.806***	***1.826***	***1.843***	***1.860***	***1.875***	***1.890***	***1.902***	***1.914***	***1.925***	***1.936***	***1.946***	***1.955***	***1.963***
	0.130	0.130	0.130	0.130	0.130	0.130	0.130	0.130	0.130	0.130	0.130	0.130	0.130	0.130
	0.503	0.525	0.547	0.567	0.585	0.602	0.618	0.632	0.645	0.658	0.670	0.680	0.690	0.701
0.40	***1.633***	***1.655***	***1.677***	***1.697***	***1.715***	***1.732***	***1.748***	***1.762***	***1.775***	***1.788***	***1.800***	***1.810***	***1.820***	***1.831***
	0.075	0.072	0.072	0.072	0.072	0.072	0.072	0.072	0.072	0.072	0.072	0.072	0.072	0.072
	0.441	0.467	0.491	0.511	0.532	0.550	0.567	0.583	0.598	0.612	0.625	0.637	0.648	0.659
0.50	***1.516***	***1.539***	***1.563***	***1.583***	***1.603***	***1.622***	***1.639***	***1.655***	***1.670***	***1.684***	***1.697***	***1.708***	***1.720***	***1.731***
	0.043	0.039	0.037	0.037	0.037	0.037	0.037	0.037	0.037	0.037	0.037	0.037	0.037	0.037
	0.375	0.405	0.430	0.453	0.475	0.495	0.513	0.530	0.547	0.562	0.576	0.589	0.603	0.614
0.60	***1.418***	***1.444***	***1.468***	***1.491***	***1.512***	***1.532***	***1.551***	***1.567***	***1.584***	***1.600***	***1.613***	***1.627***	***1.640***	***1.651***
	0.025	0.022	0.019	0.018	0.018	0.018	0.018	0.018	0.018	0.018	0.018	0.018	0.018	0.018
	0.312	0.342	0.370	0.394	0.416	0.438	0.458	0.477	0.494	0.510	0.526	0.541	0.554	0.567
0.70	***1.337***	***1.364***	***1.389***	***1.412***	***1.435***	***1.456***	***1.477***	***1.495***	***1.512***	***1.529***	***1.545***	***1.559***	***1.572***	***1.586***
	0.017	0.013	0.011	0.008	0.008	0.008	0.008	0.008	0.008	0.008	0.008	0.008	0.008	0.008
	0.251	0.282	0.310	0.337	0.360	0.382	0.402	0.422	0.441	0.459	0.475	0.491	0.506	0.520
0.80	***1.268***	***1.295***	***1.321***	***1.345***	***1.368***	***1.390***	***1.410***	***1.430***	***1.449***	***1.467***	***1.483***	***1.499***	***1.514***	***1.528***
	0.012	0.009	0.006	0.005	0.003	0.003	0.003	0.003	0.003	0.003	0.003	0.003	0.003	0.003
	0.196	0.226	0.254	0.280	0.305	0.327	0.349	0.369	0.389	0.407	0.425	0.441	0.457	0.472
0.90	***1.209***	***1.235***	***1.261***	***1.285***	***1.308***	***1.330***	***1.352***	***1.372***	***1.392***	***1.410***	***1.428***	***1.444***	***1.460***	***1.475***
	0.011	0.007	0.005	0.003	0.002	0.001	0.001	0.001	0.001	0.001	0.001	0.001	0.001	0.001
	0.148	0.177	0.204	0.229	0.253	0.277	0.298	0.319	0.339	0.358	0.376	0.394	0.410	0.426
1.00	***1.159***	***1.184***	***1.208***	***1.232***	***1.255***	***1.278***	***1.299***	***1.320***	***1.340***	***1.359***	***1.377***	***1.395***	***1.411***	***1.427***
	0.011	0.007	0.004	0.002	0.001	0.001	0.000	0.000	0.000	0.000	0.000	0.000	0.000	0.000
	0.106	0.133	0.159	0.183	0.207	0.230	0.252	0.272	0.293	0.312	0.331	0.349	0.366	0.382
1.10	***1.117***	***1.140***	***1.163***	***1.186***	***1.208***	***1.230***	***1.252***	***1.272***	***1.293***	***1.312***	***1.331***	***1.349***	***1.366***	***1.382***

Table Nine: Target at Boundary: The *Excess Costs of Production, Excess Costs of Use,* and *Effective Cost of Production and Use* when Nonconforming Units Are Scrapped

C_{pk} or P_{pk}	*Tolerance in Standard Deviation Units* 8.7	9.0	9.3	9.6	9.9	10.2	10.5	10.8	11.1	11.4	11.7	12.0
	13.93	13.93	13.93	13.93	13.93	13.93	13.93	13.93	13.93	13.93	13.93	13.93
	0.896	0.910	0.910	0.910	0.910	0.910	0.925	0.925	0.925	0.925	0.925	0.925
–0.50	***15.82***	***15.84***	***15.84***	***15.84***	***15.84***	***15.84***	***15.85***	***15.85***	***15.85***	***15.85***	***15.85***	***15.85***
	7.70	7.70	7.70	7.70	7.70	7.70	7.70	7.70	7.70	7.70	7.70	7.70
	0.896	0.896	0.904	0.904	0.904	0.913	0.913	0.913	0.913	0.922	0.922	0.922
–0.40	***9.59***	***9.59***	***9.60***	***9.60***	***9.60***	***9.61***	***9.61***	***9.61***	***9.61***	***9.62***	***9.62***	***9.62***
	4.435	4.435	4.435	4.435	4.435	4.435	4.435	4.435	4.435	4.435	4.435	4.435
	0.880	0.886	0.891	0.891	0.897	0.897	0.902	0.902	0.908	0.908	0.913	0.913
–0.30	***6.315***	***6.321***	***6.326***	***6.326***	***6.332***	***6.332***	***6.337***	***6.337***	***6.342***	***6.342***	***6.348***	***6.348***
	2.650	2.650	2.650	2.650	2.650	2.650	2.650	2.650	2.650	2.650	2.650	2.650
	0.869	0.872	0.876	0.880	0.883	0.887	0.891	0.894	0.894	0.898	0.901	0.905
–0.20	***4.518***	***4.522***	***4.526***	***4.529***	***4.533***	***4.536***	***4.540***	***4.544***	***4.544***	***4.547***	***4.551***	***4.555***
	1.618	1.618	1.618	1.618	1.618	1.618	1.618	1.618	1.618	1.618	1.618	1.618
	0.851	0.856	0.859	0.864	0.869	0.872	0.874	0.880	0.882	0.885	0.887	0.890
–0.10	***3.469***	***3.474***	***3.476***	***3.482***	***3.487***	***3.490***	***3.492***	***3.497***	***3.500***	***3.503***	***3.505***	***3.508***
	1.000	1.000	1.000	1.000	1.000	1.000	1.000	1.000	1.000	1.000	1.000	1.000
	0.830	0.836	0.840	0.846	0.850	0.854	0.858	0.862	0.864	0.868	0.872	0.874
0.00	***2.830***	***2.836***	***2.840***	***2.846***	***2.850***	***2.854***	***2.858***	***2.862***	***2.864***	***2.868***	***2.872***	***2.874***
	0.618	0.618	0.618	0.618	0.618	0.618	0.618	0.618	0.618	0.618	0.618	0.618
	0.806	0.812	0.817	0.824	0.828	0.833	0.837	0.841	0.845	0.850	0.853	0.856
0.10	***2.424***	***2.430***	***2.435***	***2.442***	***2.447***	***2.451***	***2.455***	***2.460***	***2.463***	***2.468***	***2.471***	***2.474***
	0.377	0.377	0.377	0.377	0.377	0.377	0.377	0.377	0.377	0.377	0.377	0.377
	0.778	0.785	0.791	0.798	0.803	0.809	0.813	0.818	0.822	0.826	0.831	0.835
0.20	***2.156***	***2.163***	***2.168***	***2.175***	***2.180***	***2.186***	***2.190***	***2.196***	***2.200***	***2.204***	***2.208***	***2.212***
	0.225	0.225	0.225	0.225	0.225	0.225	0.225	0.225	0.225	0.225	0.225	0.225
	0.746	0.754	0.761	0.767	0.775	0.779	0.786	0.792	0.797	0.801	0.806	0.810
0.30	***1.972***	***1.979***	***1.987***	***1.993***	***2.000***	***2.005***	***2.011***	***2.017***	***2.022***	***2.027***	***2.032***	***2.036***
	0.130	0.130	0.130	0.130	0.130	0.130	0.130	0.130	0.130	0.130	0.130	0.130
	0.710	0.718	0.727	0.733	0.741	0.748	0.754	0.760	0.766	0.772	0.777	0.782
0.40	***1.840***	***1.847***	***1.856***	***1.863***	***1.871***	***1.878***	***1.884***	***1.890***	***1.896***	***1.902***	***1.907***	***1.912***
	0.072	0.072	0.072	0.072	0.072	0.072	0.072	0.072	0.072	0.072	0.072	0.072
	0.669	0.678	0.688	0.697	0.704	0.712	0.719	0.727	0.733	0.740	0.745	0.751
0.50	***1.741***	***1.750***	***1.760***	***1.768***	***1.776***	***1.783***	***1.791***	***1.798***	***1.805***	***1.811***	***1.817***	***1.823***
	0.037	0.037	0.037	0.037	0.037	0.037	0.037	0.037	0.037	0.037	0.037	0.037
	0.626	0.636	0.646	0.656	0.665	0.673	0.682	0.689	0.697	0.703	0.711	0.717
0.60	***1.663***	***1.673***	***1.684***	***1.693***	***1.702***	***1.711***	***1.719***	***1.726***	***1.734***	***1.741***	***1.748***	***1.754***
	0.018	0.018	0.018	0.018	0.018	0.018	0.018	0.018	0.018	0.018	0.018	0.018
	0.579	0.592	0.603	0.613	0.623	0.632	0.642	0.650	0.658	0.666	0.674	0.681
0.70	***1.598***	***1.610***	***1.621***	***1.631***	***1.642***	***1.651***	***1.660***	***1.668***	***1.676***	***1.684***	***1.692***	***1.700***
	0.008	0.008	0.008	0.008	0.008	0.008	0.008	0.008	0.008	0.008	0.008	0.008
	0.533	0.545	0.557	0.570	0.580	0.591	0.600	0.610	0.619	0.627	0.635	0.643
0.80	***1.541***	***1.553***	***1.566***	***1.578***	***1.588***	***1.599***	***1.608***	***1.618***	***1.627***	***1.635***	***1.643***	***1.651***
	0.003	0.003	0.003	0.003	0.003	0.003	0.003	0.003	0.003	0.003	0.003	0.003
	0.486	0.501	0.514	0.526	0.537	0.549	0.559	0.569	0.579	0.588	0.597	0.606
0.90	***1.489***	***1.504***	***1.517***	***1.529***	***1.540***	***1.552***	***1.562***	***1.572***	***1.582***	***1.591***	***1.600***	***1.609***
	0.001	0.001	0.001	0.001	0.001	0.001	0.001	0.001	0.001	0.001	0.001	0.001
	0.441	0.455	0.469	0.482	0.495	0.507	0.519	0.530	0.540	0.550	0.560	0.569
1.00	***1.442***	***1.456***	***1.470***	***1.483***	***1.496***	***1.508***	***1.520***	***1.531***	***1.541***	***1.551***	***1.561***	***1.570***
	0.000	0.000	0.000	0.000	0.000	0.000	0.000	0.000	0.000	0.000	0.000	0.000
	0.398	0.413	0.427	0.441	0.454	0.467	0.479	0.490	0.501	0.512	0.522	0.532
1.10	***1.398***	***1.413***	***1.427***	***1.441***	***1.454***	***1.467***	***1.479***	***1.490***	***1.501***	***1.512***	***1.522***	***1.532***

Table Nine: Target at Boundary: The *Excess Costs of Production, Excess Costs of Use,* and *Effective Cost of Production and Use* when Nonconforming Units Are Scrapped

C_{pk} or P_{pk}	*Tolerance in Standard Deviation Units* 4.5	4.8	5.1	5.4	5.7	6.0	6.3	6.6	6.9	7.2	7.5	7.8	8.1	8.4
	0.012	0.007	0.004	0.002	0.001	0.001	0.000	0.000	0.000	0.000	0.000	0.000	0.000	0.000
	0.071	0.096	0.120	0.143	0.165	0.187	0.209	0.229	0.250	0.269	0.288	0.306	0.324	0.341
1.20	***1.082***	***1.104***	***1.124***	***1.145***	***1.167***	***1.188***	***1.209***	***1.230***	***1.250***	***1.269***	***1.288***	***1.306***	***1.324***	***1.341***
		0.009	0.005	0.003	0.001	0.000	0.000	0.000	0.000	0.000	0.000	0.000	0.000	0.000
		0.064	0.087	0.109	0.129	0.150	0.170	0.190	0.210	0.229	0.248	0.266	0.284	0.301
1.30		***1.073***	***1.092***	***1.111***	***1.130***	***1.150***	***1.170***	***1.190***	***1.210***	***1.229***	***1.248***	***1.266***	***1.284***	***1.301***
			0.007	0.004	0.002	0.001	0.000	0.000	0.000	0.000	0.000	0.000	0.000	0.000
			0.059	0.079	0.098	0.117	0.136	0.155	0.174	0.193	0.211	0.229	0.247	0.264
1.40 »			***1.066***	***1.083***	***1.100***	***1.118***	***1.136***	***1.155***	***1.174***	***1.193***	***1.211***	***1.229***	***1.247***	***1.264***
				0.005	0.002	0.001	0.000	0.000	0.000	0.000	0.000	0.000	0.000	0.000
				0.054	0.072	0.089	0.107	0.124	0.142	0.160	0.178	0.195	0.213	0.230
1.50 »				***1.059***	***1.074***	***1.090***	***1.107***	***1.124***	***1.142***	***1.160***	***1.178***	***1.195***	***1.213***	***1.230***
					0.004	0.002	0.001	0.000	0.000	0.000	0.000	0.000	0.000	0.000
					0.049	0.066	0.081	0.097	0.114	0.130	0.147	0.164	0.181	0.198
1.60 »					***1.054***	***1.067***	***1.082***	***1.097***	***1.114***	***1.130***	***1.147***	***1.164***	***1.181***	***1.198***
						0.003	0.001	0.000	0.000	0.000	0.000	0.000	0.000	0.000
						0.045	0.060	0.074	0.089	0.104	0.120	0.136	0.152	0.169
1.70 »						***1.049***	***1.061***	***1.075***	***1.089***	***1.104***	***1.120***	***1.136***	***1.152***	***1.169***
							0.003	0.001	0.000	0.000	0.000	0.000	0.000	0.000
							0.042	0.055	0.068	0.082	0.096	0.111	0.126	0.142
1.80 »							***1.044***	***1.056***	***1.068***	***1.082***	***1.096***	***1.111***	***1.126***	***1.142***
								0.002	0.001	0.000	0.000	0.000	0.000	0.000
								0.039	0.051	0.063	0.075	0.089	0.103	0.117
1.90 »								***1.041***	***1.051***	***1.063***	***1.075***	***1.089***	***1.103***	***1.117***
									0.002	0.000	0.000	0.000	0.000	0.000
									0.036	0.047	0.058	0.070	0.082	0.096
2.00 »									***1.037***	***1.047***	***1.058***	***1.070***	***1.082***	***1.096***
										0.001	0.000	0.000	0.000	0.000
										0.033	0.043	0.053	0.065	0.077
2.10 »										***1.034***	***1.043***	***1.053***	***1.065***	***1.077***
											0.001	0.000	0.000	0.000
											0.031	0.040	0.050	0.060
2.20 »											***1.032***	***1.040***	***1.050***	***1.060***
												0.001	0.000	0.000
												0.029	0.037	0.046
2.30 »												***1.029***	***1.037***	***1.046***
													0.001	0.000
													0.027	0.034
2.40 »													***1.027***	***1.035***
														0.001
														0.025
2.50 »														***1.025***

Table Nine: Target at Boundary: The *Excess Costs of Production, Excess Costs of Use,* and *Effective Cost of Production and Use* when Nonconforming Units Are Scrapped

C_{pk} or P_{pk}	*Tolerance in Standard Deviation Units* 8.7	9.0	9.3	9.6	9.9	10.2	10.5	10.8	11.1	11.4	11.7	12.0
	0.000	0.000	0.000	0.000	0.000	0.000	0.000	0.000	0.000	0.000	0.000	0.000
	0.357	0.372	0.387	0.401	0.415	0.428	0.441	0.453	0.464	0.476	0.486	0.497
1.20	***.357***	***1.372***	***1.387***	***1.401***	***1.415***	***1.428***	***1.441***	***1.453***	***1.464***	***1.476***	***1.486***	***1.497***
	0.000	0.000	0.000	0.000	0.000	0.000	0.000	0.000	0.000	0.000	0.000	0.000
	0.318	0.333	0.349	0.363	0.377	0.391	0.404	0.417	0.429	0.440	0.452	0.463
1.30	***.318***	***1.333***	***1.349***	***1.363***	***1.377***	***1.391***	***1.404***	***1.417***	***1.429***	***1.440***	***1.452***	***1.463***
	0.000	0.000	0.000	0.000	0.000	0.000	0.000	0.000	0.000	0.000	0.000	0.000
	0.281	0.297	0.312	0.327	0.342	0.356	0.369	0.382	0.395	0.407	0.418	0.429
1.40	***.281***	***1.297***	***1.312***	***1.327***	***1.342***	***1.356***	***1.369***	***1.382***	***1.395***	***1.407***	***1.418***	***1.429***
	0.000	0.000	0.000	0.000	0.000	0.000	0.000	0.000	0.000	0.000	0.000	0.000
	0.246	0.262	0.278	0.293	0.308	0.322	0.336	0.349	0.362	0.374	0.386	0.398
1.50	***.246***	***1.262***	***1.278***	***1.293***	***1.308***	***1.322***	***1.336***	***1.349***	***1.362***	***1.374***	***1.386***	***1.398***
	0.000	0.000	0.000	0.000	0.000	0.000	0.000	0.000	0.000	0.000	0.000	0.000
	0.214	0.230	0.246	0.261	0.276	0.290	0.304	0.317	0.330	0.343	0.355	0.367
1.60	***.214***	***1.230***	***1.246***	***1.261***	***1.276***	***1.290***	***1.304***	***1.317***	***1.330***	***1.343***	***1.355***	***1.367***
	0.000	0.000	0.000	0.000	0.000	0.000	0.000	0.000	0.000	0.000	0.000	0.000
	0.184	0.200	0.216	0.231	0.245	0.260	0.274	0.287	0.300	0.313	0.326	0.338
1.70	***.184***	***1.200***	***1.216***	***1.231***	***1.245***	***1.260***	***1.274***	***1.287***	***1.300***	***1.313***	***1.326***	***1.338***
	0.000	0.000	0.000	0.000	0.000	0.000	0.000	0.000	0.000	0.000	0.000	0.000
	0.157	0.172	0.187	0.202	0.217	0.231	0.245	0.259	0.272	0.285	0.297	0.309
1.80	***.157***	***1.172***	***1.187***	***1.202***	***1.217***	***1.231***	***1.245***	***1.259***	***1.272***	***1.285***	***1.297***	***1.309***
	0.000	0.000	0.000	0.000	0.000	0.000	0.000	0.000	0.000	0.000	0.000	0.000
	0.132	0.147	0.161	0.176	0.190	0.204	0.218	0.232	0.245	0.258	0.270	0.283
1.90	***.132***	***1.147***	***1.161***	***1.176***	***1.190***	***1.204***	***1.218***	***1.232***	***1.245***	***1.258***	***1.270***	***1.283***
	0.000	0.000	0.000	0.000	0.000	0.000	0.000	0.000	0.000	0.000	0.000	0.000
	0.110	0.123	0.137	0.151	0.165	0.179	0.193	0.206	0.219	0.232	0.245	0.257
2.00	***.110***	***1.123***	***1.137***	***1.151***	***1.165***	***1.179***	***1.193***	***1.206***	***1.219***	***1.232***	***1.245***	***1.257***
	0.000	0.000	0.000	0.000	0.000	0.000	0.000	0.000	0.000	0.000	0.000	0.000
	0.089	0.102	0.116	0.129	0.142	0.156	0.169	0.182	0.195	0.208	0.220	0.233
2.10	***.089***	***1.102***	***1.116***	***1.129***	***1.142***	***1.156***	***1.169***	***1.182***	***1.195***	***1.208***	***1.220***	***1.233***
	0.000	0.000	0.000	0.000	0.000	0.000	0.000	0.000	0.000	0.000	0.000	0.000
	0.071	0.083	0.096	0.109	0.121	0.134	0.147	0.160	0.172	0.185	0.197	0.209
2.20	***.071***	***1.083***	***1.096***	***1.109***	***1.121***	***1.134***	***1.147***	***1.160***	***1.172***	***1.185***	***1.197***	***1.209***
	0.000	0.000	0.000	0.000	0.000	0.000	0.000	0.000	0.000	0.000	0.000	0.000
	0.056	0.067	0.078	0.090	0.102	0.114	0.127	0.139	0.151	0.164	0.176	0.188
2.30	***.056***	***1.067***	***1.078***	***1.090***	***1.102***	***1.114***	***1.127***	***1.139***	***1.151***	***1.164***	***1.176***	***1.188***
	0.000	0.000	0.000	0.000	0.000	0.000	0.000	0.000	0.000	0.000	0.000	0.000
	0.043	0.052	0.063	0.073	0.085	0.096	0.108	0.120	0.132	0.143	0.155	0.167
2.40	***.043***	***1.052***	***1.063***	***1.073***	***1.085***	***1.096***	***1.108***	***1.120***	***1.132***	***1.143***	***1.155***	***1.167***
	0.000	0.000	0.000	0.000	0.000	0.000	0.000	0.000	0.000	0.000	0.000	0.000
	0.032	0.040	0.049	0.059	0.069	0.080	0.091	0.102	0.113	0.125	0.136	0.148
2.50	***.032***	***1.040***	***1.049***	***1.059***	***1.069***	***1.080***	***1.091***	***1.102***	***1.113***	***1.125***	***1.136***	***1.148***
	0.000	0.000	0.000	0.000	0.000	0.000	0.000	0.000	0.000	0.000	0.000	0.000
	0.023	0.030	0.038	0.046	0.055	0.065	0.075	0.086	0.097	0.107	0.118	0.129
2.60	***.024***	***1.030***	***1.038***	***1.046***	***1.055***	***1.065***	***1.075***	***1.086***	***1.097***	***1.107***	***1.118***	***1.129***
		0.000	0.000	0.000	0.000	0.000	0.000	0.000	0.000	0.000	0.000	0.000
		0.022	0.028	0.035	0.043	0.052	0.061	0.071	0.081	0.091	0.102	0.113
2.70		***1.022***	***1.028***	***1.035***	***1.043***	***1.052***	***1.061***	***1.071***	***1.081***	***1.091***	***1.102***	***1.113***
			0.000	0.000	0.000	0.000	0.000	0.000	0.000	0.000	0.000	0.000
			0.021	0.026	0.033	0.041	0.049	0.058	0.067	0.077	0.087	0.097
2.80 »			***1.021***	***1.026***	***1.033***	***1.041***	***1.049***	***1.058***	***1.067***	***1.077***	***1.087***	***1.097***

Table Nine: Target at Boundary: The *Excess Costs of Production, Excess Costs of Use,* and *Effective Cost of Production and Use* when Nonconforming Units Are Scrapped

C_{pk} or P_{pk}	*Tolerance in Standard Deviation Units* 9.6	9.9	10.2	10.5	10.8	11.1	11.4	11.7	12.0
	0.000	0.000	0.000	0.000	0.000	0.000	0.000	0.000	0.000
	0.019	0.025	0.031	0.038	0.046	0.055	0.064	0.073	0.083
2.90	***1.020***	***1.025***	***1.031***	***1.038***	***1.046***	***1.055***	***1.064***	***1.073***	***1.083***
		0.000	0.000	0.000	0.000	0.000	0.000	0.000	0.000
		0.018	0.023	0.029	0.036	0.044	0.052	0.061	0.069
3.00		***1.018***	***1.023***	***1.029***	***1.036***	***1.044***	***1.052***	***1.061***	***1.069***
			0.000	0.000	0.000	0.000	0.000	0.000	0.000
			0.017	0.022	0.028	0.034	0.042	0.049	0.058
3.10 »			***1.017***	***1.022***	***1.028***	***1.034***	***1.042***	***1.049***	***1.058***
				0.000	0.000	0.000	0.000	0.000	0.000
				0.016	0.021	0.026	0.033	0.040	0.047
3.20 »				***1.016***	***1.021***	***1.026***	***1.033***	***1.040***	***1.047***
					0.000	0.000	0.000	0.000	0.000
					0.015	0.020	0.025	0.031	0.038
3.30 »					***1.015***	***1.020***	***1.025***	***1.031***	***1.038***
						0.000	0.000	0.000	0.000
						0.015	0.019	0.024	0.029
3.40 »						***1.015***	***1.019***	***1.024***	***1.029***
							0.000	0.000	0.000
							0.014	0.018	0.023
3.50 »							***1.014***	***1.018***	***1.023***
								0.000	0.000
								0.013	0.017
3.60 »								***1.013***	***1.017***
									0.000
									0.013
3.70 »									***1.013***

Table Ten: Target at Boundary: The *Excess Costs of Production, Excess Costs of Use,* and *Effective Cost of Production and Use* when *All* Nonconforming Units Are Reworked

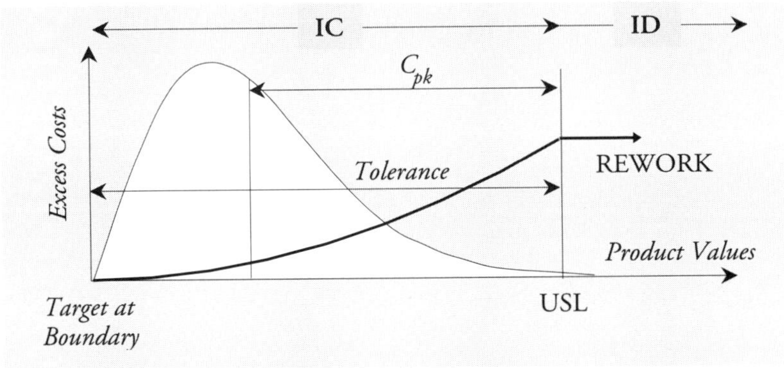

Table Ten: Excess Cost Function when Target Is at a Boundary Value

Corresponding to each pairing for *Tolerance* and C_{pk} there are three entries:

The topmost of these is the *Excess Costs of Production*, *ExCP*

The middle value is the *Excess Costs of Use*, *ExCU*

And the bottom value (in bold) is the *Effective Cost of Production and Use*, ***ECP&U***

The entries in Table Ten are for the case where all nonconforming product is reworked and the Cost of Rework, C_R, is assumed to be equal to the Nominal Cost of Production, *NC*. The values for **IC** and **ID** are given in Table A.10. The general formula for Table Ten is:

$$ECP\&U = 1.0 + \frac{C_R}{NC}\,\mathbf{ID} + \frac{C_R}{NC}\,\mathbf{IC}$$

$$= 1.0 + \textit{Excess Costs of Production} + \textit{Excess Costs of Use}$$

Since the Cost of Rework will generally be less than *NC*, the entries in Table Ten are essentially *upper bounds* for all three quantities. If greater precision is required, the entries for the *Excess Costs of Production* and the *Excess Costs of Use* may be deflated by the ratio of C_R to *NC*, as shown in the formula above, and added to 1.00 to get a customized value for the *Effective Cost of Production and Use*.

For example, with *Tolerance* = 2.7 and C_{pk} = 0.5, and with C_R = 0.50 *NC* we would convert the tabled value of 1.263 for *ECP&U* into the value 1.132. The *Excess Cost of Production* would be 0.042, while the *Excess Cost of Use* would be 0.089.

Table Ten: Target at Boundary: The *Excess Costs of Production, Excess Costs of Use,* and *Effective Cost of Production and Use* when Nonconforming Units Are Reworked

C_{pk} or P_{pk}	*Tolerance in Standard Deviation Units* 0.3	0.6	0.9	1.2	1.5	1.8	2.1	2.4	2.7	3.0	3.3	3.6	3.9	4.2
	0.973	0.955	0.944	0.937	0.932	0.933	0.933	0.933	0.933	0.933	0.933	0.933	0.933	0.933
	0.013	0.024	0.032	0.037	0.042	0.041	0.044	0.046	0.048	0.050	0.051	0.052	0.053	0.054
–0.50	***1.986***	***1.978***	***1.975***	***1.974***	***1.974***	***1.974***	***1.977***	***1.979***	***1.981***	***1.983***	***1.984***	***1.985***	***1.986***	***1.987***
	0.931	0.904	0.891	0.885	0.882	0.879	0.885	0.885	0.885	0.885	0.885	0.885	0.885	0.885
	0.030	0.046	0.057	0.064	0.070	0.074	0.072	0.077	0.080	0.083	0.085	0.088	0.089	0.091
–0.40	***1.961***	***1.950***	***1.948***	***1.949***	***1.951***	***1.954***	***1.957***	***1.962***	***1.965***	***1.968***	***1.970***	***1.973***	***1.974***	***1.976***
	0.842	0.813	0.804	0.803	0.804	0.805	0.807	0.816	0.816	0.816	0.816	0.816	0.816	0.816
	0.058	0.079	0.092	0.101	0.108	0.113	0.118	0.117	0.123	0.128	0.132	0.136	0.139	0.142
–0.30	***1.900***	***1.891***	***1.896***	***1.904***	***1.911***	***1.919***	***1.925***	***1.933***	***1.939***	***1.944***	***1.948***	***1.952***	***1.955***	***1.958***
	0.683	0.670	0.678	0.688	0.697	0.704	0.711	0.716	0.726	0.726	0.726	0.726	0.726	0.726
	0.091	0.114	0.130	0.142	0.152	0.159	0.166	0.171	0.173	0.181	0.188	0.194	0.199	0.204
–0.20	***1.774***	***1.784***	***1.808***	***1.830***	***1.848***	***1.863***	***1.876***	***1.887***	***1.899***	***1.907***	***1.914***	***1.920***	***1.925***	***1.930***
		0.486	0.519	0.546	0.566	0.581	0.593	0.602	0.610	0.618	0.618	0.618	0.618	0.618
		0.133	0.156	0.175	0.190	0.203	0.214	0.223	0.230	0.238	0.248	0.257	0.265	0.272
–0.10		***1.619***	***1.675***	***1.721***	***1.756***	***1.784***	***1.806***	***1.825***	***1.840***	***1.856***	***1.866***	***1.875***	***1.883***	***1.890***
			0.351	0.395	0.426	0.448	0.465	0.478	0.488	0.496	0.500	0.500	0.500	0.500
			0.152	0.183	0.209	0.231	0.249	0.265	0.278	0.289	0.304	0.317	0.328	0.338
	0.00 »		***1.503***	***1.577***	***1.635***	***1.679***	***1.714***	***1.743***	***1.766***	***1.786***	***1.804***	***1.817***	***1.828***	***1.838***
				0.256	0.296	0.323	0.342	0.357	0.367	0.376	0.383	0.382	0.382	0.382
				0.157	0.197	0.232	0.261	0.286	0.308	0.327	0.343	0.364	0.379	0.393
		0.10 »		***1.413***	***1.493***	***1.554***	***1.603***	***1.643***	***1.675***	***1.702***	***1.725***	***1.746***	***1.761***	***1.775***
					0.188	0.219	0.239	0.252	0.262	0.269	0.274	0.278	0.274	0.274
					0.155	0.201	0.242	0.279	0.310	0.338	0.363	0.384	0.410	0.427
			0.20 »		***1.344***	***1.420***	***1.481***	***1.531***	***1.572***	***1.607***	***1.636***	***1.662***	***1.684***	***1.701***
						0.140	0.161	0.173	0.180	0.184	0.187	0.188	0.189	0.184
						0.149	0.198	0.243	0.285	0.321	0.354	0.383	0.409	0.437
				0.30 »		***1.288***	***1.358***	***1.416***	***1.465***	***1.506***	***1.541***	***1.571***	***1.598***	***1.621***
							0.104	0.117	0.122	0.124	0.124	0.123	0.121	0.119
							0.140	0.190	0.237	0.281	0.320	0.356	0.388	0.417
					0.40 »		***1.244***	***1.307***	***1.359***	***1.405***	***1.444***	***1.479***	***1.509***	***1.537***
								0.078	0.084	0.085	0.083	0.080	0.076	0.073
								0.130	0.179	0.226	0.269	0.309	0.346	0.379
						0.50 »		***1.208***	***1.263***	***1.311***	***1.352***	***1.389***	***1.423***	***1.452***
									0.059	0.060	0.058	0.054	0.049	0.045
									0.120	0.166	0.211	0.253	0.292	0.328
							0.60 »		***1.178***	***1.227***	***1.269***	***1.307***	***1.341***	***1.372***
										0.044	0.043	0.039	0.034	0.029
										0.110	0.153	0.195	0.234	0.270
								0.70 »		***1.154***	***1.196***	***1.233***	***1.268***	***1.299***
											0.034	0.031	0.026	0.021
											0.100	0.140	0.178	0.214
									0.80 »		***1.134***	***1.171***	***1.204***	***1.234***
												0.026	0.022	0.017
												0.092	0.128	0.162
										0.90 »		***1.117***	***1.149***	***1.178***
													0.020	0.015
													0.084	0.116
											1.00 »		***1.103***	***1.131***
														0.015
														0.076
												1.10 »		***1.091***

Table Ten: Target at Boundary: The *Excess Costs of Production, Excess Costs of Use,* and *Effective Cost of Production and Use* when Nonconforming Units Are Reworked

C_{pk} or P_{pk}	*Tolerance in Standard Deviation Units* 4.5	4.8	5.1	5.4	5.7	6.0	6.3	6.6	6.9	7.2	7.5	7.8	8.1	8.4
	0.933	0.933	0.933	0.933	0.933	0.933	0.933	0.933	0.933	0.933	0.933	0.933	0.933	0.933
	0.055	0.056	0.056	0.057	0.057	0.058	0.058	0.059	0.059	0.059	0.059	0.060	0.060	0.060
–0.50	***1.988***	***1.989***	***1.989***	***1.990***	***1.990***	***1.991***	***1.991***	***1.992***	***1.992***	***1.992***	***1.992***	***1.993***	***1.993***	***1.993***
	0.885	0.885	0.885	0.885	0.885	0.885	0.885	0.885	0.885	0.885	0.885	0.885	0.885	0.885
	0.093	0.094	0.095	0.096	0.097	0.098	0.099	0.099	0.100	0.100	0.101	0.102	0.102	0.102
–0.40	***1.978***	***1.979***	***1.980***	***1.981***	***1.982***	***1.983***	***1.984***	***1.984***	***1.985***	***1.985***	***1.986***	***1.987***	***1.987***	***1.987***
	0.816	0.816	0.816	0.816	0.816	0.816	0.816	0.816	0.816	0.816	0.816	0.816	0.816	0.816
	0.144	0.146	0.148	0.150	0.152	0.153	0.155	0.156	0.157	0.158	0.159	0.160	0.161	0.162
–0.30	***1.960***	***1.962***	***1.964***	***1.966***	***1.968***	***1.969***	***1.971***	***1.972***	***1.973***	***1.974***	***1.975***	***1.976***	***1.977***	***1.978***
	0.726	0.726	0.726	0.726	0.726	0.726	0.726	0.726	0.726	0.726	0.726	0.726	0.726	0.726
	0.208	0.212	0.215	0.218	0.221	0.223	0.225	0.227	0.229	0.231	0.233	0.234	0.235	0.237
–0.20	***1.934***	***1.938***	***1.941***	***1.944***	***1.947***	***1.949***	***1.951***	***1.953***	***1.955***	***1.957***	***1.959***	***1.960***	***1.961***	***1.963***
	0.618	0.618	0.618	0.618	0.618	0.618	0.618	0.618	0.618	0.618	0.618	0.618	0.618	0.618
	0.279	0.284	0.289	0.294	0.298	0.302	0.305	0.308	0.311	0.314	0.317	0.319	0.321	0.323
–0.10	***1.897***	***1.902***	***1.907***	***1.912***	***1.916***	***1.920***	***1.923***	***1.926***	***1.929***	***1.932***	***1.935***	***1.937***	***1.939***	***1.941***
	0.500	0.500	0.500	0.500	0.500	0.500	0.500	0.500	0.500	0.500	0.500	0.500	0.500	0.500
	0.347	0.356	0.363	0.370	0.376	0.381	0.386	0.391	0.395	0.399	0.403	0.406	0.409	0.412
0.00	***1.847***	***1.856***	***1.863***	***1.870***	***1.876***	***1.881***	***1.886***	***1.891***	***1.895***	***1.899***	***1.903***	***1.906***	***1.909***	***1.912***
	0.382	0.382	0.382	0.382	0.382	0.382	0.382	0.382	0.382	0.382	0.382	0.382	0.382	0.382
	0.405	0.416	0.426	0.435	0.443	0.451	0.458	0.464	0.470	0.476	0.481	0.486	0.490	0.494
0.10	***1.787***	***1.798***	***1.808***	***1.817***	***1.825***	***1.833***	***1.840***	***1.846***	***1.852***	***1.858***	***1.863***	***1.868***	***1.872***	***1.876***
	0.274	0.274	0.274	0.274	0.274	0.274	0.274	0.274	0.274	0.274	0.274	0.274	0.274	0.274
	0.443	0.457	0.470	0.482	0.493	0.503	0.512	0.520	0.528	0.535	0.542	0.548	0.554	0.560
0.20	***1.717***	***1.731***	***1.744***	***1.756***	***1.767***	***1.777***	***1.786***	***1.794***	***1.802***	***1.809***	***1.816***	***1.822***	***1.828***	***1.834***
	0.184	0.184	0.184	0.184	0.184	0.184	0.184	0.184	0.184	0.184	0.184	0.184	0.184	0.184
	0.456	0.474	0.490	0.504	0.518	0.530	0.542	0.552	0.562	0.571	0.580	0.588	0.595	0.602
0.30	***1.640***	***1.658***	***1.674***	***1.688***	***1.702***	***1.714***	***1.726***	***1.736***	***1.746***	***1.755***	***1.764***	***1.772***	***1.779***	***1.786***
	0.115	0.115	0.115	0.115	0.115	0.115	0.115	0.115	0.115	0.115	0.115	0.115	0.115	0.115
	0.445	0.465	0.484	0.502	0.518	0.533	0.547	0.559	0.571	0.582	0.593	0.602	0.611	0.620
0.40	***1.560***	***1.580***	***1.599***	***1.617***	***1.633***	***1.648***	***1.662***	***1.674***	***1.686***	***1.697***	***1.708***	***1.717***	***1.726***	***1.735***
	0.070	0.067	0.067	0.067	0.067	0.067	0.067	0.067	0.067	0.067	0.067	0.067	0.067	0.067
	0.410	0.436	0.458	0.477	0.496	0.513	0.529	0.544	0.558	0.571	0.583	0.594	0.605	0.615
0.50	***1.480***	***1.503***	***1.525***	***1.544***	***1.563***	***1.580***	***1.596***	***1.611***	***1.625***	***1.638***	***1.650***	***1.661***	***1.672***	***1.682***
	0.041	0.037	0.036	0.036	0.036	0.036	0.036	0.036	0.036	0.036	0.036	0.036	0.036	0.036
	0.360	0.390	0.415	0.437	0.458	0.477	0.495	0.511	0.527	0.542	0.555	0.568	0.581	0.592
0.60	***1.401***	***1.428***	***1.451***	***1.473***	***1.494***	***1.513***	***1.531***	***1.547***	***1.563***	***1.578***	***1.591***	***1.604***	***1.617***	***1.628***
	0.025	0.021	0.018	0.018	0.018	0.018	0.018	0.018	0.018	0.018	0.018	0.018	0.018	0.018
	0.304	0.335	0.364	0.387	0.409	0.430	0.450	0.468	0.485	0.501	0.517	0.531	0.544	0.557
0.70	***1.329***	***1.356***	***1.382***	***1.405***	***1.427***	***1.448***	***1.468***	***1.486***	***1.503***	***1.519***	***1.535***	***1.549***	***1.562***	***1.575***
	0.016	0.013	0.010	0.008	0.008	0.008	0.008	0.008	0.008	0.008	0.008	0.008	0.008	0.008
	0.247	0.278	0.307	0.334	0.357	0.379	0.399	0.419	0.437	0.455	0.471	0.487	0.502	0.516
0.80	***1.264***	***1.291***	***1.317***	***1.342***	***1.365***	***1.387***	***1.407***	***1.427***	***1.445***	***1.463***	***1.479***	***1.495***	***1.510***	***1.524***
	0.012	0.009	0.006	0.005	0.003	0.003	0.003	0.003	0.003	0.003	0.003	0.003	0.003	0.003
	0.194	0.224	0.253	0.279	0.304	0.326	0.348	0.368	0.388	0.406	0.424	0.440	0.456	0.471
0.90	***1.206***	***1.233***	***1.259***	***1.284***	***1.307***	***1.329***	***1.351***	***1.371***	***1.391***	***1.409***	***1.427***	***1.443***	***1.459***	***1.474***
	0.011	0.007	0.005	0.003	0.002	0.001	0.001	0.001	0.001	0.001	0.001	0.001	0.001	0.001
	0.147	0.175	0.203	0.229	0.253	0.276	0.298	0.319	0.339	0.358	0.376	0.394	0.410	0.426
1.00	***1.157***	***1.182***	***1.207***	***1.231***	***1.255***	***1.278***	***1.299***	***1.320***	***1.340***	***1.359***	***1.377***	***1.395***	***1.411***	***1.427***
	0.011	0.007	0.004	0.002	0.001	0.001	0.000	0.000	0.000	0.000	0.000	0.000	0.000	0.000
	0.105	0.132	0.158	0.183	0.207	0.229	0.252	0.272	0.293	0.312	0.331	0.349	0.366	0.382
1.10	***1.116***	***1.139***	***1.162***	***1.185***	***1.208***	***1.230***	***1.252***	***1.272***	***1.293***	***1.312***	***1.331***	***1.349***	***1.366***	***1.382***

Table Ten: Target at Boundary: The *Excess Costs of Production, Excess Costs of Use,* and *Effective Cost of Production and Use* when Nonconforming Units Are Reworked

C_{pk} or P_{pk}	*Tolerance in Standard Deviation Units* 8.7	9.0	9.3	9.6	9.9	10.2	10.5	10.8	11.1	11.4	11.7	12.0
	0.933	0.933	0.933	0.933	0.933	0.933	0.933	0.933	0.933	0.933	0.933	0.933
	0.060	0.061	0.061	0.061	0.061	0.061	0.062	0.062	0.062	0.062	0.062	0.062
–0.50	***1.993***	***1.994***	***1.994***	***1.994***	***1.994***	***1.994***	***1.995***	***1.995***	***1.995***	***1.995***	***1.995***	***1.995***
	0.885	0.885	0.885	0.885	0.885	0.885	0.885	0.885	0.885	0.885	0.885	0.885
	0.103	0.103	0.104	0.104	0.104	0.105	0.105	0.105	0.105	0.106	0.106	0.106
–0.40	***1.988***	***1.988***	***1.989***	***1.989***	***1.989***	***1.990***	***1.990***	***1.990***	***1.990***	***1.991***	***1.991***	***1.991***
	0.816	0.816	0.816	0.816	0.816	0.816	0.816	0.816	0.816	0.816	0.816	0.816
	0.162	0.163	0.164	0.164	0.165	0.165	0.166	0.166	0.167	0.167	0.168	0.168
–0.30	***1.978***	***1.979***	***1.980***	***1.980***	***1.981***	***1.981***	***1.982***	***1.982***	***1.983***	***1.983***	***1.984***	***1.984***
	0.726	0.726	0.726	0.726	0.726	0.726	0.726	0.726	0.726	0.726	0.726	0.726
	0.238	0.239	0.240	0.241	0.242	0.243	0.244	0.245	0.245	0.246	0.247	0.248
–0.20	***1.964***	***1.965***	***1.966***	***1.967***	***1.968***	***1.969***	***1.970***	***1.971***	***1.971***	***1.972***	***1.973***	***1.974***
	0.618	0.618	0.618	0.618	0.618	0.618	0.618	0.618	0.618	0.618	0.618	0.618
	0.325	0.327	0.328	0.330	0.332	0.333	0.334	0.336	0.337	0.338	0.339	0.340
–0.10	***1.943***	***1.945***	***1.946***	***1.948***	***1.950***	***1.951***	***1.952***	***1.954***	***1.955***	***1.956***	***1.957***	***1.958***
	0.500	0.500	0.500	0.500	0.500	0.500	0.500	0.500	0.500	0.500	0.500	0.500
	0.415	0.418	0.420	0.423	0.425	0.427	0.429	0.431	0.432	0.434	0.436	0.437
0.00	***1.915***	***1.918***	***1.920***	***1.923***	***1.925***	***1.927***	***1.929***	***1.931***	***1.932***	***1.934***	***1.936***	***1.937***
	0.382	0.382	0.382	0.382	0.382	0.382	0.382	0.382	0.382	0.382	0.382	0.382
	0.498	0.502	0.505	0.509	0.512	0.515	0.517	0.520	0.522	0.525	0.527	0.529
0.10	***1.880***	***1.884***	***1.887***	***1.891***	***1.894***	***1.897***	***1.899***	***1.902***	***1.904***	***1.907***	***1.909***	***1.911***
	0.274	0.274	0.274	0.274	0.274	0.274	0.274	0.274	0.274	0.274	0.274	0.274
	0.565	0.570	0.574	0.579	0.583	0.587	0.590	0.594	0.597	0.600	0.603	0.606
0.20	***1.839***	***1.844***	***1.848***	***1.853***	***1.857***	***1.861***	***1.864***	***1.868***	***1.871***	***1.874***	***1.877***	***1.880***
	0.184	0.184	0.184	0.184	0.184	0.184	0.184	0.184	0.184	0.184	0.184	0.184
	0.609	0.615	0.621	0.626	0.632	0.636	0.641	0.646	0.650	0.654	0.658	0.661
0.30	***1.793***	***1.799***	***1.805***	***1.810***	***1.816***	***1.820***	***1.825***	***1.830***	***1.834***	***1.838***	***1.842***	***1.845***
	0.115	0.115	0.115	0.115	0.115	0.115	0.115	0.115	0.115	0.115	0.115	0.115
	0.628	0.635	0.643	0.649	0.656	0.662	0.667	0.673	0.678	0.683	0.688	0.692
0.40	***1.743***	***1.750***	***1.758***	***1.764***	***1.771***	***1.777***	***1.782***	***1.788***	***1.793***	***1.798***	***1.803***	***1.807***
	0.067	0.067	0.067	0.067	0.067	0.067	0.067	0.067	0.067	0.067	0.067	0.067
	0.624	0.633	0.642	0.650	0.657	0.664	0.671	0.678	0.684	0.690	0.695	0.701
0.50	***1.691***	***1.700***	***1.709***	***1.717***	***1.724***	***1.731***	***1.738***	***1.745***	***1.751***	***1.757***	***1.762***	***1.768***
	0.036	0.036	0.036	0.036	0.036	0.036	0.036	0.036	0.036	0.036	0.036	0.036
	0.603	0.613	0.623	0.632	0.641	0.649	0.657	0.664	0.672	0.678	0.685	0.691
0.60	***1.639***	***1.649***	***1.659***	***1.668***	***1.677***	***1.685***	***1.693***	***1.700***	***1.708***	***1.714***	***1.721***	***1.727***
	0.018	0.018	0.018	0.018	0.018	0.018	0.018	0.018	0.018	0.018	0.018	0.018
	0.569	0.581	0.592	0.602	0.612	0.621	0.630	0.638	0.646	0.654	0.662	0.669
0.70	***1.587***	***1.599***	***1.610***	***1.620***	***1.630***	***1.639***	***1.648***	***1.656***	***1.664***	***1.672***	***1.680***	***1.687***
	0.008	0.008	0.008	0.008	0.008	0.008	0.008	0.008	0.008	0.008	0.008	0.008
	0.529	0.541	0.553	0.565	0.575	0.586	0.595	0.605	0.614	0.622	0.630	0.638
0.80	***1.537***	***1.549***	***1.561***	***1.573***	***1.583***	***1.594***	***1.603***	***1.613***	***1.622***	***1.630***	***1.638***	***1.646***
	0.003	0.003	0.003	0.003	0.003	0.003	0.003	0.003	0.003	0.003	0.003	0.003
	0.485	0.499	0.512	0.524	0.535	0.547	0.557	0.567	0.577	0.586	0.595	0.604
0.90	***1.488***	***1.502***	***1.515***	***1.527***	***1.538***	***1.550***	***1.560***	***1.570***	***1.580***	***1.589***	***1.598***	***1.607***
	0.001	0.001	0.001	0.001	0.001	0.001	0.001	0.001	0.001	0.001	0.001	0.001
	0.441	0.455	0.469	0.482	0.495	0.506	0.518	0.529	0.539	0.549	0.559	0.568
1.00	***1.442***	***1.456***	***1.470***	***1.483***	***1.496***	***1.507***	***1.519***	***1.530***	***1.540***	***1.550***	***1.560***	***1.569***
	0.000	0.000	0.000	0.000	0.000	0.000	0.000	0.000	0.000	0.000	0.000	0.000
	0.398	0.413	0.427	0.441	0.454	0.467	0.479	0.490	0.501	0.512	0.522	0.532
1.10	***1.398***	***1.413***	***1.427***	***1.441***	***1.454***	***1.467***	***1.479***	***1.490***	***1.501***	***1.512***	***1.522***	***1.532***

Table Ten: Target at Boundary: The *Excess Costs of Production, Excess Costs of Use,* and *Effective Cost of Production and Use* when Nonconforming Units Are Reworked

C_{pk} or P_{pk}	*Tolerance in Standard Deviation Units* 4.5	4.8	5.1	5.4	5.7	6.0	6.3	6.6	6.9	7.2	7.5	7.8	8.1	8.4
	0.012	0.007	0.004	0.002	0.001	0.001	0.000	0.000	0.000	0.000	0.000	0.000	0.000	0.000
	0.070	0.096	0.120	0.143	0.165	0.187	0.209	0.229	0.250	0.269	0.288	0.306	0.324	0.341
1.20	***1.081***	***1.103***	***1.124***	***1.145***	***1.166***	***1.188***	***1.209***	***1.230***	***1.250***	***1.269***	***1.288***	***1.306***	***1.324***	***1.341***
		0.009	0.005	0.003	0.001	0.000	0.000	0.000	0.000	0.000	0.000	0.000	0.000	0.000
		0.064	0.087	0.108	0.129	0.150	0.170	0.190	0.210	0.229	0.248	0.266	0.284	0.301
1.30		***1.073***	***1.092***	***1.111***	***1.130***	***1.150***	***1.170***	***1.190***	***1.210***	***1.229***	***1.248***	***1.266***	***1.284***	***1.301***
			0.007	0.004	0.002	0.001	0.000	0.000	0.000	0.000	0.000	0.000	0.000	0.000
			0.058	0.079	0.098	0.117	0.136	0.155	0.174	0.193	0.211	0.229	0.247	0.264
1.40 »			***1.065***	***1.082***	***1.100***	***1.118***	***1.136***	***1.155***	***1.174***	***1.193***	***1.211***	***1.229***	***1.247***	***1.264***
				0.005	0.002	0.001	0.000	0.000	0.000	0.000	0.000	0.000	0.000	0.000
				0.054	0.072	0.089	0.106	0.124	0.142	0.160	0.178	0.195	0.213	0.230
1.50 »				***1.059***	***1.074***	***1.090***	***1.107***	***1.124***	***1.142***	***1.160***	***1.178***	***1.195***	***1.213***	***1.230***
					0.004	0.002	0.001	0.000	0.000	0.000	0.000	0.000	0.000	0.000
					0.049	0.066	0.081	0.097	0.114	0.130	0.147	0.164	0.181	0.198
1.60 »					***1.053***	***1.067***	***1.082***	***1.097***	***1.114***	***1.130***	***1.147***	***1.164***	***1.181***	***1.198***
						0.003	0.001	0.000	0.000	0.000	0.000	0.000	0.000	0.000
						0.045	0.060	0.074	0.089	0.104	0.120	0.136	0.152	0.169
1.70 »						***1.049***	***1.061***	***1.075***	***1.089***	***1.104***	***1.120***	***1.136***	***1.152***	***1.169***
							0.003	0.001	0.000	0.000	0.000	0.000	0.000	0.000
							0.042	0.055	0.068	0.082	0.096	0.111	0.126	0.142
1.80 »							***1.044***	***1.056***	***1.068***	***1.082***	***1.096***	***1.111***	***1.126***	***1.142***
								0.002	0.001	0.000	0.000	0.000	0.000	0.000
								0.039	0.051	0.063	0.075	0.089	0.103	0.117
1.90 »								***1.041***	***1.051***	***1.063***	***1.075***	***1.089***	***1.103***	***1.117***
									0.002	0.000	0.000	0.000	0.000	0.000
									0.036	0.047	0.058	0.070	0.082	0.096
2.00 »									***1.037***	***1.047***	***1.058***	***1.070***	***1.082***	***1.096***
										0.001	0.000	0.000	0.000	0.000
										0.033	0.043	0.053	0.065	0.077
2.10 »										***1.034***	***1.043***	***1.053***	***1.065***	***1.077***
											0.001	0.000	0.000	0.000
											0.031	0.040	0.050	0.060
2.20 »											***1.032***	***1.040***	***1.050***	***1.060***
												0.001	0.000	0.000
												0.029	0.037	0.046
2.30 »												***1.029***	***1.037***	***1.046***
													0.001	0.000
													0.027	0.034
2.40 »													***1.027***	***1.035***
														0.001
														0.025
2.50 »														***1.025***

Table Ten: Target at Boundary: The *Excess Costs of Production, Excess Costs of Use,* and *Effective Cost of Production and Use* when Nonconforming Units Are Reworked

C_{pk} or P_{pk}	*Tolerance in Standard Deviation Units* 8.7	9.0	9.3	9.6	9.9	10.2	10.5	10.8	11.1	11.4	11.7	12.0
	0.000	0.000	0.000	0.000	0.000	0.000	0.000	0.000	0.000	0.000	0.000	0.000
	0.357	0.372	0.387	0.401	0.415	0.428	0.441	0.453	0.464	0.476	0.486	0.497
1.20	***1.357***	***1.372***	***1.387***	***1.401***	***1.415***	***1.428***	***1.441***	***1.453***	***1.464***	***1.476***	***1.486***	***1.497***
	0.000	0.000	0.000	0.000	0.000	0.000	0.000	0.000	0.000	0.000	0.000	0.000
	0.318	0.333	0.349	0.363	0.377	0.391	0.404	0.417	0.429	0.440	0.452	0.463
1.30	***1.318***	***1.333***	***1.349***	***1.363***	***1.377***	***1.391***	***1.404***	***1.417***	***1.429***	***1.440***	***1.452***	***1.463***
	0.000	0.000	0.000	0.000	0.000	0.000	0.000	0.000	0.000	0.000	0.000	0.000
	0.281	0.297	0.312	0.327	0.342	0.356	0.369	0.382	0.395	0.407	0.418	0.429
1.40	***1.281***	***1.297***	***1.312***	***1.327***	***1.342***	***1.356***	***1.369***	***1.382***	***1.395***	***1.407***	***1.418***	***1.429***
	0.000	0.000	0.000	0.000	0.000	0.000	0.000	0.000	0.000	0.000	0.000	0.000
	0.246	0.262	0.278	0.293	0.308	0.322	0.336	0.349	0.362	0.374	0.386	0.398
1.50	***1.246***	***1.262***	***1.278***	***1.293***	***1.308***	***1.322***	***1.336***	***1.349***	***1.362***	***1.374***	***1.386***	***1.398***
	0.000	0.000	0.000	0.000	0.000	0.000	0.000	0.000	0.000	0.000	0.000	0.000
	0.214	0.230	0.246	0.261	0.276	0.290	0.304	0.317	0.330	0.343	0.355	0.367
1.60	***1.214***	***1.230***	***1.246***	***1.261***	***1.276***	***1.290***	***1.304***	***1.317***	***1.330***	***1.343***	***1.355***	***1.367***
	0.000	0.000	0.000	0.000	0.000	0.000	0.000	0.000	0.000	0.000	0.000	0.000
	0.184	0.200	0.216	0.231	0.245	0.260	0.274	0.287	0.300	0.313	0.326	0.338
1.70	***1.184***	***1.200***	***1.216***	***1.231***	***1.245***	***1.260***	***1.274***	***1.287***	***1.300***	***1.313***	***1.326***	***1.338***
	0.000	0.000	0.000	0.000	0.000	0.000	0.000	0.000	0.000	0.000	0.000	0.000
	0.157	0.172	0.187	0.202	0.217	0.231	0.245	0.259	0.272	0.285	0.297	0.309
1.80	***1.157***	***1.172***	***1.187***	***1.202***	***1.217***	***1.231***	***1.245***	***1.259***	***1.272***	***1.285***	***1.297***	***1.309***
	0.000	0.000	0.000	0.000	0.000	0.000	0.000	0.000	0.000	0.000	0.000	0.000
	0.132	0.147	0.161	0.176	0.190	0.204	0.218	0.232	0.245	0.258	0.270	0.283
1.90	***1.132***	***1.147***	***1.161***	***1.176***	***1.190***	***1.204***	***1.218***	***1.232***	***1.245***	***1.258***	***1.270***	***1.283***
	0.000	0.000	0.000	0.000	0.000	0.000	0.000	0.000	0.000	0.000	0.000	0.000
	0.110	0.123	0.137	0.151	0.165	0.179	0.193	0.206	0.219	0.232	0.245	0.257
2.00	***1.110***	***1.123***	***1.137***	***1.151***	***1.165***	***1.179***	***1.193***	***1.206***	***1.219***	***1.232***	***1.245***	***1.257***
	0.000	0.000	0.000	0.000	0.000	0.000	0.000	0.000	0.000	0.000	0.000	0.000
	0.089	0.102	0.116	0.129	0.142	0.156	0.169	0.182	0.195	0.208	0.220	0.233
2.10	***1.089***	***1.102***	***1.116***	***1.129***	***1.142***	***1.156***	***1.169***	***1.182***	***1.195***	***1.208***	***1.220***	***1.233***
	0.000	0.000	0.000	0.000	0.000	0.000	0.000	0.000	0.000	0.000	0.000	0.000
	0.071	0.083	0.096	0.109	0.121	0.134	0.147	0.160	0.172	0.185	0.197	0.209
2.20	***1.071***	***1.083***	***1.096***	***1.109***	***1.121***	***1.134***	***1.147***	***1.160***	***1.172***	***1.185***	***1.197***	***1.209***
	0.000	0.000	0.000	0.000	0.000	0.000	0.000	0.000	0.000	0.000	0.000	0.000
	0.056	0.067	0.078	0.090	0.102	0.114	0.127	0.139	0.151	0.164	0.176	0.188
2.30	***1.056***	***1.067***	***1.078***	***1.090***	***1.102***	***1.114***	***1.127***	***1.139***	***1.151***	***1.164***	***1.176***	***1.188***
	0.000	0.000	0.000	0.000	0.000	0.000	0.000	0.000	0.000	0.000	0.000	0.000
	0.043	0.052	0.063	0.073	0.085	0.096	0.108	0.120	0.132	0.143	0.155	0.167
2.40	***1.043***	***1.052***	***1.063***	***1.073***	***1.085***	***1.096***	***1.108***	***1.120***	***1.132***	***1.143***	***1.155***	***1.167***
	0.000	0.000	0.000	0.000	0.000	0.000	0.000	0.000	0.000	0.000	0.000	0.000
	0.032	0.040	0.049	0.059	0.069	0.080	0.091	0.102	0.113	0.125	0.136	0.148
2.50	***1.032***	***1.040***	***1.049***	***1.059***	***1.069***	***1.080***	***1.091***	***1.102***	***1.113***	***1.125***	***1.136***	***1.148***
	0.000	0.000	0.000	0.000	0.000	0.000	0.000	0.000	0.000	0.000	0.000	0.000
	0.023	0.030	0.038	0.046	0.055	0.065	0.075	0.086	0.097	0.107	0.118	0.129
2.60	***1.024***	***1.030***	***1.038***	***1.046***	***1.055***	***1.065***	***1.075***	***1.086***	***1.097***	***1.107***	***1.118***	***1.129***
		0.000	0.000	0.000	0.000	0.000	0.000	0.000	0.000	0.000	0.000	0.000
		0.022	0.028	0.035	0.043	0.052	0.061	0.071	0.081	0.091	0.102	0.113
2.70		***1.022***	***1.028***	***1.035***	***1.043***	***1.052***	***1.061***	***1.071***	***1.081***	***1.091***	***1.102***	***1.113***
			0.000	0.000	0.000	0.000	0.000	0.000	0.000	0.000	0.000	0.000
			0.021	0.026	0.033	0.041	0.049	0.058	0.067	0.077	0.087	0.097
2.80 »			***1.021***	***1.026***	***1.033***	***1.041***	***1.049***	***1.058***	***1.067***	***1.077***	***1.087***	***1.097***

Table Ten: Target at Boundary: The *Excess Costs of Production, Excess Costs of Use,* and *Effective Cost of Production and Use* when Nonconforming Units Are Reworked

C_{pk} or P_{pk}	*Tolerance in Standard Deviation Units* 9.6	9.9	10.2	10.5	10.8	11.1	11.4	11.7	12.0
	0.000	0.000	0.000	0.000	0.000	0.000	0.000	0.000	0.000
	0.019	0.025	0.031	0.038	0.046	0.055	0.064	0.073	0.083
2.90	***1.020***	***1.025***	***1.031***	***1.038***	***1.046***	***1.055***	***1.064***	***1.073***	***1.083***
		0.000	0.000	0.000	0.000	0.000	0.000	0.000	0.000
		0.018	0.023	0.029	0.036	0.044	0.052	0.061	0.069
3.00		***1.018***	***1.023***	***1.029***	***1.036***	***1.044***	***1.052***	***1.061***	***1.069***
			0.000	0.000	0.000	0.000	0.000	0.000	0.000
			0.017	0.022	0.028	0.034	0.042	0.049	0.058
	3.10 »		***1.017***	***1.022***	***1.028***	***1.034***	***1.042***	***1.049***	***1.058***
				0.000	0.000	0.000	0.000	0.000	0.000
				0.016	0.021	0.026	0.033	0.040	0.047
		3.20 »		***1.016***	***1.021***	***1.026***	***1.033***	***1.040***	***1.047***
					0.000	0.000	0.000	0.000	0.000
					0.015	0.020	0.025	0.031	0.038
			3.30 »		***1.015***	***1.020***	***1.025***	***1.031***	***1.038***
						0.000	0.000	0.000	0.000
						0.015	0.019	0.024	0.029
				3.40 »		***1.015***	***1.019***	***1.024***	***1.029***
							0.000	0.000	0.000
							0.014	0.018	0.023
					3.50 »		***1.014***	***1.018***	***1.023***
								0.000	0.000
								0.013	0.017
						3.60 »		***1.013***	***1.017***
									0.000
									0.013
							3.70 »		***1.013***

Appendix

I. Formulas for Measures of Dispersion	243
II. The Coverage of Three-Sigma Limits	244
III. The Transformation of Data	248
Table A.1 Bias Correction Factors for Measures of Dispersion	252
Table A.2 *XmR* Charts: Charts for Individual Values	254
Table A.3 Average and Range Charts Based on the Average Range	256
Table A.4 Average and Range Charts Based on the Median Range	258
Table A.5 Average and Standard Deviation Charts Based on the Average Standard Deviation Statistic	260
Table A.6 Average and Standard Deviation Charts Based on the Median Standard Deviation Statistic	262
Table A.7 Average and Root Mean Square Deviation Charts Based on the Average Root Mean Square Deviation Statistic	264
Table A.8 Average and Root Mean Square Deviation Charts Based on the Median Root Mean Square Deviation Statistic	266
Table A.9 Normal Theory Values of the Integrals **IA**, **IB**, **IC**, and **ID**	268
Table A.10 Burr and Normal Values for the Integrals **IC** and **ID**	277
Glossary of Symbols	283
Index	287

I. Formulas for Measures of Dispersion

The **Range** statistic for n values $\{X_1, X_2, \ldots, X_n\}$ will be denoted by:

$$R = \text{Max}\{X_1, X_2, \ldots, X_n\} - \text{Min}\{X_1, X_2, \ldots, X_n\}$$

The **Standard Deviation Statistic** for n values $\{X_1, X_2, \ldots, X_n\}$ will be denoted by:

$$s = \sqrt{\frac{\Sigma(X_i - \bar{X})^2}{n-1}}$$

where $\bar{X}$ denotes the average of the n values.

The **Root Mean Square Deviation** statistic for n values will be denoted by:

$$s_n = \sqrt{\frac{\Sigma(X_i - \bar{X})^2}{n}}$$

None of these measures of dispersion may be said to represent or estimate a parameter for a probability model *unless and until* those conditions exist where that parameter becomes well defined. A clear distinction between *parameters* and *statistics* is a prerequisite for effective data analysis.

All three of these statistics may be computed globally or within the subgroups. See page 20, pages 43 to 45, and page 251 for discussions of the difference between within-subgroup measures of dispersion and global measures of dispersion.

II. The Coverage of Three-Sigma Limits

Don't we need normally distributed data?

No, we don't.

Three-sigma limits are sufficiently general to work with all kinds of data. Remember, the objective is to separate the routine variation from the exceptional variation. Since the exceptional variation is, by definition, dominant, we only need to filter out the bulk of the routine variation in order to tell the difference. Thus, we need limits that will cover all, or almost all, of the routine variation.

To see how three-sigma limits do this, we can use various probability models to characterize routine variation. Figure A.1 contains six such models ranging from the uniform distribution to the exponential distribution. Each of these models was standardized so that they all have a mean of zero and a standard deviation parameter of 1.00. Figure A.1 shows the three-sigma limits and that proportion of the area under each curve that falls within those three-sigma limits.

There are four lessons that can be learned from Figure A.1.

The first lesson of Figure A.1 is that three-sigma limits will filter out virtually all of the routine variation regardless of the shape of the histogram. These six models are radically different, yet in spite of these differences, three-sigma limits cover 98 percent to 100 percent of the area under each curve.

The second lesson of Figure A.1 is that any data point that falls outside the three-sigma limits is a potential signal of a process change. Since it will be a rare event for routine variation to take you outside the three-sigma limits, it is more likely that any point that falls outside these limits is a signal of a process change.

The third lesson of Figure A.1 is that symmetric, three-sigma limits work with skewed data. Four of the six models shown are skewed. As we scan down the page we see that no matter how skewed the model becomes, the standard deviation parameter is stretched at essentially the same rate as the tail. This means that the length of the elongated tail will effectively determine the three-sigma distance in each case, and that the limits will cover the bulk of the elongated tail no matter how skewed the data become.

"But that certainly makes the other limit look silly." Yes, it does. Here we need to pause and think about those situations where we have skewed data. In most cases we have skewed data when the data pile up against a barrier or boundary condition. Whenever a boundary value falls within the computed limits, the boundary takes precedence over the computed limit, and we end up with a one-sided chart. When this happens the remaining limit covers the long tail and allows us to separate the routine variation from potential signals of deviation away from the boundary. Hence, symmetric, three-sigma limits work with skewed data.

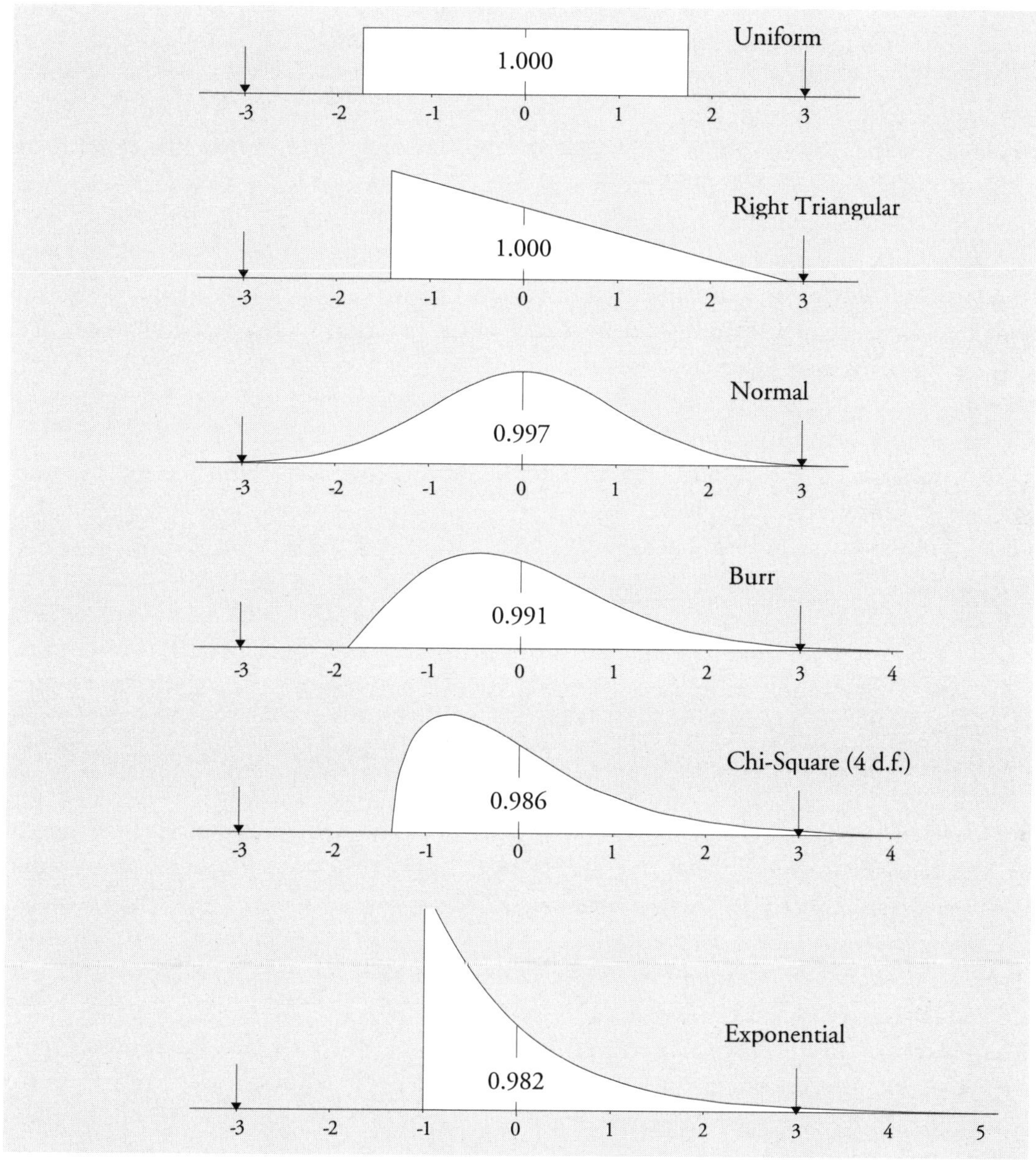

Figure A.1: The Coverage of Three-Sigma Limits for Six Probability Models

The fourth lesson of Figure A.1 is that any uncertainty in where we draw the three-sigma lines will not greatly affect the coverage of the limits. All of the curves are so flat by the time they reach the neighborhood of the three-sigma limits that any errors we may make when we estimate the limits will have, at most, a minimal impact upon how the chart works.

The six probability models in Figure A.1 are represented by white dots on the shape characterization plane in Figure A.2. The black dots define 1137 other probability models.

The black lines divide the shape characterization plane into three regions. The bottom region is for U-shaped distributions. The middle is for J-shaped distributions, and the top is for mound-shaped distributions.

The grey bands define contours wherein three-sigma limits will cover similar percentages of the distributions. For example, the light grey band along the bottom that ends with the notation of 100% shows the region where three-sigma limits will cover 100% of the distribution. The next band, shown in darker grey, defines the region where three-sigma limits will cover at least 99.5% of the distribution.

The next light grey band defines the region where three-sigma limits will cover at least 99% of the distribution. And so on, in a clockwise manner, to the last light grey zone having between 96% and 96.5% coverage.

Careful study of Figure A.2 will reveal that three-sigma limits will provide better than 98% coverage for probability models that are mound shaped.

Three-sigma limits will provide better than 97.5% coverage for all but the most extreme of the J-shaped probability models.

Of the 1143 models studied, 1112 (97.3%) had better than 97.5% coverage.

While some of the extreme U-shaped models can move out toward the Chebychev minimum, most of these models will have 100% coverage. Moreover, U-shaped histograms are almost always a mixture of two different systems. Such "two-humped camels" are a clear indication of a problem, which mitigates the weaker coverage of three-sigma limits for this class of probability models.

Since the purpose here is to have a reasonably conservative sequential analysis, the emphasis is not on having a specific, known risk of a false alarm, but rather on keeping the risk of a false alarm reasonably small. Figure A.2 shows us that generic, three-sigma limits will provide a conservative risk of less than 2.5% every time we plot a point on our process behavior chart regardless of what our histogram looks like.

Three-sigma limits work by brute force. They are sufficiently general to work with all types and shapes of histograms. They work with skewed data, and they work even when the limits are based on few data.

Therefore, we do not have to pre-qualify our data before we place them on a process behavior chart. We do not need to define a reference distribution prior to computing limits. Neither do we need to transform our data before we place them on a process behavior chart.

Anyone who tells you anything to the contrary is simply trying to complicate your life unnecessarily.

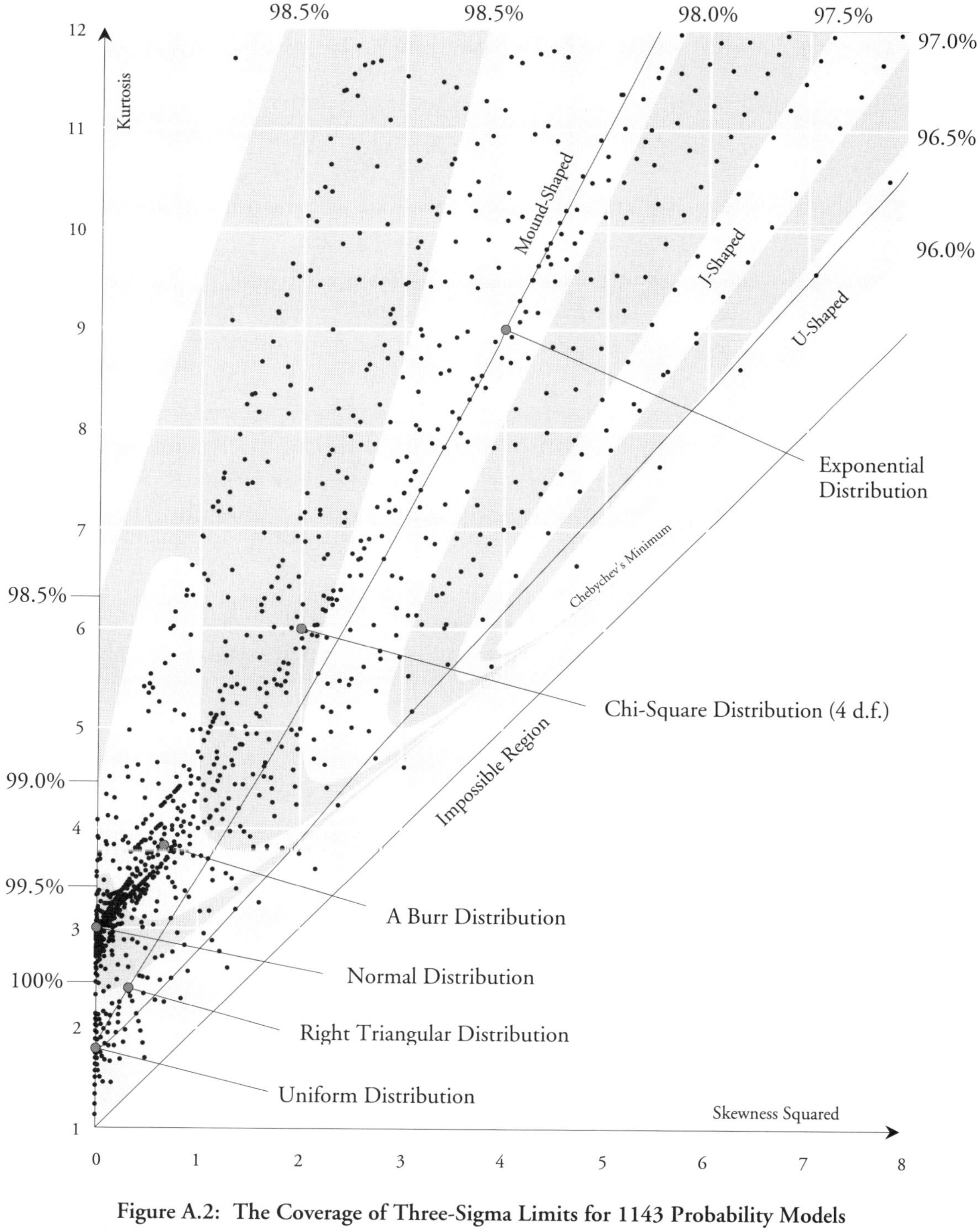

Figure A.2: The Coverage of Three-Sigma Limits for 1143 Probability Models

III. The Transformation of Data

The first principle for understanding data is that no data have meaning apart from their context. Analysis begins with context, is driven by context, and ends with the results being interpreted in the context of the original data. This principle requires that there must always be a link between what you do with the data and the original context for the data. Any transformation of the data risks breaking this linkage.

When using data in a business setting you should take care to present the data in a form that is easy to interpret. This will generally mean that you should avoid nonlinear transformations of the data such as logarithms, square roots, trigonometric functions, and probabilistic transformations.

If a transformation makes sense both in terms of the original data and the objectives of the analysis, then it will be okay to use that transformation. Transformations of this type might be things like the use of daily or weekly averages in place of hourly values or the use of proportions or rates in place of counts to take into account the differing areas of opportunity in different time periods.

Only you as the user can determine when a transformation will make sense in the context of the data. (The software cannot do this because it will never know the context.) Moreover, since these sensible transformations will tend to be fairly simple in nature, they do not tend to distort the data.

A second class of transformations would be those that rescale the data in order to achieve certain statistical properties. (These are the only type of transformations that any software can suggest.) Here the usual objective is to make the data appear to be more "normally distributed" in order to have an "estimate of dispersion that is independent of the estimate of location." Unfortunately, these transformations will tend to be very complex and nonlinear in nature, involving exponential, inverse exponential, or logarithmic functions. (And just what does the logarithm of the percentage of on-time shipments represent?) These nonlinear transformations will distort the data in two ways: At one end of the histogram, values that were originally far apart will now be close together; at the other end of the histogram, values that were originally close together will now be far apart.

To illustrate the effect of transformations to achieve statistical properties we will use the Hot Metal Transit Times (shown written in rows in the table on the next page). These values are the times (to the nearest five minutes) between the phone call alerting the steel furnace that a load of hot metal was on the way and the actual arrival of that load at the steel furnace ladle house. The histogram is shown in Figure A.3.

The Hot Metal Transit Times in Minutes

40	45	125	100	40	40	100	65	55	40	125	65	40	45	95
105	45	110	40	50	120	45	65	105	35	70	55	25	50	55
50	40	40	45	55	50	45	125	55	100	40	70	40	40	110
55	50	30	50	105	45	45	55	50	25	65	60	60	55	70
55	45	100	60	45	145	45	50	65	180	60	45	35	35	55
55	55	50	120	35	45	35	45	55	50	70	45	75	60	45
60	40	60	40	50	60	65	95	65	60	50	25	25	100	50
60	45	35	40	30	180	50	30	30	30	65	130	80	20	45
65	65	45	40	50	25	120	30	115	50	85	40	35	40	40
55	50	25	75	55	50									

Given the skewed nature of these data some software packages would suggest using a logarithmic transformation. Taking the natural logarithm of each of these transit times results in the histogram in Figure A.4. (The horizontal scales show both the original and transformed values.) Notice how the values on the left of Figure A.4 are spaced out while those on the right are crowded together. After the transformation the distance from 20 to 25 minutes is about the same size as the distance from 140 to 180 minutes. How could you begin to explain this to your boss?

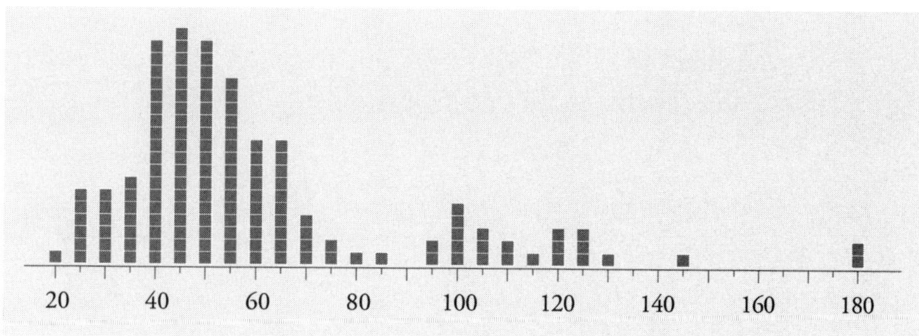

Figure A.3: Histogram of the Hot Metal Transit Times

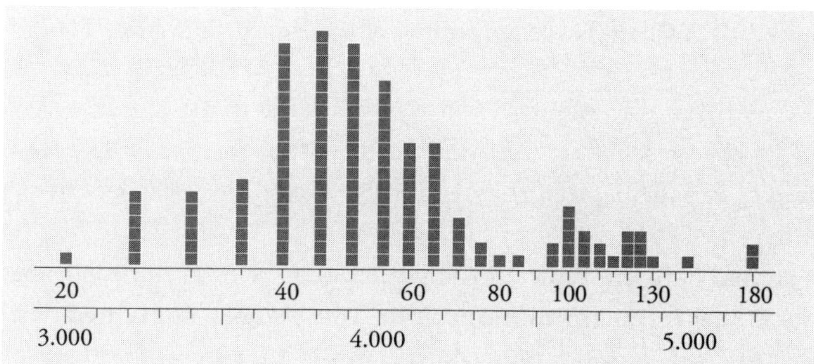

Figure A.4: Histogram of the Logarithms of Hot Metal Transit Times

By itself, this distortion of the data is sufficient to call into question the practice of transforming the data to achieve statistical properties. However, the impact of these non-linear transformations is not confined to the histograms.

Figure A.5 shows the X Chart for the original, untransformed data. The central line shown is the median of the X values, and the limits are based on the median moving range. Eleven of the 141 transit times are above the upper limit, confirming the impression given by the histogram that these data come from a mixture of at least two different processes. Even after the steel furnace gets the phone call, they still have no idea when the hot metal will arrive at the ladle house.

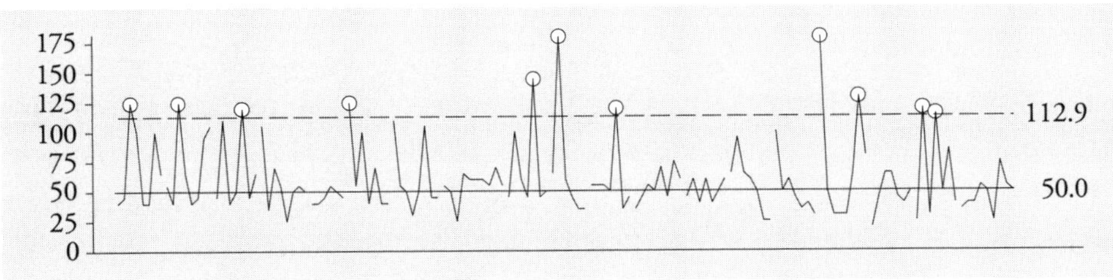

Figure A.5: ***X*** **Chart for the Hot Metal Transit Times**

However, if we transform the data before we put them on a process behavior chart we end up with Figure A.6. There we find no points outside the limits!

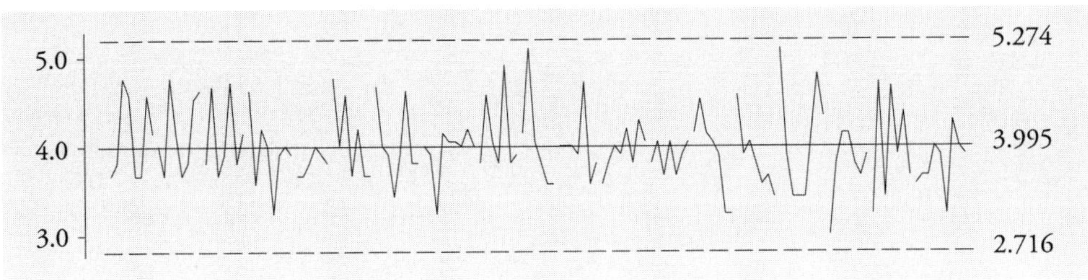

Figure A.6: ***X*** **Chart for the Logarithms of the Hot Metal Transit Times**

Clearly the logarithmic transformation has obliterated the signals. What good is a transformation that changes the message contained within the data? The transformation of the data to achieve statistical properties is simply a complex way of distorting both the data and the truth.

The results shown here are typical of what happens with nonlinear transformations of the original data. **These transformations hide the signals contained within the data simply because they are based upon computations that *presume there are no signals within the data.***

To see how the computations do this we need to pause to consider the nature of the formulas for common descriptive statistics. For a descriptive measure of location we usually use the average, which is simply based upon the sum of the data. However, once we leave the average behind the formulas become much more complex. For a descriptive measure of dispersion we commonly use the global standard deviation statistic, which is a function of the *squared deviations from the average.* For descriptive measures of shape we commonly use the skewness and kurtosis statistics which, respectively, depend upon the *third* and *fourth* powers of the deviations of the data from the average. When we aggregate the data together in this manner and use the second, third, and fourth powers of the distance between each observation and the average value, we are implicitly assuming that these computations make sense. Whether they are measures of dispersion, or skewness, or kurtosis, *any high-order descriptive statistic that is computed globally is implicitly based upon a very strong assumption that the data are homogeneous.*

When the data are not homogeneous it is not the *shape* of the histogram that is wrong, but rather the computation and use of the descriptive statistics that is erroneous. ***We do not need to distort the histogram to make the transformed values more homogeneous. Instead we need to stop and question what the lack of homogeneity means in the context of the original observations.***

So how can we determine when a data set is homogeneous? That is the purpose of the process behavior chart! Transforming the data to achieve statistical properties prior to placing those data on a process behavior chart is an example of getting everything backwards. It assumes that we need to make the data more homogeneous prior to checking them for homogeneity! Any recommendation regarding the transformation of the data *prior* to placing them on a process behavior chart reveals a fundamental lack of understanding about both the purpose of the process behavior chart and the essence of data analysis.

Shewhart's approach, with its generic limits computed empirically from the data, does not even require the specification of a probability model. As Shewhart wrote on page 54 of *Statistical Method from the Viewpoint of Quality Control, "...we are not concerned with the functional form of the universe* [i.e., the probability model], *but merely with the assumption that a universe exists."* [Italics in the original.]

When you transform the data to achieve statistical properties you deceive both yourself and everyone else who is not sophisticated enough to catch you in your deception.

Beware of computations that triumph over common sense!

Table A.1
Bias Correction Factors for Measures of Dispersion

n	c_1	c_2	c_3	c_4	c_5	c_6	d_2	d_3	d_4
2	0.477	0.5642	0.4262	0.7979	0.6028	0.675	1.128	0.8525	0.954
3	0.680	0.7236	0.3782	0.8862	0.4633	0.832	1.693	0.8884	1.588
4	0.769	0.7979	0.3367	0.9213	0.3889	0.888	2.059	0.8798	1.978
5	0.819	0.8407	0.3053	0.9400	0.3412	0.916	2.326	0.8641	2.257
6	0.852	0.8686	0.2808	0.9515	0.3076	0.933	2.534	0.8480	2.472
7	0.874	0.8882	0.2612	0.9594	0.2820	0.944	2.704	0.8332	2.645
8	0.891	0.9027	0.2452	0.9650	0.2622	0.952	2.847	0.8198	2.791
9	0.903	0.9139	0.2317	0.9693	0.2459	0.958	2.970	0.8078	2.915
10	0.913	0.9227	0.2205	0.9727	0.2321	0.963	3.078	0.7971	3.024
11	0.921	0.9300	0.2102	0.9754	0.2204	0.966	3.173	0.7873	3.121
12	0.928	0.9359	0.2019	0.9776	0.2105	0.970	3.258	0.7785	3.207
13	0.934	0.9410	0.1939	0.9794	0.2019	0.972	3.336	0.7704	3.285
14	0.939	0.9453	0.1870	0.9810	0.1940	0.974	3.407	0.7630	3.356
15	0.943	0.9490	0.1809	0.9823	0.1873	0.976	3.472	0.7562	3.422
16	0.947	0.9523	0.1750	0.9835	0.1809	0.978	3.532	0.7499	3.482
17	0.950	0.9551	0.1702	0.9845	0.1754	0.979	3.588	0.7441	3.538
18	0.953	0.9576	0.1657	0.9854	0.1703	0.980	3.640	0.7386	3.591
19	0.955	0.9599	0.1611	0.9862	0.1656	0.981	3.689	0.7335	3.640
20	0.957	0.9619	0.1573	0.9869	0.1613	0.982	3.735	0.7287	3.686
21	0.960	0.9638	0.1532	0.9876	0.1570	0.983	3.778	0.7242	3.730
22	0.962	0.9655	0.1495	0.9882	0.1532	0.984	3.819	0.7199	3.771
23	0.963	0.9670	0.1464	0.9887	0.1499	0.985	3.858	0.7159	3.811
24	0.965	0.9684	0.1433	0.9892	0.1466	0.985	3.895	0.7121	3.847
25	0.966	0.9695	0.1417	0.9896	0.1438	0.986	3.931	0.7084	3.883
30	0.972	0.9748	0.1282	0.9915	0.1301	0.988	4.086	0.6927	4.037
35	0.976	0.9784	0.1190	0.9927	0.1206	0.990	4.213	0.6799	4.166
40	0.979	0.9811	0.1115	0.9936	0.1130	0.991	4.322	0.6692	4.274
45	0.981	0.9832	0.1053	0.9943	0.1066	0.992	4.415	0.6601	4.372
50	0.983	0.9849	0.0999	0.9949	0.1009	0.993	4.498	0.6521	4.450
60	0.986	0.9874	0.0915	0.9957	0.0926	0.994	4.639	0.6389	4.591
70	0.988	0.9892	0.0848	0.9964	0.0848	0.995	4.755	0.6283	4.707
80	0.990	0.9906	0.0788	0.9968	0.0799	0.996	4.854	0.6194	4.806
90	0.991	0.9916	0.0750	0.9972	0.0748	0.996	4.939	0.6118	4.892
100	0.992	0.9925	0.0703	0.9975	0.0707	0.997	5.015	0.6052	4.968

Denote any *Adjusted Summary Within-Subgroup Measure of Dispersion* by *Sigma(X):*

$$Sigma(X) = \frac{\bar{R}}{d_2} \text{ or } \frac{\tilde{R}}{d_4} \text{ or } \frac{\bar{s}_n}{c_2} \text{ or } \frac{\tilde{s}_n}{c_1} \text{ or } \frac{\bar{s}}{c_4} \text{ or } \frac{\tilde{s}}{c_6}$$

$$Sigma(\bar{X}) = \frac{Sigma(X)}{\sqrt{n}} \qquad Sigma(R) = d_3 \; Sigma(X)$$

Table A.1
Bias Correction Factors for Measures of Dispersion

For the Range Statistic, *R*, based on *n* values:

$MEAN(R) = d_2\,\sigma$ $\qquad$ $SD(R) = d_3\,\sigma$ $\qquad$ $MEDIAN(R) = d_4\,\sigma$

where:

$$d_2 = \int_0^\infty w\, f(w)\, dw$$

$$d_3^2 = \int_0^\infty [w - d_2]^2 f(w)\, dw$$

and d_4 is defined by: $\quad 0.5000 = \int_0^{d_4} f(w)\, dw$

where

$$f(w) = \frac{n\,(n-1)}{[\sqrt{2\pi}\,]^n} \int_{-\infty}^{\infty} \left[\int_x^{x+w} e^{-(u^2/2)}\, du \right]^{n-2} e^{-((x+w)^2/2)}\; e^{-(x^2/2)}\, dx$$

For the Standard Deviation Statistic, *s*, based on *n* values:

$MEAN(s) = c_4\,\sigma$ $\qquad$ $SD(s) = c_5\,\sigma$ $\qquad$ $MEDIAN(s) = c_6\,\sigma$

where:

$$c_4 = \sqrt{\frac{2}{n-1}}\;\frac{\Gamma(\frac{n}{2})}{\Gamma(\frac{n-1}{2})} \qquad\qquad c_5 = \sqrt{1 - c_4^2}$$

and c_6 is defined by: $\quad 0.5\,\Gamma(\frac{n-1}{2}) = \int_0^{c_6^2/2(n-1)} t^{(n-3)/2}\, e^{-t}\, dt$

For the Root Mean Square Deviation Statistic, s_n, based on *n* values:

$MEAN(s_n) = c_2\,\sigma$ $\qquad$ $SD(s_n) = c_3\,\sigma$ $\qquad$ $MEDIAN(s_n) = c_1\,\sigma$

where:

$$c_2 = \sqrt{\frac{n-1}{n}}\;c_4 \qquad c_3 = \sqrt{\frac{n-1}{n}}\;c_5 \qquad c_1 = \sqrt{\frac{n-1}{n}}\;c_6$$

Table A.2
XmR Charts: Charts for Individual Values

When a *time series consisting of logically comparable values* has a logical subgroup size of $n = 1$ we can plot the individual values and use a two-point moving range to measure the dispersion. The average moving range or the median moving range may then be used to compute the Natural Process Limits for the individual values and the Upper Range Limit for the moving ranges according to the formulas:

$$UNPL = \bar{X} + E_2\,\bar{R} \quad \text{or} \quad \bar{X} + E_5\,\tilde{R}$$
$$Central\ Line = \bar{X}$$
$$LNPL = \bar{X} - E_2\,\bar{R} \quad \text{or} \quad \bar{X} - E_5\,\tilde{R}$$

$$URL = D_4\,\bar{R} \quad \text{or} \quad D_6\,\tilde{R}$$
$$Central\ Line = \bar{R} \quad \text{or} \quad \tilde{R}$$

For two-period moving ranges these constants are:

$$E_2 = \frac{3}{d_2} = 2.660 \qquad E_5 = \frac{3}{d_4} = 3.145$$

$$D_4 = \left[\,1 + \frac{3\,d_3}{d_2}\,\right] = 3.268 \qquad D_6 = \frac{d_2 + 3\,d_3}{d_4} = 3.865$$

where d_2, d_3 and d_4 are the bias correction factors given in Table A.1 for $n = 2$.

The average of (k–1) two-point moving ranges will have approximately 0.62 (k–1) degrees of freedom.

The median of (k–1) two-point moving ranges will have approximately 0.32 (k–1) degrees of freedom.

Verification Table for Individual Values: 16 Resistances in Megohms from Table 2.4

	X	mR		X	mR		X	mR		X	mR
-1-	5045		-5-	4290	315	-9-	3980	305	-13-	3300	460
-2-	4350	695	-6-	4430	140	-10-	3925	55	-14-	3685	385
-3-	4350	0	-7-	4485	55	-11-	3645	280	-15-	3463	222
-4-	3975	375	-8-	4285	200	-12-	3760	115	-16-	5200	1737

Table A.2
XmR Charts: Charts for Individual Values

- Moving ranges which fall above the Upper Range Limit may be taken as indications of a potential break in the time series—some sudden shift which is so large that it is unlikely to have occurred by chance.
- There is no advantage to using any other measure of dispersion besides a two-point moving range. Moving ranges based on any $n > 2$ will be harder to compute and less efficient than the two-point moving ranges. Mean square successive differences will be harder to compute and more prone to being inflated by extreme values while being no more efficient than the two-point moving range. Global measures of dispersion beg the question of homogeneity by making a strong assumption that the data are completely homogeneous, and so they can never be properly used to compute limits for individual values.
- If the data do not come from a time series, i.e., if the individual values do not possess a definite and known time order, then the use of a moving range to measure the dispersion of the data is essentially arbitrary because the order of the data is arbitrary.
- Moreover, if the individual values are *known* to come from different cause systems, as would be the case in a sequence of experimental runs, then the moving ranges may not be used to construct Natural Process Limits even when the data constitute a time series.
- Since most software packages contain options that compute limits incorrectly you should verify the computations in your software. To determine which option in a software package computes the limits for an X Chart correctly you may use the software with the verification data set on the preceding page.

  Using the average of 4135.50 and the average moving range of 355.93 the correct limits are:

  $LNPL$ = 3188.72, $UNPL$ = 5082.28, and URL = 1163.19.

  Or, using the median moving range of 280, the correct limits are:

  $LNPL$ = 3254.90, $UNPL$ = 5016.10, and URL = 1082.20.

  Alternately, using the median for x of 4132.5 and the median moving range of 280, the correct limits are:

  $LNPL$ = 3251.90, $UNPL$ = 5013.10, and URL = 1082.20.

  If an option in a software package is used to compute limits for Individual Values using the 16 values from the verification data set, and if those limits do not agree with one of the three sets of limits given above, then that option is incorrect and should not be used in practice.

Table A.3

Average and Range Charts Based on the Average Range

Given k subgroups each with n observations
with Grand Average, $\bar{\bar{X}}$ and Average Range, $\bar{R}$
use the tabled constants with the formulas to obtain:

- limits for Subgroup Averages,
- limits for Subgroup Ranges, or
- Natural Process Limits for X, and
- the approximate Degrees of Freedom for these limits.

n	A_2	D_3	D_4	E_2	*Approximate Degrees of Freedom*
2	1.880	—	3.268	2.660	.88 k (n–1)
3	1.023	—	2.574	1.772	.91 k (n–1)
4	0.729	—	2.282	1.457	.91 k (n–1)
5	0.577	—	2.114	1.290	.91 k (n–1)
6	0.483	—	2.004	1.184	.89 k (n–1)
7	0.419	0.076	1.924	1.109	.88 k (n–1)
8	0.373	0.136	1.864	1.054	.86 k (n–1)
9	0.337	0.184	1.816	1.010	.85 k (n–1)
10	0.308	0.223	1.777	0.975	.83 k (n–1)
11	0.285	0.256	1.744	0.945	.81 k (n–1)
12	0.266	0.283	1.717	0.921	.80 k (n–1)
13	0.249	0.307	1.693	0.899	.78 k (n–1)
14	0.235	0.328	1.672	0.881	.77 k (n–1)
15	0.223	0.347	1.653	0.864	.75 k (n–1)

Since most software packages contain options that compute limits incorrectly, you should verify the computations in your software. To see if your software uses the *Average Range* to correctly compute limits for an Average and Range Chart use the 10 subgroups below. The grand average is 4468.45, the average range is 687.0, and the correct limits are: *LAL* = 3967.63, *UAL* = 4969.27, *URL* = 1567.73, *LNPL* = 3467.49, and *UNPL* = 5469.41. These limits have approximately 27 degrees of freedom.

Verification Table for Subgrouped Data: 40 Resistances in Megohms from Table 2.2

Subgroup	-1-	-2-	-3-	-4-	-5-	-6-	-7-	-8-	-9-	-10-
Value 1	5045	4290	3980	3300	5100	4635	4410	4725	4790	4110
Value 2	4350	4430	3925	3685	4635	4720	4065	4640	4845	4410
Value 3	4350	4485	3645	3463	5100	4810	4565	4640	4700	4180
Value 4	3975	4285	3760	5200	5450	4565	5190	4895	4600	4790
Average	***4430.0***	***4372.5***	***3827.5***	***3912.0***	***5071.25***	***4682.5***	***4557.5***	***4725.0***	***4733.75***	***4372.5***
Range	***1070***	***200***	***335***	***1900***	***815***	***245***	***1125***	***255***	***245***	***680***
Std Dev.	***446.49***	***100.71***	***153.43***	***873.04***	***334.37***	***106.03***	***470.61***	***120.21***	***107.34***	***306.42***
RMS Dev	***386.67***	***87.21***	***132.88***	***756.08***	***289.58***	***91.82***	***407.56***	***104.10***	***92.96***	***265.37***

Table A.3
Average and Range Charts Based on the Average Range

Limits for ***Subgroup Averages*** *will be:*

$$UAL = \bar{\bar{X}} + A_2 \bar{R}$$
$$\text{Central Line} = \bar{\bar{X}}$$
$$LAL = \bar{\bar{X}} - A_2 \bar{R}$$

Limits for ***Subgroup Ranges*** *will be:*

$$URL = D_4 \bar{R}$$
$$\text{Central Line} = \bar{R}$$
$$LRL = D_3 \bar{R}$$

While ***Natural Process Limits for individual values*** *may be obtained from:*

$$UNPL = \bar{\bar{X}} + E_2 \bar{R}$$
$$\text{Central Line} = \bar{\bar{X}}$$
$$LNPL = \bar{\bar{X}} - E_2 \bar{R}$$

where

$$A_2 = \frac{3}{d_2 \sqrt{n}}$$

$$D_3 = \left[1 - \frac{3\, d_3}{d_2} \right] \qquad D_4 = \left[1 + \frac{3\, d_3}{d_2} \right]$$

$$E_2 = \frac{3}{d_2}$$

If an option in a software package is used to compute limits for the Average Chart using the verification data set on the preceding page, and if those limits do not match one of the sets of limits given in Tables A.3 to A.8, then that option uses an incorrect computation and should be avoided in practice.

Table A.4
Average and Range Charts Based on the Median Range

Given k subgroups each with n observations
with Grand Average, $\bar{\bar{X}}$ and Median Range, $\tilde{R}$
use the tabled constants with the formulas to obtain:

- limits for Subgroup Averages,
- limits for Subgroup Ranges, or
- Natural Process Limits for X, and
- the approximate Degrees of Freedom for these limits.

n	A_4	D_5	D_6	E_5	*Approximate* * *Degrees of Freedom*
2	2.224	—	3.865	3.145	$(1.56 + .023\,k)\,(n-1)\,\sqrt{k}$
3	1.091	—	2.745	1.889	$(1.61 + .031\,k)\,(n-1)\,\sqrt{k}$
4	0.758	—	2.375	1.517	$(1.61 + .032\,k)\,(n-1)\,\sqrt{k}$
5	0.594	—	2.179	1.329	$(1.61 + .028\,k)\,(n-1)\,\sqrt{k}$
6	0.495	—	2.055	1.214	$(1.57 + .025\,k)\,(n-1)\,\sqrt{k}$
7	0.429	0.078	1.967	1.134	$(1.56 + .023\,k)\,(n-1)\,\sqrt{k}$
8	0.380	0.139	1.901	1.075	$(1.52 + .020\,k)\,(n-1)\,\sqrt{k}$
9	0.343	0.187	1.850	1.029	$(1.50 + .018\,k)\,(n-1)\,\sqrt{k}$
10	0.314	0.227	1.809	0.992	$(1.47 + .016\,k)\,(n-1)\,\sqrt{k}$
11	0.290	0.260	1.773	0.961	$(1.43 + .015\,k)\,(n-1)\,\sqrt{k}$
12	0.270	0.288	1.744	0.935	$(1.42 + .014\,k)\,(n-1)\,\sqrt{k}$
13	0.253	0.312	1.719	0.913	$(1.38 + .013\,k)\,(n-1)\,\sqrt{k}$
14	0.239	0.333	1.697	0.894	$(1.36 + .012\,k)\,(n-1)\,\sqrt{k}$
15	0.226	0.352	1.678	0.877	$(1.33 + .010\,k)\,(n-1)\,\sqrt{k}$

* formulas for even k— *for odd values of k use (k–1) in place of k in formula,*

Since most software packages contain options that compute limits incorrectly, you should verify the computations in your software. To see if your software uses the *Median Range* to correctly compute limits for an Average and Range Chart use the data from the Verification Table on page 256. The grand average is 4468.45, the median range is 507.5, and the correct limits are:

LAL = 4083.77, *UAL* = 4853.14, *URL* = 1205.31, *LNPL* = 3698.57, and *UNPL* = 5238.33
These limits have approximately 18 degrees of freedom.

Table A.4
Average and Range Charts Based on the Median Range

Limits for ***Subgroup Averages*** *will be:*

$$UAL = \bar{\bar{X}} + A_4 \tilde{R}$$
$$Central\ Line = \bar{\bar{X}}$$
$$LAL = \bar{\bar{X}} - A_4 \tilde{R}$$

Limits for ***Subgroup Ranges*** *will be:*

$$URL = D_6 \tilde{R}$$
$$Central\ Line = \tilde{R}$$
$$LRL = D_5 \tilde{R}$$

While ***Natural Process Limits for individual values*** *may be obtained from:*

$$UNPL = \bar{\bar{X}} + E_5 \tilde{R}$$
$$Central\ Line = \bar{\bar{X}}$$
$$LNPL = \bar{\bar{X}} - E_5 \tilde{R}$$

where

$$A_4 = \frac{3}{d_4 \sqrt{n}}$$

$$D_5 = \frac{d_2 - 3\,d_3}{d_4} \qquad D_6 = \frac{d_2 + 3\,d_3}{d_4}$$

$$E_5 = \frac{3}{d_4}$$

If an option in a software package is used to compute limits for the Average Chart using the verification data set on page 256, and if those limits do not match one of the sets of limits given in Tables A.3 to A.8, then that option uses an incorrect computation and should be avoided in practice.

Table A.5

Average and Standard Deviation Charts Based on the Average Standard Deviation Statistic

Given k subgroups each with n observations
with Grand Average, $\bar{\bar{X}}$, and Average Standard Deviation Statistic, $\bar{s}$
use the tabled constants with the formulas to obtain:

- limits for Subgroup Averages,
- limits for Subgroup Standard Deviations, or
- Natural Process Limits for individual values, and
- the approximate Degrees of Freedom for these limits.

n	A_3	B_3	B_4	E_3	*Approximate Degrees of Freedom*
2	2.659	–	3.267	3.760	.88 k $(n-1)$
3	1.954	–	2.568	3.385	.92 k $(n-1)$
4	1.628	–	2.266	3.256	.93 k $(n-1)$
5	1.427	–	2.089	3.191	.95 k $(n-1)$
6	1.287	0.030	1.970	3.153	.96 k $(n-1)$
7	1.182	0.118	1.882	3.127	.97 k $(n-1)$
8	1.099	0.185	1.815	3.109	.97 k $(n-1)$
9	1.032	0.239	1.761	3.095	.98 k $(n-1)$
10	0.975	0.284	1.716	3.084	.98 k $(n-1)$
11	0.927	0.322	1.678	3.076	.98 k $(n-1)$
12	0.886	0.354	1.646	3.069	.98 k $(n-1)$
13	0.850	0.382	1.619	3.063	.98 k $(n-1)$
14	0.817	0.407	1.593	3.058	.98 k $(n-1)$
15	0.789	0.428	1.572	3.054	.98 k $(n-1)$
20	0.680	0.510	1.490	3.040	.98 k $(n-1)$
25	0.606	0.564	1.436	3.032	.98 k $(n-1)$
30	0.552	0.606	1.394	3.026	.98 k $(n-1)$
35	0.511	0.636	1.364	3.022	.98 k $(n-1)$
40	0.477	0.659	1.341	3.019	.98 k $(n-1)$
45	0.450	0.678	1.322	3.017	.98 k $(n-1)$
50	0.426	0.696	1.304	3.015	.98 k $(n-1)$

Since most software packages contain options that compute limits incorrectly, you should verify the computations in your software. To see if your software uses the *Average Standard Deviation* to correctly compute limits for an Average and Standard Deviation Chart use the data from the Verification Table on page 256. The grand average is 4468.45, the average standard deviation is 301.865, and the correct limits are:

LAL = 3977.01, UAL = 4959.89, $USDL$ = 684.03, $LNPL$ = 3485.58, and $UNPL$ = 5451.32

These limits have approximately 28 degrees of freedom.

Table A.5
Average and Standard Deviation Charts
Based on the Average Standard Deviation Statistic

*Limits for **Subgroup Averages** will be:*

$$UAL = \bar{\bar{X}} + A_3 \, \bar{s}$$

$$Central\ Line = \bar{\bar{X}}$$

$$LAL = \bar{\bar{X}} - A_3 \, \bar{s}$$

*Limits for **Subgroup Standard Deviation Statistics** will be:*

$$USDL = B_4 \, \bar{s}$$

$$Central\ Line = \bar{s}$$

$$LSDL = B_3 \, \bar{s}$$

*While **Natural Process Limits for individual values** may be obtained from:*

$$UNPL = \bar{\bar{X}} + E_3 \, \bar{s}$$

$$Central\ Line = \bar{\bar{X}}$$

$$LNPL = \bar{\bar{X}} - E_3 \, \bar{s}$$

where

$$A_3 = \frac{3}{c_4 \sqrt{n}}$$

$$B_3 = 1 - \frac{3}{c_4}\sqrt{1-(c_4)^2} \qquad B_4 = 1 + \frac{3}{c_4}\sqrt{1-(c_4)^2}$$

$$E_3 = \frac{3}{c_4}$$

If an option in a software package is used to compute limits for the Average Chart using the verification data set on page 256, and if those limits do not match one of the sets of limits given in Tables A.3 to A.8, then that option uses an incorrect computation and should be avoided in practice.

Table A.6
Average and Standard Deviation Charts
Based on the Median Standard Deviation Statistic

Given k subgroups each with n observations
with Grand Average, $\bar{\bar{X}}$, and Median Standard Deviation, $\tilde{s}$
use the tabled constants with the formulas to obtain:

- limits for Subgroup Averages,
- limits for Subgroup Standard Deviations, or
- Natural Process Limits for individual values, and
- the approximate Degrees of Freedom for these limits.

n	A_{10}	B_9	B_{10}	E_6	*Approximate * Degrees of Freedom*
2	3.143	–	3.864	4.444	$(1.55 + .023\, k)\,(n–1)\,\sqrt{k}$
3	2.082	–	2.733	3.606	$(1.74 + .034\, k)\,(n–1)\,\sqrt{k}$
4	1.689	–	2.351	3.378	$(1.64 + .033\, k)\,(n–1)\,\sqrt{k}$
5	1.465	–	2.145	3.275	$(1.67 + .029\, k)\,(n–1)\,\sqrt{k}$
6	1.313	0.031	2.008	3.215	$(1.67 + .027\, k)\,(n–1)\,\sqrt{k}$
7	1.201	0.120	1.913	3.178	$(1.68 + .025\, k)\,(n–1)\,\sqrt{k}$
8	1.114	0.188	1.839	3.151	$(1.68 + .022\, k)\,(n–1)\,\sqrt{k}$
9	1.044	0.242	1.782	3.132	$(1.68 + .020\, k)\,(n–1)\,\sqrt{k}$
10	0.985	0.287	1.735	3.115	$(1.69 + .018\, k)\,(n–1)\,\sqrt{k}$
11	0.936	0.324	1.695	3.106	$(1.69 + .018\, k)\,(n–1)\,\sqrt{k}$
12	0.893	0.357	1.661	3.093	$(1.68 + .016\, k)\,(n–1)\,\sqrt{k}$
13	0.856	0.384	1.631	3.086	$(1.65 + .016\, k)\,(n–1)\,\sqrt{k}$
14	0.823	0.409	1.604	3.080	$(1.64 + .014\, k)\,(n–1)\,\sqrt{k}$
15	0.794	0.431	1.582	3.074	$(1.63 + .014\, k)\,(n–1)\,\sqrt{k}$

* formulas for even k— *for odd values of k use (k–1) in place of k in formula,*

Since most software packages contain options that compute limits incorrectly, you should verify the computations in your software. To see if your software uses the *Median Standard Deviation* to correctly compute limits for an Average and Standard Deviation Chart use the data from the Verification Table on page 256. The grand average is 4468.45, the median standard deviation is 229.925, and the correct limits are:

LAL = 4080.11, *UAL* = 4856.79, *USDL* = 540.55, *LNPL* = 3691.76, and *UNPL* = 5245.14
These limits have approximately 19 degrees of freedom.

Table A.6
Average and Standard Deviation Charts
Based on the Median Standard Deviation Statistic

*Limits for **Subgroup Averages** will be:*

$$UAL = \bar{\bar{X}} + A_{10}\,\tilde{s}$$
$$\text{Central Line} = \bar{\bar{X}}$$
$$LAL = \bar{\bar{X}} - A_{10}\,\tilde{s}$$

*Limits for **Subgroup Standard Deviation Statistics** will be:*

$$USDL = B_{10}\,\tilde{s}$$
$$\text{Central Line} = \tilde{s}$$
$$LSDL = B_9\,\tilde{s}$$

*While **Natural Process Limits for individual values** may be obtained from:*

$$UNPL = \bar{\bar{X}} + E_6\,\tilde{s}$$
$$\text{Central Line} = \bar{\bar{X}}$$
$$LNPL = \bar{\bar{X}} - E_6\,\tilde{s}$$

where

$$A_{10} = \frac{3}{c_6\sqrt{n}}$$

$$B_9 = \frac{c_4 - 3\sqrt{1-(c_4)^2}}{c_6} \qquad B_{10} = \frac{c_4 + 3\sqrt{1-(c_4)^2}}{c_6}$$

$$E_6 = \frac{3}{c_6}$$

If an option in a software package is used to compute limits for the Average Chart using the verification data set on page 256, and if those limits do not match one of the sets of limits given in Tables A.3 to A.8, then that option uses an incorrect computation and should be avoided in practice.

Table A.7

Average and Root Mean Square Deviation Charts Based on the Average Root Mean Square Deviation Statistic

Given k subgroups each with n observations
with Grand Average, $\bar{\bar{X}}$, and Average Root Mean Square Deviation Statistic, $\bar{s}_n$
use the tabled constants with the formulas to obtain:

- limits for Subgroup Averages,
- limits for Subgroup Root Mean Square Deviations, or
- Natural Process Limits for individual values, and
- the approximate Degrees of Freedom for these limits.

n	A_1	B_3	B_4	E_1	*Approximate Degrees of Freedom*
2	3.760	–	3.267	5.317	.88 $k\ (n-1)$
3	2.393	–	2.568	4.146	.92 $k\ (n-1)$
4	1.880	–	2.266	3.760	.93 $k\ (n-1)$
5	1.595	–	2.089	3.568	.95 $k\ (n-1)$
6	1.410	0.030	1.970	3.454	.96 $k\ (n-1)$
7	1.277	0.118	1.882	3.378	.97 $k\ (n-1)$
8	1.175	0.185	1.815	3.323	.97 $k\ (n-1)$
9	1.095	0.239	1.761	3.283	.98 $k\ (n-1)$
10	1.028	0.284	1.716	3.251	.98 $k\ (n-1)$
11	0.973	0.322	1.678	3.226	.98 $k\ (n-1)$
12	0.925	0.354	1.646	3.205	.98 $k\ (n-1)$
13	0.884	0.382	1.619	3.188	.98 $k\ (n-1)$
14	0.848	0.407	1.593	3.174	.98 $k\ (n-1)$
15	0.816	0.428	1.572	3.161	.98 $k\ (n-1)$
20	0.697	0.510	1.490	3.117	.98 $k\ (n-1)$
25	0.619	0.564	1.436	3.095	.98 $k\ (n-1)$
30	0.562	0.606	1.394	3.078	.98 $k\ (n-1)$
35	0.518	0.636	1.364	3.065	.98 $k\ (n-1)$
40	0.483	0.659	1.341	3.055	.98 $k\ (n-1)$
45	0.455	0.678	1.322	3.052	.98 $k\ (n-1)$
50	0.431	0.696	1.304	3.048	.98 $k\ (n-1)$

Since most software packages contain options that compute limits incorrectly, you should verify the computations in your software. To see if your software uses the *Average Root Mean Square Deviation* to correctly compute limits for an Average and Root Mean Square Deviation Chart use the data from page 256. The grand average is 4468.45, the average root mean square deviation is 261.423, and the correct limits are:

LAL = 3976.97, *UAL* = 4959.93, *URMSL* = 592.38, *LNPL* = 3485.50, and *UNPL* = 5451.40
These limits have approximately 28 degrees of freedom.

Table A.7
Average and Root Mean Square Deviation Charts
Based on the Average Root Mean Square Deviation Statistic

Limits for ***Subgroup Averages*** *will be:*

$$UAL = \bar{\bar{X}} + A_1\ \bar{s}_n$$
$$\text{Central Line} = \bar{\bar{X}}$$
$$LAL = \bar{\bar{X}} - A_1\ \bar{s}_n$$

Limits for ***Subgroup Root Mean Square Deviation Statistics*** *will be:*

$$URMSL = B_4\ \bar{s}_n$$
$$\text{Central Line} = \bar{s}_n$$
$$LRMSL = B_3\ \bar{s}_n$$

While ***Natural Process Limits for individual values*** *may be obtained from:*

$$UNPL = \bar{\bar{X}} + E_1\ \bar{s}_n$$
$$\text{Central Line} = \bar{\bar{X}}$$
$$LNPL = \bar{\bar{X}} - E_1\ \bar{s}_n$$

where

$$A_1 = \frac{3}{c_2\sqrt{n}}$$

$$B_3 = 1 - \frac{3}{c_4}\sqrt{1-(c_4)^2} \qquad B_4 = 1 + \frac{3}{c_4}\sqrt{1-(c_4)^2}$$

$$E_1 = \frac{3}{c_2}$$

If an option in a software package is used to compute limits for the Average Chart using the verification data set on page 256, and if those limits do not match one of the sets of limits given in Tables A.3 to A.8, then that option uses an incorrect computation and should be avoided in practice.

Table A.8

Average and Root Mean Square Deviation Charts Based on the Median Root Mean Square Deviation Statistic

Given k subgroups each with n observations
with Grand Average, $\bar{\bar{X}}$, and Median Root Mean Square Deviation, $\tilde{s}_n$
use the tabled constants with the formulas to obtain:

- limits for Subgroup Averages,
- limits for Subgroup Root Mean Square Deviations, or
- Natural Process Limits for individual values, and
- the approximate Degrees of Freedom for these limits.

n	A_5	B_9	B_{10}	E_4	*Approximate Degrees of Freedom* *
2	4.447	–	3.864	6.289	$(1.55 + .023\,k)\,(n{-}1)\,\sqrt{k}$
3	2.547	–	2.733	4.412	$(1.74 + .034\,k)\,(n{-}1)\,\sqrt{k}$
4	1.951	–	2.351	4.115	$(1.64 + .033\,k)\,(n{-}1)\,\sqrt{k}$
5	1.638	–	2.145	3.663	$(1.67 + .029\,k)\,(n{-}1)\,\sqrt{k}$
6	1.438	0.031	2.008	3.521	$(1.67 + .027\,k)\,(n{-}1)\,\sqrt{k}$
7	1.297	0.120	1.913	3.432	$(1.68 + .025\,k)\,(n{-}1)\,\sqrt{k}$
8	1.190	0.188	1.839	3.367	$(1.68 + .022\,k)\,(n{-}1)\,\sqrt{k}$
9	1.107	0.242	1.782	3.322	$(1.68 + .020\,k)\,(n{-}1)\,\sqrt{k}$
10	1.039	0.287	1.735	3.286	$(1.69 + .018\,k)\,(n{-}1)\,\sqrt{k}$
11	0.982	0.324	1.695	3.257	$(1.69 + .018\,k)\,(n{-}1)\,\sqrt{k}$
12	0.933	0.357	1.661	3.233	$(1.68 + .016\,k)\,(n{-}1)\,\sqrt{k}$
13	0.891	0.384	1.631	3.212	$(1.65 + .016\,k)\,(n{-}1)\,\sqrt{k}$
14	0.854	0.409	1.604	3.195	$(1.64 + .014\,k)\,(n{-}1)\,\sqrt{k}$
15	0.821	0.431	1.582	3.181	$(1.63 + .014\,k)\,(n{-}1)\,\sqrt{k}$

* formulas for even k— *for odd values of k use (k–1) in place of k in formula,*

Since most software packages contain options that compute limits incorrectly, you should verify the computations in your software. To see if your software uses the *Median Root Mean Square Deviation* to correctly compute limits for an Average and Root Mean Square Deviation Chart use the data from page 256. The grand average is 4468.45, the median root mean square deviation is 199.121, and the correct limits are:

LAL = 4079.96, *UAL* = 4856.94, *URMSL* = 468.13, *LNPL* = 3649.07, and *UNPL* = 5287.83
These limits have approximately 19 degrees of freedom.

Table A.8
Average and Root Mean Square Deviation Charts Based on the Median Root Mean Square Deviation Statistic

Limits for ***Subgroup Averages*** *will be:*

$$UAL = \bar{\bar{X}} + A_5 \tilde{s}_n$$
$$\text{Central Line} = \bar{\bar{X}}$$
$$LAL = \bar{\bar{X}} - A_5 \tilde{s}_n$$

Limits for ***Subgroup Root Mean Square Deviation Statistics*** *will be:*

$$URMSL = B_{10} \tilde{s}_n$$
$$\text{Central Line} = \tilde{s}_n$$
$$LRMSL = B_9 \tilde{s}_n$$

While ***Natural Process Limits for individual values*** *may be obtained from:*

$$UNPL = \bar{\bar{X}} + E_4 \tilde{s}_n$$
$$\text{Central Line} = \bar{\bar{X}}$$
$$LNPL = \bar{\bar{X}} - E_4 \tilde{s}_n$$

where

$$A_5 = \frac{3}{c_1 \sqrt{n}}$$

$$B_9 = \frac{c_4 - 3\sqrt{1-(c_4)^2}}{c_6} \qquad B_{10} = \frac{c_4 + 3\sqrt{1-(c_4)^2}}{c_6}$$

$$E_4 = \frac{3}{c_1}$$

If an option in a software package is used to compute limits for the Average Chart using the verification data set on page 256, and if those limits do not match one of the sets of limits given in Tables A.3 to A.8, then that option uses an incorrect computation and should be avoided in practice.

Table A.9: Normal Theory Values of the Integrals IA, IB, IC, and ID

C_{pk}		C_p values 0.10	0.20	0.30	0.40	0.50	0.60	0.70	0.80	0.90	1.00	1.10	1.20	1.30	1.40
-0.5	**IA**	0.018	0.003	0.000	0.000	0.000	0.000	0.000	0.000	0.000	0.000	0.000	0.000	0.000	0.000
	IB	0.005	0.003	0.001	0.000	0.000	0.000	0.000	0.000	0.000	0.000	0.000	0.000	0.000	0.000
	IC	0.012	0.021	0.028	0.033	0.038	0.041	0.044	0.046	0.048	0.050	0.051	0.052	0.053	0.054
	ID	0.933	0.933	0.933	0.933	0.933	0.933	0.933	0.933	0.933	0.933	0.933	0.933	0.933	0.933
-0.4	**IA**	0.036	0.008	0.001	0.000	0.000	0.000	0.000	0.000	0.000	0.000	0.000	0.000	0.000	0.000
	IB	0.009	0.007	0.003	0.001	0.000	0.000	0.000	0.000	0.000	0.000	0.000	0.000	0.000	0.000
	IC	0.018	0.032	0.044	0.054	0.061	0.067	0.072	0.077	0.080	0.083	0.085	0.088	0.089	0.091
	ID	0.885	0.885	0.885	0.885	0.885	0.885	0.885	0.885	0.885	0.885	0.885	0.885	0.885	0.885
-0.3	**IA**	0.067	0.018	0.003	0.000	0.000	0.000	0.000	0.000	0.000	0.000	0.000	0.000	0.000	0.000
	IB	0.014	0.012	0.006	0.002	0.001	0.000	0.000	0.000	0.000	0.000	0.000	0.000	0.000	0.000
	IC	0.025	0.046	0.064	0.079	0.091	0.101	0.110	0.117	0.123	0.128	0.132	0.136	0.139	0.142
	ID	0.816	0.816	0.816	0.816	0.816	0.816	0.816	0.816	0.816	0.816	0.816	0.816	0.816	0.816
-0.2	**IA**	0.115	0.036	0.008	0.001	0.000	0.000	0.000	0.000	0.000	0.000	0.000	0.000	0.000	0.000
	IB	0.021	0.021	0.012	0.005	0.002	0.000	0.000	0.000	0.000	0.000	0.000	0.000	0.000	0.000
	IC	0.032	0.060	0.085	0.106	0.124	0.140	0.153	0.164	0.173	0.181	0.188	0.194	0.199	0.204
	ID	0.726	0.726	0.726	0.726	0.726	0.726	0.726	0.726	0.726	0.726	0.726	0.726	0.726	0.726
-0.1	**IA**	0.184	0.067	0.018	0.003	0.000	0.000	0.000	0.000	0.000	0.000	0.000	0.000	0.000	0.000
	IB	0.028	0.032	0.021	0.010	0.004	0.001	0.000	0.000	0.000	0.000	0.000	0.000	0.000	0.000
	IC	0.037	0.072	0.103	0.131	0.156	0.178	0.196	0.212	0.226	0.238	0.248	0.257	0.265	0.272
	ID	0.618	0.618	0.618	0.618	0.618	0.618	0.618	0.618	0.618	0.618	0.618	0.618	0.618	0.618
0.0	**IA**	0.274	0.115	0.036	0.008	0.001	0.000	0.000	0.000	0.000	0.000	0.000	0.000	0.000	0.000
	IB	0.035	0.046	0.035	0.019	0.008	0.003	0.001	0.000	0.000	0.000	0.000	0.000	0.000	0.000
	IC	0.040	0.078	0.115	0.149	0.180	0.208	0.233	0.254	0.273	0.290	0.304	0.317	0.328	0.338
	ID	0.500	0.500	0.500	0.500	0.500	0.500	0.500	0.500	0.500	0.500	0.500	0.500	0.500	0.500
0.1	**IA**	0.382	0.184	0.067	0.018	0.003	0.000	0.000	0.000	0.000	0.000	0.000	0.000	0.000	0.000
	IB	0.039	0.060	0.053	0.033	0.016	0.006	0.002	0.001	0.000	0.000	0.000	0.000	0.000	0.000
	IC	0.039	0.078	0.118	0.155	0.191	0.224	0.255	0.282	0.306	0.328	0.347	0.364	0.379	0.393
	ID	0.382	0.382	0.382	0.382	0.382	0.382	0.382	0.382	0.382	0.382	0.382	0.382	0.382	0.382
0.2	**IA**		0.274	0.115	0.036	0.008	0.001	0.000	0.000	0.000	0.000	0.000	0.000	0.000	0.000
	IB		0.072	0.074	0.053	0.029	0.013	0.005	0.002	0.001	0.000	0.000	0.000	0.000	0.000
	IC	0.035	0.072	0.110	0.149	0.187	0.223	0.258	0.290	0.319	0.345	0.369	0.390	0.410	0.427
	ID	0.274	0.274	0.274	0.274	0.274	0.274	0.274	0.274	0.274	0.274	0.274	0.274	0.274	0.274
0.3	**IA**			0.184	0.067	0.018	0.003	0.000	0.000	0.000	0.000	0.000	0.000	0.000	0.000
	IB			0.094	0.078	0.049	0.024	0.010	0.004	0.001	0.000	0.000	0.000	0.000	0.000
	IC	0.028	0.060	0.094	0.131	0.168	0.205	0.242	0.276	0.309	0.339	0.367	0.393	0.416	0.437
	ID	0.184	0.184	0.184	0.184	0.184	0.184	0.184	0.184	0.184	0.184	0.184	0.184	0.184	0.184
0.4	**IA**				0.115	0.036	0.008	0.001	0.000	0.000	0.000	0.000	0.000	0.000	0.000
	IB				0.105	0.075	0.042	0.019	0.008	0.003	0.001	0.000	0.000	0.000	0.000
	IC	0.021	0.046	0.074	0.106	0.139	0.174	0.210	0.245	0.280	0.312	0.343	0.372	0.398	0.422
	ID	0.115	0.115	0.115	0.115	0.115	0.115	0.115	0.115	0.115	0.115	0.115	0.115	0.115	0.115
0.5	**IA**					0.067	0.018	0.003	0.000	0.000	0.000	0.000	0.000	0.000	0.000
	IB					0.106	0.067	0.035	0.016	0.006	0.003	0.001	0.000	0.000	0.000
	IC	0.014	0.032	0.054	0.078	0.106	0.137	0.169	0.203	0.237	0.270	0.302	0.332	0.361	0.388
	ID	0.067	0.067	0.067	0.067	0.067	0.067	0.067	0.067	0.067	0.067	0.067	0.067	0.067	0.067
0.6	**IA**						0.036	0.008	0.001	0.000	0.000	0.000	0.000	0.000	0.000
	IB						0.099	0.057	0.028	0.013	0.005	0.002	0.001	0.000	0.000
	IC	0.009	0.021	0.035	0.053	0.075	0.099	0.127	0.156	0.188	0.219	0.251	0.282	0.312	0.340
	ID	0.036	0.036	0.036	0.036	0.036	0.036	0.036	0.036	0.036	0.036	0.036	0.036	0.036	0.036

Table A.9: Normal Theory Values of the Integrals IA, IB, IC, and ID

C_{pk}		C_p values 1.50	1.60	1.70	1.80	1.90	2.00	2.10	2.20	2.30	2.40	2.50	2.60	2.70	2.80
-0.5	**IA**	0.000	0.000	0.000	0.000	0.000	0.000	0.000	0.000	0.000	0.000	0.000	0.000	0.000	0.000
	IB	0.000	0.000	0.000	0.000	0.000	0.000	0.000	0.000	0.000	0.000	0.000	0.000	0.000	0.000
	IC	0.055	0.056	0.056	0.057	0.057	0.058	0.058	0.059	0.059	0.059	0.059	0.060	0.060	0.060
	ID	0.933	0.933	0.933	0.933	0.933	0.933	0.933	0.933	0.933	0.933	0.933	0.933	0.933	0.933
-0.4	**IA**	0.000	0.000	0.000	0.000	0.000	0.000	0.000	0.000	0.000	0.000	0.000	0.000	0.000	0.000
	IB	0.000	0.000	0.000	0.000	0.000	0.000	0.000	0.000	0.000	0.000	0.000	0.000	0.000	0.000
	IC	0.093	0.094	0.095	0.096	0.097	0.098	0.099	0.099	0.100	0.100	0.101	0.102	0.102	0.102
	ID	0.885	0.885	0.885	0.885	0.885	0.885	0.885	0.885	0.885	0.885	0.885	0.885	0.885	0.885
-0.3	**IA**	0.000	0.000	0.000	0.000	0.000	0.000	0.000	0.000	0.000	0.000	0.000	0.000	0.000	0.000
	IB	0.000	0.000	0.000	0.000	0.000	0.000	0.000	0.000	0.000	0.000	0.000	0.000	0.000	0.000
	IC	0.144	0.146	0.148	0.150	0.152	0.153	0.155	0.156	0.157	0.158	0.159	0.160	0.161	0.162
	ID	0.816	0.816	0.816	0.816	0.816	0.816	0.816	0.816	0.816	0.816	0.816	0.816	0.816	0.816
-0.2	**IA**	0.000	0.000	0.000	0.000	0.000	0.000	0.000	0.000	0.000	0.000	0.000	0.000	0.000	0.000
	IB	0.000	0.000	0.000	0.000	0.000	0.000	0.000	0.000	0.000	0.000	0.000	0.000	0.000	0.000
	IC	0.208	0.212	0.215	0.218	0.221	0.223	0.225	0.227	0.229	0.231	0.233	0.234	0.235	0.237
	ID	0.726	0.726	0.726	0.726	0.726	0.726	0.726	0.726	0.726	0.726	0.726	0.726	0.726	0.726
-0.1	**IA**	0.000	0.000	0.000	0.000	0.000	0.000	0.000	0.000	0.000	0.000	0.000	0.000	0.000	0.000
	IB	0.000	0.000	0.000	0.000	0.000	0.000	0.000	0.000	0.000	0.000	0.000	0.000	0.000	0.000
	IC	0.279	0.284	0.289	0.294	0.298	0.302	0.305	0.308	0.311	0.314	0.317	0.319	0.321	0.323
	ID	0.618	0.618	0.618	0.618	0.618	0.618	0.618	0.618	0.618	0.618	0.618	0.618	0.618	0.618
0.0	**IA**	0.000	0.000	0.000	0.000	0.000	0.000	0.000	0.000	0.000	0.000	0.000	0.000	0.000	0.000
	IB	0.000	0.000	0.000	0.000	0.000	0.000	0.000	0.000	0.000	0.000	0.000	0.000	0.000	0.000
	IC	0.347	0.356	0.363	0.370	0.376	0.381	0.386	0.391	0.395	0.399	0.403	0.406	0.409	0.412
	ID	0.500	0.500	0.500	0.500	0.500	0.500	0.500	0.500	0.500	0.500	0.500	0.500	0.500	0.500
0.1	**IA**	0.000	0.000	0.000	0.000	0.000	0.000	0.000	0.000	0.000	0.000	0.000	0.000	0.000	0.000
	IB	0.000	0.000	0.000	0.000	0.000	0.000	0.000	0.000	0.000	0.000	0.000	0.000	0.000	0.000
	IC	0.405	0.416	0.426	0.435	0.443	0.451	0.458	0.464	0.470	0.476	0.481	0.486	0.490	0.494
	ID	0.382	0.382	0.382	0.382	0.382	0.382	0.382	0.382	0.382	0.382	0.382	0.382	0.382	0.382
0.2	**IA**	0.000	0.000	0.000	0.000	0.000	0.000	0.000	0.000	0.000	0.000	0.000	0.000	0.000	0.000
	IB	0.000	0.000	0.000	0.000	0.000	0.000	0.000	0.000	0.000	0.000	0.000	0.000	0.000	0.000
	IC	0.443	0.457	0.470	0.482	0.493	0.503	0.512	0.520	0.528	0.535	0.542	0.548	0.554	0.560
	ID	0.274	0.274	0.274	0.274	0.274	0.274	0.274	0.274	0.274	0.274	0.274	0.274	0.274	0.274
0.3	**IA**	0.000	0.000	0.000	0.000	0.000	0.000	0.000	0.000	0.000	0.000	0.000	0.000	0.000	0.000
	IB	0.000	0.000	0.000	0.000	0.000	0.000	0.000	0.000	0.000	0.000	0.000	0.000	0.000	0.000
	IC	0.456	0.474	0.490	0.504	0.518	0.530	0.542	0.552	0.562	0.571	0.580	0.588	0.595	0.602
	ID	0.184	0.184	0.184	0.184	0.184	0.184	0.184	0.184	0.184	0.184	0.184	0.184	0.184	0.184
0.4	**IA**	0.000	0.000	0.000	0.000	0.000	0.000	0.000	0.000	0.000	0.000	0.000	0.000	0.000	0.000
	IB	0.000	0.000	0.000	0.000	0.000	0.000	0.000	0.000	0.000	0.000	0.000	0.000	0.000	0.000
	IC	0.445	0.465	0.484	0.502	0.518	0.533	0.547	0.559	0.571	0.582	0.593	0.602	0.611	0.620
	ID	0.115	0.115	0.115	0.115	0.115	0.115	0.115	0.115	0.115	0.115	0.115	0.115	0.115	0.115
0.5	**IA**	0.000	0.000	0.000	0.000	0.000	0.000	0.000	0.000	0.000	0.000	0.000	0.000	0.000	0.000
	IB	0.000	0.000	0.000	0.000	0.000	0.000	0.000	0.000	0.000	0.000	0.000	0.000	0.000	0.000
	IC	0.413	0.436	0.458	0.477	0.496	0.513	0.529	0.544	0.558	0.571	0.583	0.594	0.605	0.615
	ID	0.067	0.067	0.067	0.067	0.067	0.067	0.067	0.067	0.067	0.067	0.067	0.067	0.067	0.067
0.6	**IA**	0.000	0.000	0.000	0.000	0.000	0.000	0.000	0.000	0.000	0.000	0.000	0.000	0.000	0.000
	IB	0.000	0.000	0.000	0.000	0.000	0.000	0.000	0.000	0.000	0.000	0.000	0.000	0.000	0.000
	IC	0.367	0.392	0.415	0.437	0.458	0.477	0.495	0.511	0.527	0.542	0.555	0.568	0.581	0.592
	ID	0.036	0.036	0.036	0.036	0.036	0.036	0.036	0.036	0.036	0.036	0.036	0.036	0.036	0.036

Table A.9: Normal Theory Values of the Integrals IA, IB, IC, and ID

C_{pk}		C_p values 2.90	3.00	3.20	3.40	3.60	3.80	4.00	4.50	5.00	5.50	6.0	8.0	12.0	20.0
-0.5	IA	0.000	0.000	0.000	0.000	0.000	0.000	0.000	0.000	0.000	0.000	0.000	0.000	0.000	0.000
	IB	0.000	0.000	0.000	0.000	0.000	0.000	0.000	0.000	0.000	0.000	0.000	0.000	0.000	0.000
	IC	0.060	0.061	0.061	0.061	0.062	0.062	0.062	0.063	0.063	0.063	0.064	0.065	0.065	0.066
	ID	0.933	0.933	0.933	0.933	0.933	0.933	0.933	0.933	0.933	0.933	0.933	0.933	0.933	0.933
-0.4	IA	0.000	0.000	0.000	0.000	0.000	0.000	0.000	0.000	0.000	0.000	0.000	0.000	0.000	0.000
	IB	0.000	0.000	0.000	0.000	0.000	0.000	0.000	0.000	0.000	0.000	0.000	0.000	0.000	0.000
	IC	0.103	0.103	0.104	0.105	0.105	0.106	0.106	0.107	0.108	0.109	0.109	0.111	0.112	0.114
	ID	0.885	0.885	0.885	0.885	0.885	0.885	0.885	0.885	0.885	0.885	0.885	0.885	0.885	0.885
-0.3	IA	0.000	0.000	0.000	0.000	0.000	0.000	0.000	0.000	0.000	0.000	0.000	0.000	0.000	0.000
	IB	0.000	0.000	0.000	0.000	0.000	0.000	0.000	0.000	0.000	0.000	0.000	0.000	0.000	0.000
	IC	0.162	0.163	0.164	0.165	0.166	0.167	0.168	0.170	0.171	0.173	0.173	0.176	0.179	0.182
	ID	0.816	0.816	0.816	0.816	0.816	0.816	0.816	0.816	0.816	0.816	0.816	0.816	0.816	0.816
-0.2	IA	0.000	0.000	0.000	0.000	0.000	0.000	0.000	0.000	0.000	0.000	0.000	0.000	0.000	0.000
	IB	0.000	0.000	0.000	0.000	0.000	0.000	0.000	0.000	0.000	0.000	0.000	0.000	0.000	0.000
	IC	0.238	0.239	0.241	0.243	0.245	0.246	0.248	0.250	0.253	0.255	0.256	0.261	0.266	0.270
	ID	0.726	0.726	0.726	0.726	0.726	0.726	0.726	0.726	0.726	0.726	0.726	0.726	0.726	0.726
-0.1	IA	0.000	0.000	0.000	0.000	0.000	0.000	0.000	0.000	0.000	0.000	0.000	0.000	0.000	0.000
	IB	0.000	0.000	0.000	0.000	0.000	0.000	0.000	0.000	0.000	0.000	0.000	0.000	0.000	0.000
	IC	0.325	0.327	0.330	0.333	0.336	0.338	0.340	0.345	0.348	0.351	0.354	0.361	0.368	0.375
	ID	0.618	0.618	0.618	0.618	0.618	0.618	0.618	0.618	0.618	0.618	0.618	0.618	0.618	0.618
0.0	IA	0.000	0.000	0.000	0.000	0.000	0.000	0.000	0.000	0.000	0.000	0.000	0.000	0.000	0.000
	IB	0.000	0.000	0.000	0.000	0.000	0.000	0.000	0.000	0.000	0.000	0.000	0.000	0.000	0.000
	IC	0.415	0.418	0.423	0.427	0.431	0.434	0.437	0.444	0.449	0.454	0.458	0.468	0.479	0.488
	ID	0.500	0.500	0.500	0.500	0.500	0.500	0.500	0.500	0.500	0.500	0.500	0.500	0.500	0.500
0.1	IA	0.000	0.000	0.000	0.000	0.000	0.000	0.000	0.000	0.000	0.000	0.000	0.000	0.000	0.000
	IB	0.000	0.000	0.000	0.000	0.000	0.000	0.000	0.000	0.000	0.000	0.000	0.000	0.000	0.000
	IC	0.498	0.502	0.509	0.515	0.520	0.525	0.529	0.539	0.546	0.553	0.558	0.573	0.588	0.601
	ID	0.382	0.382	0.382	0.382	0.382	0.382	0.382	0.382	0.382	0.382	0.382	0.382	0.382	0.382
0.2	IA	0.000	0.000	0.000	0.000	0.000	0.000	0.000	0.000	0.000	0.000	0.000	0.000	0.000	0.000
	IB	0.000	0.000	0.000	0.000	0.000	0.000	0.000	0.000	0.000	0.000	0.000	0.000	0.000	0.000
	IC	0.565	0.570	0.579	0.587	0.594	0.600	0.606	0.619	0.629	0.637	0.644	0.664	0.685	0.702
	ID	0.274	0.274	0.274	0.274	0.274	0.274	0.274	0.274	0.274	0.274	0.274	0.274	0.274	0.274
0.3	IA	0.000	0.000	0.000	0.000	0.000	0.000	0.000	0.000	0.000	0.000	0.000	0.000	0.000	0.000
	IB	0.000	0.000	0.000	0.000	0.000	0.000	0.000	0.000	0.000	0.000	0.000	0.000	0.000	0.000
	IC	0.609	0.615	0.626	0.636	0.646	0.654	0.661	0.677	0.690	0.701	0.710	0.736	0.762	0.784
	ID	0.184	0.184	0.184	0.184	0.184	0.184	0.184	0.184	0.184	0.184	0.184	0.184	0.184	0.184
0.4	IA	0.000	0.000	0.000	0.000	0.000	0.000	0.000	0.000	0.000	0.000	0.000	0.000	0.000	0.000
	IB	0.000	0.000	0.000	0.000	0.000	0.000	0.000	0.000	0.000	0.000	0.000	0.000	0.000	0.000
	IC	0.628	0.635	0.649	0.662	0.673	0.683	0.692	0.712	0.728	0.742	0.753	0.785	0.817	0.844
	ID	0.115	0.115	0.115	0.115	0.115	0.115	0.115	0.115	0.115	0.115	0.115	0.115	0.115	0.115
0.5	IA	0.000	0.000	0.000	0.000	0.000	0.000	0.000	0.000	0.000	0.000	0.000	0.000	0.000	0.000
	IB	0.000	0.000	0.000	0.000	0.000	0.000	0.000	0.000	0.000	0.000	0.000	0.000	0.000	0.000
	IC	0.624	0.633	0.650	0.664	0.678	0.690	0.701	0.724	0.744	0.760	0.773	0.812	0.851	0.884
	ID	0.067	0.067	0.067	0.067	0.067	0.067	0.067	0.067	0.067	0.067	0.067	0.067	0.067	0.067
0.6	IA	0.000	0.000	0.000	0.000	0.000	0.000	0.000	0.000	0.000	0.000	0.000	0.000	0.000	0.000
	IB	0.000	0.000	0.000	0.000	0.000	0.000	0.000	0.000	0.000	0.000	0.000	0.000	0.000	0.000
	IC	0.603	0.613	0.632	0.649	0.664	0.678	0.691	0.719	0.741	0.760	0.776	0.820	0.867	0.905
	ID	0.036	0.036	0.036	0.036	0.036	0.036	0.036	0.036	0.036	0.036	0.036	0.036	0.036	0.036

Table A.9: Normal Theory Values of the Integrals IA, IB, IC, and ID

C_{pk}		C_p values 0.10	0.20	0.30	0.40	0.50	0.60	0.70	0.80	0.90	1.00	1.10	1.20	1.30	1.40
0.7	IA							0.018	0.003	0.000	0.000	0.000	0.000	0.000	0.000
	IB							0.088	0.048	0.023	0.010	0.004	0.002	0.001	0.000
	IC	0.005	0.012	0.021	0.033	0.049	0.067	0.088	0.113	0.140	0.168	0.198	0.227	0.257	0.285
	ID	0.018	0.018	0.018	0.018	0.018	0.018	0.018	0.018	0.018	0.018	0.018	0.018	0.018	0.018
0.8	IA								0.008	0.001	0.000	0.000	0.000	0.000	0.000
	IB								0.076	0.040	0.019	0.009	0.004	0.002	0.001
	IC	0.003	0.007	0.012	0.019	0.029	0.042	0.057	0.076	0.098	0.122	0.148	0.175	0.202	0.230
	ID	0.008	0.008	0.008	0.008	0.008	0.008	0.008	0.008	0.008	0.008	0.008	0.008	0.008	0.008
0.9	IA									0.003	0.000	0.000	0.000	0.000	0.000
	IB									0.064	0.033	0.016	0.007	0.003	0.001
	IC	0.001	0.003	0.006	0.010	0.016	0.024	0.035	0.048	0.064	0.083	0.105	0.128	0.153	0.179
	ID	0.003	0.003	0.003	0.003	0.003	0.003	0.003	0.003	0.003	0.003	0.003	0.003	0.003	0.003
1.0	IA										0.001	0.000	0.000	0.000	0.000
	IB										0.054	0.028	0.013	0.006	0.003
	IC	0.001	0.001	0.003	0.005	0.008	0.013	0.019	0.028	0.040	0.054	0.071	0.090	0.111	0.134
	ID	0.001	0.001	0.001	0.001	0.001	0.001	0.001	0.001	0.001	0.001	0.001	0.001	0.001	0.001
1.1	IA											0.000	0.000	0.000	0.000
	IB											0.045	0.023	0.011	0.005
	IC	0.000	0.001	0.001	0.002	0.004	0.006	0.010	0.016	0.023	0.033	0.045	0.060	0.077	0.097
	ID	0.000	0.000	0.000	0.000	0.000	0.000	0.000	0.000	0.000	0.000	0.000	0.000	0.000	0.000
1.2	IA												0.000	0.000	0.000
	IB												0.038	0.020	0.010
	IC			0.000	0.001	0.002	0.003	0.005	0.008	0.013	0.019	0.028	0.038	0.052	0.067
	ID			0.000	0.000	0.000	0.000	0.000	0.000	0.000	0.000	0.000	0.000	0.000	0.000
1.3	IA													0.000	0.000
	IB													0.033	0.017
	IC				0.000	0.001	0.001	0.002	0.004	0.006	0.010	0.016	0.023	0.033	0.045
	ID				0.000	0.000	0.000	0.000	0.000	0.000	0.000	0.000	0.000	0.000	0.000
1.4	IA														0.000
	IB														0.028
	IC						0.000	0.001	0.002	0.003	0.005	0.009	0.013	0.020	0.028
	ID						0.000	0.000	0.000	0.000	0.000	0.000	0.000	0.000	0.000
1.5	IA														
	IB														
	IC							0.000	0.001	0.001	0.003	0.004	0.007	0.011	0.017
	ID							0.000	0.000	0.000	0.000	0.000	0.000	0.000	0.000
1.6	IA														
	IB														
	IC								0.000	0.001	0.001	0.002	0.004	0.006	0.010
	ID								0.000	0.000	0.000	0.000	0.000	0.000	0.000
1.7	IA														
	IB														
	IC										0.000	0.001	0.002	0.003	0.005
	ID										0.000	0.000	0.000	0.000	0.000
1.8	IA														
	IB														
	IC											0.000	0.001	0.002	0.003
	ID											0.000	0.000	0.000	0.000

Table A.9: Normal Theory Values of the Integrals IA, IB, IC, and ID

C_{pk}		C_p values 1.50	1.60	1.70	1.80	1.90	2.00	2.10	2.20	2.30	2.40	2.50	2.60	2.70	2.80
0.7	**IA**	0.000	0.000	0.000	0.000	0.000	0.000	0.000	0.000	0.000	0.000	0.000	0.000	0.000	0.000
	IB	0.000	0.000	0.000	0.000	0.000	0.000	0.000	0.000	0.000	0.000	0.000	0.000	0.000	0.000
	IC	0.313	0.339	0.364	0.387	0.409	0.430	0.450	0.468	0.485	0.501	0.517	0.531	0.544	0.557
	ID	0.018	0.018	0.018	0.018	0.018	0.018	0.018	0.018	0.018	0.018	0.018	0.018	0.018	0.018
0.8	**IA**	0.000	0.000	0.000	0.000	0.000	0.000	0.000	0.000	0.000	0.000	0.000	0.000	0.000	0.000
	IB	0.000	0.000	0.000	0.000	0.000	0.000	0.000	0.000	0.000	0.000	0.000	0.000	0.000	0.000
	IC	0.257	0.284	0.309	0.334	0.357	0.379	0.399	0.419	0.437	0.455	0.471	0.487	0.502	0.516
	ID	0.008	0.008	0.008	0.008	0.008	0.008	0.008	0.008	0.008	0.008	0.008	0.008	0.008	0.008
0.9	**IA**	0.000	0.000	0.000	0.000	0.000	0.000	0.000	0.000	0.000	0.000	0.000	0.000	0.000	0.000
	IB	0.001	0.000	0.000	0.000	0.000	0.000	0.000	0.000	0.000	0.000	0.000	0.000	0.000	0.000
	IC	0.205	0.231	0.256	0.280	0.304	0.326	0.348	0.368	0.388	0.406	0.424	0.440	0.456	0.471
	ID	0.003	0.003	0.003	0.003	0.003	0.003	0.003	0.003	0.003	0.003	0.003	0.003	0.003	0.003
1.0	**IA**	0.000	0.000	0.000	0.000	0.000	0.000	0.000	0.000	0.000	0.000	0.000	0.000	0.000	0.000
	IB	0.001	0.000	0.000	0.000	0.000	0.000	0.000	0.000	0.000	0.000	0.000	0.000	0.000	0.000
	IC	0.158	0.182	0.206	0.230	0.254	0.276	0.298	0.319	0.339	0.358	0.376	0.394	0.410	0.426
	ID	0.001	0.001	0.001	0.001	0.001	0.001	0.001	0.001	0.001	0.001	0.001	0.001	0.001	0.001
1.1	**IA**	0.000	0.000	0.000	0.000	0.000	0.000	0.000	0.000	0.000	0.000	0.000	0.000	0.000	0.000
	IB	0.002	0.001	0.000	0.000	0.000	0.000	0.000	0.000	0.000	0.000	0.000	0.000	0.000	0.000
	IC	0.118	0.140	0.162	0.185	0.207	0.230	0.251	0.272	0.293	0.312	0.331	0.349	0.366	0.382
	ID	0.000	0.000	0.000	0.000	0.000	0.000	0.000	0.000	0.000	0.000	0.000	0.000	0.000	0.000
1.2	**IA**	0.000	0.000	0.000	0.000	0.000	0.000	0.000	0.000	0.000	0.000	0.000	0.000	0.000	0.000
	IB	0.005	0.002	0.001	0.000	0.000	0.000	0.000	0.000	0.000	0.000	0.000	0.000	0.000	0.000
	IC	0.085	0.104	0.124	0.145	0.166	0.188	0.209	0.229	0.250	0.269	0.288	0.306	0.324	0.341
	ID	0.000	0.000	0.000	0.000	0.000	0.000	0.000	0.000	0.000	0.000	0.000	0.000	0.000	0.000
1.3	**IA**	0.000	0.000	0.000	0.000	0.000	0.000	0.000	0.000	0.000	0.000	0.000	0.000	0.000	0.000
	IB	0.009	0.004	0.002	0.001	0.000	0.000	0.000	0.000	0.000	0.000	0.000	0.000	0.000	0.000
	IC	0.059	0.074	0.092	0.111	0.130	0.150	0.170	0.190	0.210	0.229	0.248	0.266	0.284	0.301
	ID	0.000	0.000	0.000	0.000	0.000	0.000	0.000	0.000	0.000	0.000	0.000	0.000	0.000	0.000
1.4	**IA**	0.000	0.000	0.000	0.000	0.000	0.000	0.000	0.000	0.000	0.000	0.000	0.000	0.000	0.000
	IB	0.015	0.008	0.004	0.002	0.001	0.000	0.000	0.000	0.000	0.000	0.000	0.000	0.000	0.000
	IC	0.039	0.052	0.066	0.082	0.099	0.117	0.136	0.155	0.174	0.193	0.211	0.229	0.247	0.264
	ID	0.000	0.000	0.000	0.000	0.000	0.000	0.000	0.000	0.000	0.000	0.000	0.000	0.000	0.000
1.5	**IA**	0.000	0.000	0.000	0.000	0.000	0.000	0.000	0.000	0.000	0.000	0.000	0.000	0.000	0.000
	IB	0.025	0.013	0.007	0.003	0.001	0.001	0.000	0.000	0.000	0.000	0.000	0.000	0.000	0.000
	IC	0.025	0.034	0.046	0.059	0.074	0.090	0.107	0.124	0.142	0.160	0.178	0.195	0.213	0.230
	ID	0.000	0.000	0.000	0.000	0.000	0.000	0.000	0.000	0.000	0.000	0.000	0.000	0.000	0.000
1.6	**IA**		0.000	0.000	0.000	0.000	0.000	0.000	0.000	0.000	0.000	0.000	0.000	0.000	0.000
	IB		0.022	0.012	0.006	0.003	0.001	0.001	0.000	0.000	0.000	0.000	0.000	0.000	0.000
	IC	0.015	0.022	0.030	0.041	0.053	0.066	0.081	0.097	0.114	0.130	0.147	0.164	0.181	0.198
	ID	0.000	0.000	0.000	0.000	0.000	0.000	0.000	0.000	0.000	0.000	0.000	0.000	0.000	0.000
1.7	**IA**			0.000	0.000	0.000	0.000	0.000	0.000	0.000	0.000	0.000	0.000	0.000	0.000
	IB			0.019	0.010	0.005	0.003	0.001	0.001	0.000	0.000	0.000	0.000	0.000	0.000
	IC	0.009	0.013	0.019	0.027	0.037	0.048	0.060	0.074	0.089	0.104	0.120	0.136	0.152	0.169
	ID	0.000	0.000	0.000	0.000	0.000	0.000	0.000	0.000	0.000	0.000	0.000	0.000	0.000	0.000
1.8	**IA**				0.000	0.000	0.000	0.000	0.000	0.000	0.000	0.000	0.000	0.000	0.000
	IB				0.017	0.009	0.005	0.002	0.001	0.000	0.000	0.000	0.000	0.000	0.000
	IC	0.005	0.008	0.012	0.017	0.024	0.033	0.043	0.055	0.068	0.082	0.096	0.111	0.126	0.142
	ID	0.000	0.000	0.000	0.000	0.000	0.000	0.000	0.000	0.000	0.000	0.000	0.000	0.000	0.000

Table A.9: Normal Theory Values of the Integrals IA, IB, IC, and ID

C_{pk}		C_p values 2.90	3.00	3.20	3.40	3.60	3.80	4.00	4.50	5.00	5.50	6.0	8.0	12.0	20.0
0.7	IA	0.000	0.000	0.000	0.000	0.000	0.000	0.000	0.000	0.000	0.000	0.000	0.000	0.000	0.000
	IB	0.000	0.000	0.000	0.000	0.000	0.000	0.000	0.000	0.000	0.000	0.000	0.000	0.000	0.000
	IC	0.569	0.581	0.602	0.621	0.638	0.654	0.669	0.700	0.725	0.747	0.765	0.816	0.869	0.914
	ID	0.018	0.018	0.018	0.018	0.018	0.018	0.018	0.018	0.018	0.018	0.018	0.018	0.018	0.018
0.8	IA	0.000	0.000	0.000	0.000	0.000	0.000	0.000	0.000	0.000	0.000	0.000	0.000	0.000	0.000
	IB	0.000	0.000	0.000	0.000	0.000	0.000	0.000	0.000	0.000	0.000	0.000	0.000	0.000	0.000
	IC	0.529	0.541	0.565	0.586	0.605	0.622	0.638	0.673	0.701	0.725	0.746	0.803	0.864	0.914
	ID	0.008	0.008	0.008	0.008	0.008	0.008	0.008	0.008	0.008	0.008	0.008	0.008	0.008	0.008
0.9	IA	0.000	0.000	0.000	0.000	0.000	0.000	0.000	0.000	0.000	0.000	0.000	0.000	0.000	0.000
	IB	0.000	0.000	0.000	0.000	0.000	0.000	0.000	0.000	0.000	0.000	0.000	0.000	0.000	0.000
	IC	0.485	0.499	0.524	0.547	0.567	0.586	0.604	0.642	0.673	0.700	0.722	0.786	0.853	0.909
	ID	0.003	0.003	0.003	0.003	0.003	0.003	0.003	0.003	0.003	0.003	0.003	0.003	0.003	0.003
1.0	IA	0.000	0.000	0.000	0.000	0.000	0.000	0.000	0.000	0.000	0.000	0.000	0.000	0.000	0.000
	IB	0.000	0.000	0.000	0.000	0.000	0.000	0.000	0.000	0.000	0.000	0.000	0.000	0.000	0.000
	IC	0.441	0.455	0.482	0.506	0.529	0.549	0.568	0.609	0.643	0.672	0.696	0.766	0.840	0.901
	ID	0.001	0.001	0.001	0.001	0.001	0.001	0.001	0.001	0.001	0.001	0.001	0.001	0.001	0.001
1.1	IA	0.000	0.000	0.000	0.000	0.000	0.000	0.000	0.000	0.000	0.000	0.000	0.000	0.000	0.000
	IB	0.000	0.000	0.000	0.000	0.000	0.000	0.000	0.000	0.000	0.000	0.000	0.000	0.000	0.000
	IC	0.398	0.413	0.441	0.467	0.490	0.512	0.532	0.576	0.612	0.643	0.670	0.745	0.825	0.893
	ID	0.000	0.000	0.000	0.000	0.000	0.000	0.000	0.000	0.000	0.000	0.000	0.000	0.000	0.000
1.2	IA	0.000	0.000	0.000	0.000	0.000	0.000	0.000	0.000	0.000	0.000	0.000	0.000	0.000	0.000
	IB	0.000	0.000	0.000	0.000	0.000	0.000	0.000	0.000	0.000	0.000	0.000	0.000	0.000	0.000
	IC	0.357	0.372	0.401	0.428	0.453	0.476	0.497	0.543	0.582	0.615	0.643	0.724	0.811	0.884
	ID	0.000	0.000	0.000	0.000	0.000	0.000	0.000	0.000	0.000	0.000	0.000	0.000	0.000	0.000
1.3	IA	0.000	0.000	0.000	0.000	0.000	0.000	0.000	0.000	0.000	0.000	0.000	0.000	0.000	0.000
	IB	0.000	0.000	0.000	0.000	0.000	0.000	0.000	0.000	0.000	0.000	0.000	0.000	0.000	0.000
	IC	0.318	0.333	0.363	0.391	0.417	0.440	0.463	0.511	0.552	0.587	0.617	0.703	0.796	0.874
	ID	0.000	0.000	0.000	0.000	0.000	0.000	0.000	0.000	0.000	0.000	0.000	0.000	0.000	0.000
1.4	IA	0.000	0.000	0.000	0.000	0.000	0.000	0.000	0.000	0.000	0.000	0.000	0.000	0.000	0.000
	IB	0.000	0.000	0.000	0.000	0.000	0.000	0.000	0.000	0.000	0.000	0.000	0.000	0.000	0.000
	IC	0.281	0.297	0.327	0.356	0.382	0.407	0.429	0.480	0.523	0.559	0.591	0.682	0.781	0.865
	ID	0.000	0.000	0.000	0.000	0.000	0.000	0.000	0.000	0.000	0.000	0.000	0.000	0.000	0.000
1.5	IA	0.000	0.000	0.000	0.000	0.000	0.000	0.000	0.000	0.000	0.000	0.000	0.000	0.000	0.000
	IB	0.000	0.000	0.000	0.000	0.000	0.000	0.000	0.000	0.000	0.000	0.000	0.000	0.000	0.000
	IC	0.246	0.262	0.293	0.322	0.349	0.374	0.398	0.450	0.494	0.533	0.566	0.662	0.766	0.856
	ID	0.000	0.000	0.000	0.000	0.000	0.000	0.000	0.000	0.000	0.000	0.000	0.000	0.000	0.000
1.6	IA	0.000	0.000	0.000	0.000	0.000	0.000	0.000	0.000	0.000	0.000	0.000	0.000	0.000	0.000
	IB	0.000	0.000	0.000	0.000	0.000	0.000	0.000	0.000	0.000	0.000	0.000	0.000	0.000	0.000
	IC	0.214	0.230	0.261	0.290	0.317	0.343	0.367	0.421	0.467	0.506	0.541	0.642	0.752	0.847
	ID	0.000	0.000	0.000	0.000	0.000	0.000	0.000	0.000	0.000	0.000	0.000	0.000	0.000	0.000
1.7	IA	0.000	0.000	0.000	0.000	0.000	0.000	0.000	0.000	0.000	0.000	0.000	0.000	0.000	0.000
	IB	0.000	0.000	0.000	0.000	0.000	0.000	0.000	0.000	0.000	0.000	0.000	0.000	0.000	0.000
	IC	0.184	0.200	0.231	0.260	0.287	0.313	0.338	0.393	0.440	0.481	0.517	0.622	0.738	0.838
	ID	0.000	0.000	0.000	0.000	0.000	0.000	0.000	0.000	0.000	0.000	0.000	0.000	0.000	0.000
1.8	IA	0.000	0.000	0.000	0.000	0.000	0.000	0.000	0.000	0.000	0.000	0.000	0.000	0.000	0.000
	IB	0.000	0.000	0.000	0.000	0.000	0.000	0.000	0.000	0.000	0.000	0.000	0.000	0.000	0.000
	IC	0.157	0.172	0.202	0.231	0.259	0.285	0.309	0.365	0.414	0.456	0.493	0.602	0.723	0.828
	ID	0.000	0.000	0.000	0.000	0.000	0.000	0.000	0.000	0.000	0.000	0.000	0.000	0.000	0.000

Table A.9: Normal Theory Values of the Integrals IA, IB, IC, and ID

C_{pk}		C_p values 1.50	1.60	1.70	1.80	1.90	2.00	2.10	2.20	2.30	2.40	2.50	2.60	2.70	2.80
1.9	IA					0.000	0.000	0.000	0.000	0.000	0.000	0.000	0.000	0.000	0.000
	IB					0.015	0.008	0.004	0.002	0.001	0.000	0.000	0.000	0.000	0.000
	IC	0.002	0.004	0.007	0.010	0.015	0.022	0.030	0.039	0.050	0.062	0.075	0.089	0.103	0.117
	ID	0.000	0.000	0.000	0.000	0.000	0.000	0.000	0.000	0.000	0.000	0.000	0.000	0.000	0.000
2.0	IA						0.000	0.000	0.000	0.000	0.000	0.000	0.000	0.000	0.000
	IB						0.014	0.008	0.004	0.002	0.001	0.000	0.000	0.000	0.000
	IC	0.001	0.002	0.004	0.006	0.009	0.014	0.020	0.027	0.036	0.046	0.057	0.070	0.082	0.096
	ID	0.000	0.000	0.000	0.000	0.000	0.000	0.000	0.000	0.000	0.000	0.000	0.000	0.000	0.000
2.1	IA							0.000	0.000	0.000	0.000	0.000	0.000	0.000	0.000
	IB							0.013	0.007	0.004	0.002	0.001	0.000	0.000	0.000
	IC	0.001	0.001	0.002	0.003	0.005	0.008	0.013	0.018	0.025	0.033	0.043	0.053	0.064	0.077
	ID	0.000	0.000	0.000	0.000	0.000	0.000	0.000	0.000	0.000	0.000	0.000	0.000	0.000	0.000
2.2	IA								0.000	0.000	0.000	0.000	0.000	0.000	0.000
	IB								0.011	0.006	0.003	0.002	0.001	0.000	0.000
	IC		0.000	0.001	0.002	0.003	0.005	0.008	0.011	0.017	0.023	0.031	0.039	0.049	0.060
	ID		0.000	0.000	0.000	0.000	0.000	0.000	0.000	0.000	0.000	0.000	0.000	0.000	0.000
2.3	IA									0.000	0.000	0.000	0.000	0.000	0.000
	IB									0.011	0.006	0.003	0.002	0.001	0.000
	IC			0.000	0.001	0.001	0.003	0.004	0.007	0.011	0.015	0.021	0.028	0.036	0.046
	ID			0.000	0.000	0.000	0.000	0.000	0.000	0.000	0.000	0.000	0.000	0.000	0.000
2.4	IA										0.000	0.000	0.000	0.000	0.000
	IB										0.010	0.005	0.003	0.001	0.001
	IC				0.000	0.001	0.001	0.000	0.004	0.006	0.010	0.014	0.020	0.026	0.034
	ID				0.000	0.000	0.000	0.000	0.000	0.000	0.000	0.000	0.000	0.000	0.000
2.5	IA											0.000	0.000	0.000	0.000
	IB											0.009	0.005	0.003	0.001
	IC					0.000	0.001	0.001	0.002	0.004	0.006	0.009	0.013	0.018	0.024
	ID					0.000	0.000	0.000	0.000	0.000	0.000	0.000	0.000	0.000	0.000
2.6	IA												0.000	0.000	0.000
	IB												0.008	0.005	0.002
	IC						0.000	0.001	0.001	0.002	0.003	0.005	0.008	0.012	0.017
	ID						0.000	0.000	0.000	0.000	0.000	0.000	0.000	0.000	0.000
2.7	IA													0.000	0.000
	IB													0.008	0.004
	IC								0.000	0.000	0.002	0.003	0.005	0.008	0.011
	ID								0.000	0.000	0.000	0.000	0.000	0.000	0.000
2.8	IA														0.000
	IB														0.007
	IC									0.000	0.001	0.002	0.003	0.005	0.007
	ID									0.000	0.000	0.000	0.000	0.000	0.000
2.9	IA														
	IB														
	IC										0.000	0.001	0.002	0.003	0.004
	ID										0.000	0.000	0.000	0.000	0.000
3.0	IA														
	IB														
	IC											0.000	0.001	0.001	0.002
	ID											0.000	0.000	0.000	0.000

Table A.9: Normal Theory Values of the Integrals IA, IB, IC, and ID

C_{pk}		C_p values 2.90	3.00	3.20	3.40	3.60	3.80	4.00	4.50	5.00	5.50	6.0	8.0	12.0	20.0
1.9	**IA**	0.000	0.000	0.000	0.000	0.000	0.000	0.000	0.000	0.000	0.000	0.000	0.000	0.000	0.000
	IB	0.000	0.000	0.000	0.000	0.000	0.000	0.000	0.000	0.000	0.000	0.000	0.000	0.000	0.000
	IC	0.132	0.147	0.176	0.204	0.232	0.258	0.283	0.339	0.389	0.432	0.470	0.583	0.709	0.819
	ID	0.000	0.000	0.000	0.000	0.000	0.000	0.000	0.000	0.000	0.000	0.000	0.000	0.000	0.000
2.0	**IA**	0.000	0.000	0.000	0.000	0.000	0.000	0.000	0.000	0.000	0.000	0.000	0.000	0.000	0.000
	IB	0.000	0.000	0.000	0.000	0.000	0.000	0.000	0.000	0.000	0.000	0.000	0.000	0.000	0.000
	IC	0.110	0.123	0.151	0.179	0.206	0.232	0.257	0.314	0.364	0.409	0.448	0.564	0.695	0.810
	ID	0.000	0.000	0.000	0.000	0.000	0.000	0.000	0.000	0.000	0.000	0.000	0.000	0.000	0.000
2.1	**IA**	0.000	0.000	0.000	0.000	0.000	0.000	0.000	0.000	0.000	0.000	0.000	0.000	0.000	0.000
	IB	0.000	0.000	0.000	0.000	0.000	0.000	0.000	0.000	0.000	0.000	0.000	0.000	0.000	0.000
	IC	0.089	0.102	0.129	0.156	0.182	0.208	0.233	0.290	0.341	0.386	0.426	0.546	0.681	0.801
	ID	0.000	0.000	0.000	0.000	0.000	0.000	0.000	0.000	0.000	0.000	0.000	0.000	0.000	0.000
2.2	**IA**	0.000	0.000	0.000	0.000	0.000	0.000	0.000	0.000	0.000	0.000	0.000	0.000	0.000	0.000
	IB	0.000	0.000	0.000	0.000	0.000	0.000	0.000	0.000	0.000	0.000	0.000	0.000	0.000	0.000
	IC	0.071	0.083	0.109	0.134	0.160	0.185	0.209	0.267	0.318	0.364	0.404	0.527	0.668	0.792
	ID	0.000	0.000	0.000	0.000	0.000	0.000	0.000	0.000	0.000	0.000	0.000	0.000	0.000	0.000
2.3	**IA**	0.000	0.000	0.000	0.000	0.000	0.000	0.000	0.000	0.000	0.000	0.000	0.000	0.000	0.000
	IB	0.000	0.000	0.000	0.000	0.000	0.000	0.000	0.000	0.000	0.000	0.000	0.000	0.000	0.000
	IC	0.056	0.067	0.090	0.114	0.139	0.164	0.188	0.244	0.296	0.342	0.383	0.509	0.654	0.784
	ID	0.000	0.000	0.000	0.000	0.000	0.000	0.000	0.000	0.000	0.000	0.000	0.000	0.000	0.000
2.4	**IA**	0.000	0.000	0.000	0.000	0.000	0.000	0.000	0.000	0.000	0.000	0.000	0.000	0.000	0.000
	IB	0.000	0.000	0.000	0.000	0.000	0.000	0.000	0.000	0.000	0.000	0.000	0.000	0.000	0.000
	IC	0.043	0.052	0.073	0.096	0.120	0.143	0.167	0.223	0.275	0.321	0.363	0.492	0.641	0.775
	ID	0.000	0.000	0.000	0.000	0.000	0.000	0.000	0.000	0.000	0.000	0.000	0.000	0.000	0.000
2.5	**IA**	0.000	0.000	0.000	0.000	0.000	0.000	0.000	0.000	0.000	0.000	0.000	0.000	0.000	0.000
	IB	0.001	0.000	0.000	0.000	0.000	0.000	0.000	0.000	0.000	0.000	0.000	0.000	0.000	0.000
	IC	0.032	0.040	0.059	0.080	0.102	0.125	0.148	0.203	0.254	0.301	0.343	0.474	0.628	0.766
	ID	0.000	0.000	0.000	0.000	0.000	0.000	0.000	0.000	0.000	0.000	0.000	0.000	0.000	0.000
2.6	**IA**	0.000	0.000	0.000	0.000	0.000	0.000	0.000	0.000	0.000	0.000	0.000	0.000	0.000	0.000
	IB	0.001	0.001	0.000	0.000	0.000	0.000	0.000	0.000	0.000	0.000	0.000	0.000	0.000	0.000
	IC	0.023	0.030	0.046	0.065	0.086	0.107	0.129	0.184	0.235	0.282	0.324	0.457	0.614	0.757
	ID	0.000	0.000	0.000	0.000	0.000	0.000	0.000	0.000	0.000	0.000	0.000	0.000	0.000	0.000
2.7	**IA**	0.000	0.000	0.000	0.000	0.000	0.000	0.000	0.000	0.000	0.000	0.000	0.000	0.000	0.000
	IB	0.002	0.001	0.000	0.000	0.000	0.000	0.000	0.000	0.000	0.000	0.000	0.000	0.000	0.000
	IC	0.016	0.021	0.035	0.052	0.071	0.091	0.113	0.165	0.216	0.263	0.306	0.441	0.601	0.749
	ID	0.000	0.000	0.000	0.000	0.000	0.000	0.000	0.000	0.000	0.000	0.000	0.000	0.000	0.000
2.8	**IA**	0.000	0.000	0.000	0.000	0.000	0.000	0.000	0.000	0.000	0.000	0.000	0.000	0.000	0.000
	IB	0.004	0.002	0.001	0.000	0.000	0.000	0.000	0.000	0.000	0.000	0.000	0.000	0.000	0.000
	IC	0.010	0.015	0.026	0.041	0.058	0.077	0.097	0.148	0.198	0.245	0.288	0.424	0.589	0.740
	ID	0.000	0.000	0.000	0.000	0.000	0.000	0.000	0.000	0.000	0.000	0.000	0.000	0.000	0.000
2.9	**IA**	0.000	0.000	0.000	0.000	0.000	0.000	0.000	0.000	0.000	0.000	0.000	0.000	0.000	0.000
	IB	0.007	0.004	0.001	0.000	0.000	0.000	0.000	0.000	0.000	0.000	0.000	0.000	0.000	0.000
	IC	0.007	0.010	0.019	0.031	0.046	0.064	0.083	0.132	0.181	0.227	0.270	0.408	0.576	0.731
	ID	0.000	0.000	0.000	0.000	0.000	0.000	0.000	0.000	0.000	0.000	0.000	0.000	0.000	0.000
3.0	**IA**		0.000	0.000	0.000	0.000	0.000	0.000	0.000	0.000	0.000	0.000	0.000	0.000	0.000
	IB		0.006	0.002	0.000	0.000	0.000	0.000	0.000	0.000	0.000	0.000	0.000	0.000	0.000
	IC	0.004	0.006	0.013	0.023	0.036	0.052	0.069	0.117	0.164	0.210	0.253	0.392	0.563	0.723
	ID	0.000	0.000	0.000	0.000	0.000	0.000	0.000	0.000	0.000	0.000	0.000	0.000	0.000	0.000

Table A.9: Normal Theory Values of the Integrals IA, IB, IC, and ID

C_{pk}		C_p values 3.20	3.40	3.60	3.80	4.00	4.50	5.00	5.50	6.0	8.0	12.0	20.0
3.2	**IA**	0.000	0.000	0.000	0.000	0.000	0.000	0.000	0.000	0.000	0.000	0.000	0.000
	IB	0.005	0.002	0.000	0.000	0.000	0.000	0.000	0.000	0.000	0.000	0.000	0.000
	IC	0.005	0.011	0.021	0.033	0.047	0.089	0.134	0.179	0.221	0.362	0.539	0.706
	ID	0.000	0.000	0.000	0.000	0.000	0.000	0.000	0.000	0.000	0.000	0.000	0.000
3.4	**IA**		0.000	0.000	0.000	0.000	0.000	0.000	0.000	0.000	0.000	0.000	0.000
	IB		0.005	0.001	0.000	0.000	0.000	0.000	0.000	0.000	0.000	0.000	0.000
	IC		0.005	0.010	0.018	0.029	0.065	0.107	0.149	0.191	0.332	0.514	0.689
	ID		0.000	0.000	0.000	0.000	0.000	0.000	0.000	0.000	0.000	0.000	0.000
3.6	**IA**			0.000	0.000	0.000	0.000	0.000	0.000	0.000	0.000	0.000	0.000
	IB			0.004	0.001	0.000	0.000	0.000	0.000	0.000	0.000	0.000	0.000
	IC			0.004	0.009	0.017	0.045	0.083	0.123	0.163	0.304	0.491	0.673
	ID			0.000	0.000	0.000	0.000	0.000	0.000	0.000	0.000	0.000	0.000
3.8	**IA**				0.000	0.000	0.000	0.000	0.000	0.000	0.000	0.000	0.000
	IB				0.004	0.001	0.000	0.000	0.000	0.000	0.000	0.000	0.000
	IC				0.004	0.008	0.030	0.062	0.099	0.138	0.277	0.468	0.656
	ID				0.000	0.000	0.000	0.000	0.000	0.000	0.000	0.000	0.000
4.0	**IA**					0.000	0.000	0.000	0.000	0.000	0.000	0.000	0.000
	IB					0.003	0.000	0.000	0.000	0.000	0.000	0.000	0.000
	IC					0.003	0.018	0.044	0.078	0.114	0.252	0.445	0.640
	ID					0.000	0.000	0.000	0.000	0.000	0.000	0.000	0.000
4.5	**IA**						0.000	0.000	0.000	0.000	0.000	0.000	0.000
	IB						0.003	0.000	0.000	0.000	0.000	0.000	0.000
	IC						0.003	0.014	0.037	0.066	0.193	0.391	0.601
	ID						0.000	0.000	0.000	0.000	0.000	0.000	0.000
5.0	**IA**							0.000	0.000	0.000	0.000	0.000	0.000
	IB							0.002	0.000	0.000	0.000	0.000	0.000
	IC							0.002	0.012	0.031	0.142	0.341	0.563
	ID							0.000	0.000	0.000	0.000	0.000	0.000
5.5	**IA**								0.000	0.000	0.000	0.000	0.000
	IB								0.002	0.000	0.000	0.000	0.000
	IC								0.002	0.010	0.099	0.294	0.526
	ID								0.000	0.000	0.000	0.000	0.000
6.0	**IA**									0.000	0.000	0.000	0.000
	IB									0.002	0.000	0.000	0.000
	IC									0.002	0.064	0.251	0.490
	ID									0.000	0.000	0.000	0.000
8.0	**IA**										0.000	0.000	0.000
	IB										0.001	0.000	0.000
	IC										0.001	0.112	0.360
	ID										0.000	0.000	0.000
12.0	**IA**											0.000	0.000
	IB											0.000	0.000
	IC											0.000	0.160
	ID											0.000	0.000
20.0	**IA**												0.000
	IB												0.000
	IC												0.000
	ID												0.000

Table A.10: *Burr* and Normal Values for the Integrals IC and ID

C_{pk}		*Tolerance values*													
		0.3	*0.6*	*0.9*	*1.2*	*1.5*	*1.8*	*2.1*	*2.4*	*2.7*	*3.0*	*3.3*	*3.6*	*3.9*	*4.2*
-0.5	**IC**	*0.013*	*0.024*	*0.032*	*0.037*	*0.042*	0.041	0.044	0.046	0.048	0.050	0.051	0.052	0.053	0.054
	ID	*0.973*	*0.955*	*0.944*	*0.937*	*0.932*	0.933	0.933	0.933	0.933	0.933	0.933	0.933	0.933	0.933
-0.4	**IC**	*0.030*	*0.046*	*0.057*	*0.064*	*0.070*	*0.074*	0.072	0.077	0.080	0.083	0.085	0.088	0.089	0.091
	ID	*0.931*	*0.904*	*0.891*	*0.885*	*0.882*	*0.879*	0.885	0.885	0.885	0.885	0.885	0.885	0.885	0.885
-0.3	**IC**	*0.058*	*0.079*	*0.092*	*0.101*	*0.108*	*0.113*	*0.118*	0.117	0.123	0.128	0.132	0.136	0.139	0.142
	ID	*0.842*	*0.813*	*0.804*	*0.803*	*0.804*	*0.805*	*0.807*	0.816	0.816	0.816	0.816	0.816	0.816	0.816
-0.2	**IC**	*0.091*	*0.114*	*0.130*	*0.142*	*0.152*	*0.159*	*0.166*	*0.171*	0.173	0.181	0.188	0.194	0.199	0.204
	ID	*0.683*	*0.670*	*0.678*	*0.688*	*0.697*	*0.704*	*0.711*	*0.716*	0.726	0.726	0.726	0.726	0.726	0.726
-0.1	**IC**		*0.133*	*0.156*	*0.175*	*0.190*	*0.203*	*0.214*	*0.223*	*0.230*	0.238	0.248	0.257	0.265	0.272
	ID		*0.486*	*0.519*	*0.546*	*0.566*	*0.581*	*0.593*	*0.602*	*0.610*	0.618	0.618	0.618	0.618	0.618
0	**IC**			*0.152*	*0.183*	*0.209*	*0.231*	*0.249*	*0.265*	*0.278*	*0.289*	0.304	0.317	0.328	0.338
	ID			*0.351*	*0.395*	*0.426*	*0.448*	*0.465*	*0.478*	*0.488*	*0.496*	0.500	0.500	0.500	0.500
0.1	**IC**				*0.157*	*0.197*	*0.232*	*0.261*	*0.286*	*0.308*	*0.327*	*0.343*	0.364	0.379	0.393
	ID				*0.256*	*0.296*	*0.323*	*0.342*	*0.357*	*0.367*	*0.376*	*0.383*	0.382	0.382	0.382
0.2	**IC**					*0.155*	*0.201*	*0.242*	*0.279*	*0.310*	*0.338*	*0.363*	*0.384*	0.410	0.427
	ID					*0.188*	*0.219*	*0.239*	*0.252*	*0.262*	*0.269*	*0.274*	*0.278*	0.274	0.274
0.3	**IC**						*0.149*	*0.198*	*0.243*	*0.285*	*0.321*	*0.354*	*0.383*	*0.409*	0.437
	ID						*0.140*	*0.161*	*0.173*	*0.180*	*0.184*	*0.187*	*0.188*	*0.189*	0.184
0.4	**IC**							*0.140*	*0.190*	*0.237*	*0.281*	*0.320*	*0.356*	*0.388*	*0.417*
	ID							*0.104*	*0.117*	*0.122*	*0.124*	*0.124*	*0.123*	*0.121*	*0.119*
0.5	**IC**								*0.130*	*0.179*	*0.226*	*0.269*	*0.309*	*0.346*	*0.379*
	ID								*0.078*	*0.084*	*0.085*	*0.083*	*0.080*	*0.076*	*0.073*
0.6	**IC**									*0.120*	*0.166*	*0.211*	*0.253*	*0.292*	*0.328*
	ID									*0.059*	*0.060*	*0.058*	*0.054*	*0.049*	*0.045*
0.7	**IC**										*0.110*	*0.153*	*0.195*	*0.234*	*0.270*
	ID										*0.044*	*0.043*	*0.039*	*0.034*	*0.029*
0.8	**IC**											*0.100*	*0.140*	*0.178*	*0.214*
	ID											*0.034*	*0.031*	*0.026*	*0.021*
0.9	**IC**												*0.092*	*0.128*	*0.162*
	ID												*0.026*	*0.022*	*0.017*
1.0	**IC**													*0.084*	*0.116*
	ID													*0.020*	*0.015*
1.1	**IC**														*0.076*
	ID														*0.015*

Italicized values obtained from a Burr distribution.
Non-italicized values obtained from the normal distribution.
Blanks correspond to situations that are either not feasible or impossible.

Table A.10: *Burr* and Normal Values for the Integrals IC and ID

C_{pk}		*Tolerance values* 4.5	4.8	5.1	5.4	5.7	6.0	6.3	6.6	6.9	7.2	7.5	7.8	8.1	8.4
-0.5	**IC**	0.055	0.056	0.056	0.057	0.057	0.058	0.058	0.059	0.059	0.059	0.059	0.060	0.060	0.060
	ID	0.933	0.933	0.933	0.933	0.933	0.933	0.933	0.933	0.933	0.933	0.933	0.933	0.933	0.933
-0.4	**IC**	0.093	0.094	0.095	0.096	0.097	0.098	0.099	0.099	0.100	0.100	0.101	0.102	0.102	0.102
	ID	0.885	0.885	0.885	0.885	0.885	0.885	0.885	0.885	0.885	0.885	0.885	0.885	0.885	0.885
-0.3	**IC**	0.144	0.146	0.148	0.150	0.152	0.153	0.155	0.156	0.157	0.158	0.159	0.160	0.161	0.162
	ID	0.816	0.816	0.816	0.816	0.816	0.816	0.816	0.816	0.816	0.816	0.816	0.816	0.816	0.816
-0.2	**IC**	0.208	0.212	0.215	0.218	0.221	0.223	0.225	0.227	0.229	0.231	0.233	0.234	0.235	0.237
	ID	0.726	0.726	0.726	0.726	0.726	0.726	0.726	0.726	0.726	0.726	0.726	0.726	0.726	0.726
-0.1	**IC**	0.279	0.284	0.289	0.294	0.298	0.302	0.305	0.308	0.311	0.314	0.317	0.319	0.321	0.323
	ID	0.618	0.618	0.618	0.618	0.618	0.618	0.618	0.618	0.618	0.618	0.618	0.618	0.618	0.618
0	**IC**	0.347	0.356	0.363	0.370	0.376	0.381	0.386	0.391	0.395	0.399	0.403	0.406	0.409	0.412
	ID	0.500	0.500	0.500	0.500	0.500	0.500	0.500	0.500	0.500	0.500	0.500	0.500	0.500	0.500
0.1	**IC**	0.405	0.416	0.426	0.435	0.443	0.451	0.458	0.464	0.470	0.476	0.481	0.486	0.490	0.494
	ID	0.382	0.382	0.382	0.382	0.382	0.382	0.382	0.382	0.382	0.382	0.382	0.382	0.382	0.382
0.2	**IC**	0.443	0.457	0.470	0.482	0.493	0.503	0.512	0.520	0.528	0.535	0.542	0.548	0.554	0.560
	ID	0.274	0.274	0.274	0.274	0.274	0.274	0.274	0.274	0.274	0.274	0.274	0.274	0.274	0.274
0.3	**IC**	0.456	0.474	0.490	0.504	0.518	0.530	0.542	0.552	0.562	0.571	0.580	0.588	0.595	0.602
	ID	0.184	0.184	0.184	0.184	0.184	0.184	0.184	0.184	0.184	0.184	0.184	0.184	0.184	0.184
0.4	**IC**	0.445	0.465	0.484	0.502	0.518	0.533	0.547	0.559	0.571	0.582	0.593	0.602	0.611	0.620
	ID	0.115	0.115	0.115	0.115	0.115	0.115	0.115	0.115	0.115	0.115	0.115	0.115	0.115	0.115
0.5	**IC**	*0.410*	0.436	0.458	0.477	0.496	0.513	0.529	0.544	0.558	0.571	0.583	0.594	0.605	0.615
	ID	*0.070*	0.067	0.067	0.067	0.067	0.067	0.067	0.067	0.067	0.067	0.067	0.067	0.067	0.067
0.6	**IC**	*0.360*	*0.390*	0.415	0.437	0.458	0.477	0.495	0.511	0.527	0.542	0.555	0.568	0.581	0.592
	ID	*0.041*	*0.037*	0.036	0.036	0.036	0.036	0.036	0.036	0.036	0.036	0.036	0.036	0.036	0.036
0.7	**IC**	*0.304*	*0.335*	*0.364*	0.387	0.409	0.430	0.450	0.468	0.485	0.501	0.517	0.531	0.544	0.557
	ID	*0.025*	*0.021*	*0.018*	0.018	0.018	0.018	0.018	0.018	0.018	0.018	0.018	0.018	0.018	0.018
0.8	**IC**	*0.247*	*0.278*	*0.307*	*0.334*	0.357	0.379	0.399	0.419	0.437	0.455	0.471	0.487	0.502	0.516
	ID	*0.016*	*0.013*	*0.010*	*0.008*	0.008	0.008	0.008	0.008	0.008	0.008	0.008	0.008	0.008	0.008
0.9	**IC**	*0.194*	*0.224*	*0.253*	*0.279*	*0.304*	0.326	0.348	0.368	0.388	0.406	0.424	0.440	0.456	0.471
	ID	*0.012*	*0.009*	*0.006*	*0.005*	*0.003*	0.003	0.003	0.003	0.003	0.003	0.003	0.003	0.003	0.003
1.0	**IC**	*0.147*	*0.175*	*0.203*	*0.229*	*0.253*	*0.276*	0.298	0.319	0.339	0.358	0.376	0.394	0.410	0.426
	ID	*0.011*	*0.007*	*0.005*	*0.003*	*0.002*	*0.001*	0.001	0.001	0.001	0.001	0.001	0.001	0.001	0.001
1.1	**IC**	*0.105*	*0.132*	*0.158*	*0.183*	*0.207*	*0.229*	*0.252*	0.272	0.293	0.312	0.331	0.349	0.366	0.382
	ID	*0.011*	*0.007*	*0.004*	*0.002*	*0.001*	*0.001*	*0.000*	0.000	0.000	0.000	0.000	0.000	0.000	0.000

Italicized values obtained from a Burr distribution.
Non-italicized values obtained from the normal distribution.

Table A.10: *Burr* and Normal Values for the Integrals IC and ID

C_{pk}		*8.7*	*9.0*	*9.3*	*9.6*	*9.9*	*10.2*	*10.5*	*10.8*	*11.1*	*11.4*	*11.7*	*12.0*
		Tolerance values											
-0.5	**IC**	0.060	0.061	0.061	0.061	0.061	0.061	0.062	0.062	0.062	0.062	0.062	0.062
	ID	0.933	0.933	0.933	0.933	0.933	0.933	0.933	0.933	0.933	0.933	0.933	0.933
-0.4	**IC**	0.103	0.103	0.104	0.104	0.104	0.105	0.105	0.105	0.105	0.106	0.106	0.106
	ID	0.885	0.885	0.885	0.885	0.885	0.885	0.885	0.885	0.885	0.885	0.885	0.885
-0.3	**IC**	0.162	0.163	0.164	0.164	0.165	0.165	0.166	0.166	0.167	0.167	0.168	0.168
	ID	0.816	0.816	0.816	0.816	0.816	0.816	0.816	0.816	0.816	0.816	0.816	0.816
-0.2	**IC**	0.238	0.239	0.240	0.241	0.242	0.243	0.244	0.245	0.245	0.246	0.247	0.248
	ID	0.726	0.726	0.726	0.726	0.726	0.726	0.726	0.726	0.726	0.726	0.726	0.726
-0.1	**IC**	0.325	0.327	0.328	0.330	0.332	0.333	0.334	0.336	0.337	0.338	0.339	0.340
	ID	0.618	0.618	0.618	0.618	0.618	0.618	0.618	0.618	0.618	0.618	0.618	0.618
0	**IC**	0.415	0.418	0.420	0.423	0.425	0.427	0.429	0.431	0.432	0.434	0.436	0.437
	ID	0.500	0.500	0.500	0.500	0.500	0.500	0.500	0.500	0.500	0.500	0.500	0.500
0.1	**IC**	0.498	0.502	0.505	0.509	0.512	0.515	0.517	0.520	0.522	0.525	0.527	0.529
	ID	0.382	0.382	0.382	0.382	0.382	0.382	0.382	0.382	0.382	0.382	0.382	0.382
0.2	**IC**	0.565	0.570	0.574	0.579	0.583	0.587	0.590	0.594	0.597	0.600	0.603	0.606
	ID	0.274	0.274	0.274	0.274	0.274	0.274	0.274	0.274	0.274	0.274	0.274	0.274
0.3	**IC**	0.609	0.615	0.621	0.626	0.632	0.636	0.641	0.646	0.650	0.654	0.658	0.661
	ID	0.184	0.184	0.184	0.184	0.184	0.184	0.184	0.184	0.184	0.184	0.184	0.184
0.4	**IC**	0.628	0.635	0.643	0.649	0.656	0.662	0.667	0.673	0.678	0.683	0.688	0.692
	ID	0.115	0.115	0.115	0.115	0.115	0.115	0.115	0.115	0.115	0.115	0.115	0.115
0.5	**IC**	0.624	0.633	0.642	0.650	0.657	0.664	0.671	0.678	0.684	0.690	0.695	0.701
	ID	0.067	0.067	0.067	0.067	0.067	0.067	0.067	0.067	0.067	0.067	0.067	0.067
0.6	**IC**	0.603	0.613	0.623	0.632	0.641	0.649	0.657	0.664	0.672	0.678	0.685	0.691
	ID	0.036	0.036	0.036	0.036	0.036	0.036	0.036	0.036	0.036	0.036	0.036	0.036
0.7	**IC**	0.569	0.581	0.592	0.602	0.612	0.621	0.630	0.638	0.646	0.654	0.662	0.669
	ID	0.018	0.018	0.018	0.018	0.018	0.018	0.018	0.018	0.018	0.018	0.018	0.018
0.8	**IC**	0.529	0.541	0.553	0.565	0.575	0.586	0.595	0.605	0.614	0.622	0.630	0.638
	ID	0.008	0.008	0.008	0.008	0.008	0.008	0.008	0.008	0.008	0.008	0.008	0.008
0.9	**IC**	0.485	0.499	0.512	0.524	0.535	0.547	0.557	0.567	0.577	0.586	0.595	0.604
	ID	0.003	0.003	0.003	0.003	0.003	0.003	0.003	0.003	0.003	0.003	0.003	0.003
1.0	**IC**	0.441	0.455	0.469	0.482	0.495	0.506	0.518	0.529	0.539	0.549	0.559	0.568
	ID	0.001	0.001	0.001	0.001	0.001	0.001	0.001	0.001	0.001	0.001	0.001	0.001
1.1	**IC**	0.398	0.413	0.427	0.441	0.454	0.467	0.479	0.490	0.501	0.512	0.522	0.532
	ID	0.000	0.000	0.000	0.000	0.000	0.000	0.000	0.000	0.000	0.000	0.000	0.000

Non-italicized values obtained from the normal distribution.

Table A.10: *Burr* and Normal Values for the Integrals IC and ID

C_{pk}		*Tolerance values* 4.5	4.8	5.1	5.4	5.7	6.0	6.3	6.6	6.9	7.2	7.5	7.8	8.1	8.4
1.2	**IC**	*0.070*	*0.096*	*0.120*	*0.143*	*0.165*	*0.187*	*0.209*	*0.229*	0.250	0.269	0.288	0.306	0.324	0.341
	ID	*0.012*	*0.007*	*0.004*	*0.002*	*0.001*	*0.001*	*0.000*	*0.000*	0.000	0.000	0.000	0.000	0.000	0.000
1.3	**IC**		*0.064*	*0.087*	*0.108*	*0.129*	*0.150*	*0.170*	*0.190*	*0.210*	0.229	0.248	0.266	0.284	0.301
	ID		*0.009*	*0.005*	*0.003*	*0.001*	*0.000*	*0.000*	*0.000*	*0.000*	0.000	0.000	0.000	0.000	0.000
1.4	**IC**			*0.058*	*0.079*	*0.098*	*0.117*	*0.136*	*0.155*	*0.174*	*0.193*	0.211	0.229	0.247	0.264
	ID			*0.007*	*0.004*	*0.002*	*0.001*	*0.000*	*0.000*	*0.000*	*0.000*	0.000	0.000	0.000	0.000
1.5	**IC**				*0.054*	*0.072*	*0.089*	*0.106*	*0.124*	*0.142*	*0.160*	*0.178*	0.195	0.213	0.230
	ID				*0.005*	*0.002*	*0.001*	*0.000*	*0.000*	*0.000*	*0.000*	*0.000*	0.000	0.000	0.000
1.6	**IC**					*0.049*	*0.066*	*0.081*	*0.097*	*0.114*	*0.130*	*0.147*	*0.164*	0.181	0.198
	ID					*0.004*	*0.002*	*0.001*	*0.000*	*0.000*	*0.000*	*0.000*	*0.000*	0.000	0.000
1.7	**IC**						*0.045*	*0.060*	*0.074*	*0.089*	*0.104*	*0.120*	*0.136*	*0.152*	0.169
	ID						*0.003*	*0.001*	*0.000*	*0.000*	*0.000*	*0.000*	*0.000*	*0.000*	0.000
1.8	**IC**							*0.042*	*0.055*	*0.068*	*0.082*	*0.096*	*0.111*	*0.126*	*0.142*
	ID							*0.003*	*0.001*	*0.000*	*0.000*	*0.000*	*0.000*	*0.000*	*0.000*
1.9	**IC**								*0.039*	*0.051*	*0.063*	*0.075*	*0.089*	*0.103*	*0.117*
	ID								*0.002*	*0.001*	*0.000*	*0.000*	*0.000*	*0.000*	*0.000*
2.0	**IC**									*0.036*	*0.047*	*0.058*	*0.070*	*0.082*	*0.096*
	ID									*0.002*	*0.000*	*0.000*	*0.000*	*0.000*	*0.000*
2.1	**IC**										*0.033*	*0.043*	*0.053*	*0.065*	*0.077*
	ID										*0.001*	*0.000*	*0.000*	*0.000*	*0.000*
2.2	**IC**											*0.031*	*0.040*	*0.050*	*0.060*
	ID											*0.001*	*0.000*	*0.000*	*0.000*
2.3	**IC**												*0.029*	*0.037*	*0.046*
	ID												*0.001*	*0.000*	*0.000*
2.4	**IC**													*0.027*	*0.034*
	ID													*0.001*	*0.000*
2.5	**IC**														*0.025*
	ID														*0.001*

Italicized values obtained from a Burr distribution.
Non-italicized values obtained from the normal distribution.
Blanks correspond to situations that are either not feasible or impossible.

Table A.10: *Burr* and Normal Values for the Integrals IC and ID

C_{pk}		*Tolerance values*											
		8.7	*9.0*	*9.3*	*9.6*	*9.9*	*10.2*	*10.5*	*10.8*	*11.1*	*11.4*	*11.7*	*12.0*
1.2	**IC**	0.357	0.372	0.387	0.401	0.415	0.428	0.441	0.453	0.464	0.476	0.486	0.497
	ID	0.000	0.000	0.000	0.000	0.000	0.000	0.000	0.000	0.000	0.000	0.000	0.000
1.3	**IC**	0.318	0.333	0.349	0.363	0.377	0.391	0.404	0.417	0.429	0.440	0.452	0.463
	ID	0.000	0.000	0.000	0.000	0.000	0.000	0.000	0.000	0.000	0.000	0.000	0.000
1.4	**IC**	0.281	0.297	0.312	0.327	0.342	0.356	0.369	0.382	0.395	0.407	0.418	0.429
	ID	0.000	0.000	0.000	0.000	0.000	0.000	0.000	0.000	0.000	0.000	0.000	0.000
1.5	**IC**	0.246	0.262	0.278	0.293	0.308	0.322	0.336	0.349	0.362	0.374	0.386	0.398
	ID	0.000	0.000	0.000	0.000	0.000	0.000	0.000	0.000	0.000	0.000	0.000	0.000
1.6	**IC**	0.214	0.230	0.246	0.261	0.276	0.290	0.304	0.317	0.330	0.343	0.355	0.367
	ID	0.000	0.000	0.000	0.000	0.000	0.000	0.000	0.000	0.000	0.000	0.000	0.000
1.7	**IC**	0.184	0.200	0.216	0.231	0.245	0.260	0.274	0.287	0.300	0.313	0.326	0.338
	ID	0.000	0.000	0.000	0.000	0.000	0.000	0.000	0.000	0.000	0.000	0.000	0.000
1.8	**IC**	0.157	0.172	0.187	0.202	0.217	0.231	0.245	0.259	0.272	0.285	0.297	0.309
	ID	0.000	0.000	0.000	0.000	0.000	0.000	0.000	0.000	0.000	0.000	0.000	0.000
1.9	**IC**	*0.132*	0.147	0.161	0.176	0.190	0.204	0.218	0.232	0.245	0.258	0.270	0.283
	ID	*0.000*	0.000	0.000	0.000	0.000	0.000	0.000	0.000	0.000	0.000	0.000	0.000
2.0	**IC**	*0.110*	*0.123*	0.137	0.151	0.165	0.179	0.193	0.206	0.219	0.232	0.245	0.257
	ID	*0.000*	*0.000*	0.000	0.000	0.000	0.000	0.000	0.000	0.000	0.000	0.000	0.000
2.1	**IC**	*0.089*	*0.102*	*0.116*	0.129	0.142	0.156	0.169	0.182	0.195	0.208	0.220	0.233
	ID	*0.000*	*0.000*	*0.000*	0.000	0.000	0.000	0.000	0.000	0.000	0.000	0.000	0.000
2.2	**IC**	*0.071*	*0.083*	*0.096*	*0.109*	0.121	0.134	0.147	0.160	0.172	0.185	0.197	0.209
	ID	*0.000*	*0.000*	*0.000*	*0.000*	0.000	0.000	0.000	0.000	0.000	0.000	0.000	0.000
2.3	**IC**	*0.056*	*0.067*	*0.078*	*0.090*	*0.102*	0.114	0.127	0.139	0.151	0.164	0.176	0.188
	ID	*0.000*	*0.000*	*0.000*	*0.000*	*0.000*	0.000	0.000	0.000	0.000	0.000	0.000	0.000
2.4	**IC**	*0.043*	*0.052*	*0.063*	*0.073*	*0.085*	*0.096*	0.108	0.120	0.132	0.143	0.155	0.167
	ID	*0.000*	*0.000*	*0.000*	*0.000*	*0.000*	*0.000*	0.000	0.000	0.000	0.000	0.000	0.000
2.5	**IC**	*0.032*	*0.040*	*0.049*	*0.059*	*0.069*	*0.080*	*0.091*	0.102	0.113	0.125	0.136	0.148
	ID	*0.000*	*0.000*	*0.000*	*0.000*	*0.000*	*0.000*	*0.000*	0.000	0.000	0.000	0.000	0.000

Italicized values obtained from a Burr distribution.
Non-italicized values obtained from the normal distribution.

Table A.10: *Burr* and Normal Values for the Integrals IC and ID

C_{pk}		*Tolerance values*											
		8.7	*9.0*	*9.3*	*9.6*	*9.9*	*10.2*	*10.5*	*10.8*	*11.1*	*11.4*	*11.7*	*12.0*
2.6	**IC**	*0.023*	*0.030*	*0.038*	*0.046*	*0.055*	*0.065*	*0.075*	*0.086*	0.097	0.107	0.118	0.129
	ID	*0.000*	*0.000*	*0.000*	*0.000*	*0.000*	*0.000*	*0.000*	*0.000*	0.000	0.000	0.000	0.000
2.7	**IC**		*0.022*	*0.028*	*0.035*	*0.043*	*0.052*	*0.061*	*0.071*	*0.081*	0.091	0.102	0.113
	ID		*0.000*	*0.000*	*0.000*	*0.000*	*0.000*	*0.000*	*0.000*	*0.000*	0.000	0.000	0.000
2.8	**IC**			*0.021*	*0.026*	*0.033*	*0.041*	*0.049*	*0.058*	*0.067*	*0.077*	0.087	0.097
	ID			*0.000*	*0.000*	*0.000*	*0.000*	*0.000*	*0.000*	*0.000*	*0.000*	0.000	0.000
2.9	**IC**				*0.019*	*0.025*	*0.031*	*0.038*	*0.046*	*0.055*	*0.064*	*0.073*	0.083
	ID				*0.000*	*0.000*	*0.000*	*0.000*	*0.000*	*0.000*	*0.000*	*0.000*	0.000
3.0	**IC**					*0.018*	*0.023*	*0.029*	*0.036*	*0.044*	*0.052*	*0.061*	*0.069*
	ID					*0.000*	*0.000*	*0.000*	*0.000*	*0.000*	*0.000*	*0.000*	*0.000*
3.1	**IC**						*0.017*	*0.022*	*0.028*	*0.034*	*0.042*	*0.049*	*0.058*
	ID						*0.000*	*0.000*	*0.000*	*0.000*	*0.000*	*0.000*	*0.000*
3.2	**IC**							*0.016*	*0.021*	*0.026*	*0.033*	*0.040*	*0.047*
	ID							*0.000*	*0.000*	*0.000*	*0.000*	*0.000*	*0.000*
3.3	**IC**								*0.015*	*0.020*	*0.025*	*0.031*	*0.038*
	ID								*0.000*	*0.000*	*0.000*	*0.000*	*0.000*
3.4	**IC**									*0.015*	*0.019*	*0.024*	*0.029*
	ID									*0.000*	*0.000*	*0.000*	*0.000*
3.5	**IC**										*0.014*	*0.018*	*0.023*
	ID										*0.000*	*0.000*	*0.000*
3.6	**IC**											*0.013*	*0.017*
	ID											*0.000*	*0.000*
3.7	**IC**												*0.013*
	ID												*0.000*

Italicized values obtained from a Burr distribution.
Non-italicized values obtained from the normal distribution.
Blanks correspond to situations that are either not feasible or impossible.

Glossary of Symbols

A_1 a scaling factor for average chart limits, p. 265
A_2 a scaling factor for average chart limits, p. 257
A_3 a scaling factor for average chart limits, p. 261
A_4 a scaling factor for average chart limits, p. 259
A_5 a scaling factor for average chart limits, p. 267
A_{10} a scaling factor for average chart limits, p. 263
ACP *Actual Cost of Production*, p. 132
ACP&U *Actual Cost of Production and Use*, p. 137
Average(X) a generic reference to any measure of location for *X*, p. 26
Average($\bar{X}$) a generic reference to any measure of location for $\bar{X}$, p. 26
Average(R) a generic reference to any measure of location for *R*, p. 26

B_3 a scaling factor for standard deviation chart limits, p. 261
B_4 a scaling factor for standard deviation chart limits, p. 261
B_9 a scaling factor for standard deviation chart limits, p. 263
B_{10} a scaling factor for standard deviation chart limits, p. 263

c a parameter for a Burr distribution, p. 163
c_1 a bias correction factor for the root mean square deviation statistic, p. 253
c_2 a bias correction factor for the root mean square deviation statistic, p. 253
c_3 a bias correction factor for the root mean square deviation statistic, p. 253
c_4 a bias correction factor for the standard deviation statistic, p. 253
c_5 a bias correction factor for the standard deviation statistic, p. 253
c_6 a bias correction factor for the standard deviation statistic, p. 253
C_p the Capability Ratio, p. 45
C_{pk} the Centered Capability Ratio, p. 46
C_R the cost of reworking an item, p. 76
c_R the excess cost of reworking an item, p. 147
C_S the cost of scrapping an item, p. 76
c_S the excess cost of scrapping an item, p. 147

d_2	a bias correction factor for the range statistic, p. 253
d_3	a bias correction factor for the range statistic, p. 253
d_4	a bias correction factor for the range statistic, p. 253
D_3	a scaling factor for range chart limits, p. 257
D_4	a scaling factor for range chart limits, p. 257
D_5	a scaling factor for range chart limits, p. 259
D_6	a scaling factor for range chart limits, p. 259
DNS	the distance to the nearest specification, p. 46
E_1	a scaling factor for natural process limits, p. 265
E_2	a scaling factor for natural process limits, p. 257
E_3	a scaling factor for natural process limits, p. 261
E_3	a scaling factor for natural process limits, p. 261
E_4	a scaling factor for natural process limits, p. 267
E_6	a scaling factor for natural process limits, p. 263
ECP&U	*Effective Cost of Production and Use*, p. 95
ExCP	*Excess Cost of Production*, p. 93
ExCU	*Excess Cost of Use*, p. 94
$f(x)$	a probability density function, p. 78
IA	an integral needed to evaluate the Effective Cost of Production, p. 156
IB	an integral needed to evaluate the Effective Cost of Production, p. 156
IC	an integral needed to evaluate the Effective Cost of Production, p. 156
ID	an integral needed to evaluate the Effective Cost of Production, p. 156
k	a parameter for a Burr distribution, p. 163
K	a constant that defines a quadratic Excess Cost Function, p. 79
$\mathcal{L}(x)$	the Excess Cost Function, p. 77
LAL	the lower average limit, pp. 257, 259
LNPL	the lower natural process limit, p. 254
LRL	the lower range limit, pp. 257, 259
LRMSL	the lower *RMS* deviation limit, pp. 265, 267
LSL	the lower specification limit, p. 45
LSDL	the lower standard deviation limit, pp. 261, 263
μ	the mean parameter for a probability model, p. 78
MEAN(R)	the mean of the distribution of the ranges, a parameter, p. 253
MEDIAN(R)	the median of the distribution of the ranges, a parameter, p. 253

NC	the nominal cost of production, p. 133
p_2	the proportion of second quality material, p. 148
p_3	the proportion of third quality material, p. 148
P_p	the Performance Ratio, p. 47
P_{pk}	the Centered Performance Ratio, p. 47
p_R	the proportion of reworked material, p. 147
p_S	the proportion of scrap material, p. 147
PD_2	the price difference between first and second quality material, p. 148
PD_3	the price difference between first and third quality material, p. 148
R	the range statistic, p. 243
$\bar{R}$	the average range, p. 21
s	the standard deviation statistic, p. 243
σ	the standard deviation parameter for a probability model, p. 78
s_n	the root mean square deviation statistic, p. 243
$SD(R)$	the standard deviation parameter of the distribution of the ranges, p. 253
$Sigma(X)$	an adjusted summary within-subgroup measure of dispersion, p. 252
$Sigma(\bar{X})$	an adjusted summary within-subgroup measure of dispersion, p. 252
$Sigma(R)$	an adjusted summary within-subgroup measure of dispersion, p. 252
SL_R	the rework specification limit, p. 76
SL_S	the scrap specification limit, p. 76
τ	the target value for a process, p. 78
UAL	the upper average limit, pp. 257, 259
$UNPL$	the upper natural process limit, p. 254
LRL	the upper range limit, pp. 257, 259
$URMSL$	the upper *RMS* deviation limit, pp. 265, 267
USL	the upper specification limit, p. 45
$USDL$	the upper standard deviation limit, pp. 261, 263
$\bar{\bar{X}}$	the grand average, p. 21
XmR	chart for individual values and moving ranges, p. 24

Index

1.5 sigma shift 99

analysis, conservative or traditional 16
art 7
assembly 57-73
assignable causes 8-14, 18-40
average and range charts 20-23
average excess costs 75-79
average moving range 26
average range 21
axles 57-73

baseline costs 81-83
bearings 57-73
behavior of process 12-13
brink of chaos 53-54
burn the toast and scrape it 7
Burr distributions 163-167

capability indexes 45-47
capability ratio 45, 93-98
cause-and-effect relationships 1-4
  new 10
center-cut approach 60-61
centered capability ratio 46
centered costs 81-84
centered performance ratio 48
chance causes 23-24
chaos 53 56
chart for individual values 24-29
chart for moving ranges 24-29
common causes 8-14, 17-19, 23
conservative analysis 16
continual improvement 31-40
control charts 19
control factors 4-14, 23
  fine tuning of 8
controlling dominant causes 4-14
conventional wisdom 12
costs of using conforming product 57-73
cycle of despair 56

data 1
data sets
  brake hose cord 102
  batch weights 43, 82
  bead board 41
  compound M 101
  contamination data 116, 126, 138, 142
  fill volumes 106
  methanol data 28, 87
  nonfunctional stampings 146
  resistivity data 21, 91
  run-out data 109, 115, 124, 132
  Tokai Rika 31, 78, 89
Deming, W. E. 7
descriptive tolerance 113
diminishing returns 9
dominant causes 2-14

Economic Control of Quality 19-24
economic equilibrium 9
economic operation 93-98
economic zone 98
effective cost of prod & use 75-168
effective cost of production 158
effective cost of use 159
entropy 56
equipment, new 10
estimates of process average 22
exceptional variation 18-20
excess cost function 77-79
excess costs of production 78, 93-94, 158
excess costs of use 78, 94-95, 158
experimental studies 14-16

four possibilities for a process 51-56
full potential 10-14

gaps
  between axle and bearing 58
  between performance and potential 49, 80-82
global variation 20, 43-45, 251
grand average 21

ideal for process 29-30
ideal state 51-52
identifying control factors 5

Jennett, W. J. 25
judgment of process 31

magic 7
meeting requirements 11
methodology for moving toward ideal 30
minimum costs 82-86
moving ranges 24-29

natural process limits 32
new equipment 10
new technology 10
numbers 1

observational studies 14-40
one-sided specifications 101-119
on-target with minimum variance 79, 90
operating at full potential 10-14
operating predictably 13
operating to meet requirements 11

Pareto diagram 2
performance 41-45
performance indexes 47-48
performance ratio 47
potential 41-45
power function 96
predictable costs 81-84
predictable tolerance 113
pre-sorting the parts 65-66
problem solvers 56
process average, estimates of 22
process behavior 12-14
process behavior charts 13-40, 73-74
process dispersion, estimates of 22
process improvements 11-14
process performance 41-45
process potential 41-45
process specifications 104-107, 145-147
process upgrade 9-11
product specifications,
  one-sided 108-119
  two-sided 92

random assembly 61
range charts 20-23
reducing variation at the source 68-73
reengineer process 9-13
research 5
rework 76
routine variation 8, 17-20

safety specifications 101
scrap 76
sequential analysis 16
Shewhart, Walter 17, 19, 21, 24
software 29
spurious 1.5 sigma shift 99
starting point for research 5
state of chaos 54-56
Statistical Method 29
subgroup size of one 24
successive differences 25
system optimization 8

targets at boundaries 121-128
technology, new 10
three-sigma limits 19, 23
threshold state 53
Tokai Rika data 31-41
tolerance 113
traditional analysis 16
traditional process upgrades 56
trouble, definitions 50-51
two-sided specifications 92

uncontrolled factors 7-14
upgrade process 9-13, 56

variation 1
  routine 8
voices of the process 41-45

wisdom, conventional 12
within-subgroup variation 20, 43-45
world class quality 90

XmR chart 24-29